Praise for
The Language Puzzle

"How humans acquired their most important and mysterious mental skill remains a fascinating mystery. Steven Mithen describes the leading clues from diverse sources so clearly that *The Language Puzzle* is a sleuth's equivalent to one-stop shopping. The origin of language is beginning to look like a solvable problem." —Richard Wrangham, author of *Catching Fire*

"An authoritative, dense, yet accessible synthesis, *The Language Puzzle* is a superbly up-to-date guide to the complex and variegated evolution of language. Encompassing a huge and multidisciplinary scope of knowledge and covering some five million years, this fascinating book shows that asking how and why we came to speak also means exploring what it is to be human." —Rebecca Wragg Sykes, author of *Kindred*

"A fascinating history of ideas and a masterful synthesis of the latest insights from linguistics, archaeology, genetics, neuroscience, and AI—providing us with a compelling theory of the evolution of language. *The Language Puzzle* is a tour de force." —Alice Roberts, author of *Ancestors*

"Relating the evolution of the human lineage while attempting to integrate linguistics, genetics, archeology, and semiotics in proposing a holistic explanation for language evolution is no small task. However, in this remarkably accessible narrative, Mithen weaves a thoughtful and engaging account across time, bodies, places, and materials. Whether or not one agrees, in total or in part, with the assumptions and assertions in the book, it offers a bounty of valuable insights and has much to teach us all." —Agustín Fuentes, author of *The Creative Spark*

"A remarkably comprehensive biography of the single most important thing we all share—language—written with Mithen's wonderful ability to combine deep insights with a story engagingly told." —Robin Dunbar, author of *Grooming, Gossip, and the Evolution of Language*

"Mithen's book *The Language Puzzle* addresses when, why, and in which hominins nascent linguistic abilities emerged and how, over vast amounts of time, they evolved into language as we know it. Chapter by chapter, Mithen gleans relevant clues—he calls them fragments or puzzle pieces—from fossil hominins, genetics, comparative anatomy, brain evolution, evolutionary developmental biology (evo-devo), the archaeological record of stone tools and art, and much more. Fragments about linguistics, for example, suggest that the evolution of language had its roots in apelike vocalizations that paved the way for the subsequent invention of various kinds of words, grammar, and, ultimately, abstract language and thought. Intriguingly, Mithen uses the puzzle pieces to attribute different levels of linguistic development to various hominin species and then draws parallels with the cognitive skills each species likely used to produce their kinds of stone tools and (where applicable) art. However, Mithen keeps the reader deliciously hanging on until the end, when he finally assembles all of the puzzle pieces into a big picture that spells out the different levels of linguistic accomplishment of australopithecines, *Homo habilis*, *Homo erectus*, *Homo heidelbergensis*, *Homo neanderthalensis*, and *Homo sapiens*. (You'll need to read the book to find out which one is credited with developing the first true language.) Not only is *The Language Puzzle* extremely readable, it is also an epic achievement that, more than any other book out there, rises to the challenge of elucidating the immense complexity that underpinned the emergence and evolution of human language." —Dean Falk, author of *Finding Our Tongues*

The Language Puzzle

PIECING TOGETHER THE SIX-MILLION-YEAR STORY OF HOW WORDS EVOLVED

STEVEN MITHEN

BASIC BOOKS
New York

To Sue Mithen
for all her wise words

———————————————

Basic Books
Hachette Book Group
1290 Avenue of the Americas, New York, NY 10104
www.basicbooks.com

Printed in the United States of America

Originally published in 2024 by Profile Books Ltd in Great Britain
First US Edition: June 2024

Published by Basic Books, an imprint of Hachette Book Group, Inc. The Basic
Books name and logo is a registered trademark of the Hachette Book Group.

The Hachette Speakers Bureau provides a wide range of authors for speaking
events. To find out more, go to hachettespeakersbureau.com or email
HachetteSpeakers@hbgusa.com.

Basic Books copies may be purchased in bulk for business, educational,
or promotional use. For more information, please contact your local
bookseller or the Hachette Book Group Special Markets Department
at special.markets@hbgusa.com.

The publisher is not responsible for websites (or their content) that are not
owned by the publisher.

Typeset in Dante by MacGuru Ltd

Library of Congress Control Number: 2023949069

ISBNs: 9781541605381 (hardcover), 9781541605398 (ebook)

LSC-C

Printing 2, 2025

Contents

Preface

How, when and why language evolved have been enduring questions since the time of Plato, and most probably long before. Although always of interest, I avoided addressing these questions directly in my two previous books about the evolution of the human mind, *The Prehistory of the Mind* (1996) and *The Singing Neanderthals* (2005). Several other scholars were putting forward fascinating ideas and theories about language but they were neglecting other aspects of the evolving mind that I wanted to address, notably creative thought and music. As much as I tried to avoid language, however, I kept being drawn towards it as the most fundamental aspect of the modern mind.

Proposals for how, when and why language evolved continued to be published throughout the last two decades. While I read and applauded many accounts, none appeared satisfactory. Some drew primarily on evidence from one discipline, such as linguistics or anthropology, but could be readily discounted by evidence from another, such as archaeology or psychology. Others dealt with one aspect of language while neglecting others or provided elegant scenarios for how language evolved but entirely lacked a chronology for when that occurred. Hypotheses came and went with considerable speed, often reflecting the pace of new discoveries about the past, the brain and language itself. I suspected any contribution I could make would be of similar transient value. But I continued to think about the language questions, discussed them with my colleagues and students, and read in as many subject areas as I could manage. The questions were never far from my mind as I undertook my excavations to find Stone Age artefacts, the makers of which had been silenced by the passage of time.

Around 2020, I began to suspect that embedded in the recent research of linguists and archaeologists, of computer scientists and anthropologists, of philosophers, psychologists and geneticists were the fragments of a comprehensive account for language evolution. An account that could build the necessary bridges between disciplines and would stand the test of time despite the inevitability of new discoveries and new ideas. Finding those fragments from within so many disciplines was only half the challenge. The other half was working out how they join together. That was a puzzle and my solution has become *The Language Puzzle*.

INTRODUCTION: THE
PUZZLE OF LANGUAGE

By choosing to read this book, I suspect you know at least 50,000 words and say around 16,000 words a day. Thousands more will pass through your mind, either heard from others, as you are reading, thinking what to say or musing to yourself.[1] You are good at words, speaking between 120 and 200 words a minute and reading them at twice that speed.[2] When speaking, writing or using sign language, you effortlessly create unique sequences of words. These convey meanings beyond those of the individual words themselves, meanings that others can understand with equivalent ease despite never having heard or seen that string of words before. You might even be able to do this in another language, perhaps several. How so? How can you remember and manipulate so many words? That is a puzzle.

We have a love of words. Think crosswords, Scrabble and texting. Think chatting to a friend, listening to a story, sharing a joke or hearing a speech by the orator of your choice – Churchill, Obama or Mandela. Moreover, we are never satisfied with the words we have, frequently changing their meanings and inventing new ones. Think tablets, clouds and surfing. Think Covid, Brexit and, if you can, trequartista. That was one of the 2,000 new entries to the *Oxford English Dictionary* in 2022. In case you didn't know, it means an attacking football player who operates in the space between the midfielders and the strikers and whose primary role is to create opportunities for teammates to score.[3]

Now you have at least 50,001 words. Where do you keep them all? How do you know which ones to use, and how to combine them to make a statement or ask a question that someone else will understand?

Just as we love words, we know their power and may fear their consequences. We know how a few ill-thought-out words can damage a relationship and flunk an interview; how eloquent politicians can sway a crowd; how words can abuse and offend; how words can rouse people to hatred, violence and war. We tolerate and suffer the consequences of such words because of our unbounded desire to talk and listen to what others have to say.

Your lexicon love affair began in childhood. Before reaching the age of one, you were likely saying your first words and knew the meaning of several hundred. Within your second year, you had started combining words into simple sentences while learning new ones at an average rate of nine a day.[4] That rate continued unabated into your adolescence, maybe even learning two or more languages at once. How were you able to acquire language at such pace?

The answer is that you had your parents, carers, family and friends for help. You inherited a genetic predisposition to acquire language from your biological parents, which was realised by growing up amid people who were continuously using words, whether spoken or signed. Your parents had done likewise, helped by their own parents, family, friends and wider community. And so on, back through the generations. But how did it begin?

And when?

A long time ago. It must have been after 6 million years ago, the date when we shared a common ancestor with the chimpanzee. Although there are word-like qualities to chimpanzee barks and grunts, these are insufficient to characterise their vocal communication as a form of language. Unlike tool making, walking on two legs and complex patterns of social relations, language has remained stubbornly aloof from the primate world, becoming the last bastion of human uniqueness. With no antecedent in the animal world, explaining how language began has become the mother of all puzzles.

We need to solve that puzzle to explain language today – how you can extract meaning from this sentence and (hopefully) tell others about the interesting book you are reading. Equally, we need to solve the language puzzle to know about our past. I suspect you have heard about the Neanderthals of the Ice Age, and Lucy who left her footprints in Tanzania 3.7 million years ago. Anthropologists describe their bones, archaeologists their tools and biologists can tell us about their genes.[5] As fascinating as all that is, without knowledge about their language our ancestors will always remain ill defined, providing us with little understanding of the past. Did Lucy and the Neanderthals have words? If so, did they also have rules for how they could be strung together to make meaningful utterances? Or did they merely mumble and howl? We need to know. Otherwise, they will forever remain as nothing more than objects for scientific study, rather than acknowledged as sentient beings from our distant past.

Whenever language of the type we have today emerged, my proposition is that it enabled the most fundamental social, economic and cultural event of the human past: the origin of farming at *c*.10,000 years ago. That put an end to millions of years of hunting and gathering and was effectively the end of the Stone Age because metallurgy was soon discovered within the new farming communities.[6] The beginning of agriculture was not just the turning point of human history but also the crossroads for planet Earth. Farming rapidly led to towns and cities; ancient civilisations and empires soon followed; then came the industrial and digital revolutions, followed by globalisation. Marvellous things have been achieved – the music of Bach and men on the Moon. But the first farmers also ignited the slow-burning fuse of our present-day climate crisis and agriculture is responsible for extensive environmental degradation and loss of biodiversity.

Although the first farming communities are dated to 10,000 years ago, they were the outcome of a long, slow process of change in the way people thought about and acted in the world. That process began as soon as fully modern language evolved and was spurred on by climatic events that followed the peak of the last glaciation at 20,000 years ago. While archaeologists have focused on the impacts of climate change, they provide only half the story for the origin of farming.[7] The other half is language, when it evolved and its impact on the human mind and behaviour. Without that we would still be living as Stone Age hunter-gatherers.

Solving the puzzle

Understanding the origin of language has been described as the hardest problem in science.[8] Attempts to solve it began when Plato asked about the origin of names, and possibly long before. Today we have a plethora of theories, hypotheses and ideas. But there is no agreement.[9]

Some argue for a sudden emergence of language from a genetic mutation at 100,000 years ago, while others suggest phases of 'protolanguage' or a slow emergence of language over millions of years;[10] some propose language evolved from singing, while others promote social bonding, storytelling, tool making and hunting;[11] some cherry-pick a feature of language and claim its evolution was the transformative event, such as 'displacement' (the ability to talk about the future and the past) or 'recursion' (the way in which we can embed multiple clauses into a single utterance).[12] No one seems to agree with anyone else.

There have been two constraints on reaching consensus.

The first is the sheer complexity of the task, because language is such an all-encompassing, brain, body, social and cultural phenomenon. The second is that critical pieces of the evidence have been missing.

With regard to the first constraint, many academic disciplines are required to explain how we create and use language today, and even more to explain how this remarkable capacity evolved.

Linguistics is essential because it defines the nature of language, as is psychology because language is a product of the human mind, drawing on a host of mental processes including memory, perception and attention. Neuroscience digs deeper by examining how language is generated by the brain, while genetics considers how inherited genes interact with our environment to enable linguistic capabilities to develop and evolve. Anthropology is required because language users must be placed into their social and cultural context. Palaeoanthropology does likewise for our human ancestors, along with reconstructing their anatomy and its linguistic implications from skeletal remains. Archaeology is essential for inferring linguistic capabilities from stone artefacts and other human debris. Ethology is also required because studying chimpanzees and other non-human primates in captive and wild settings provides insights into the pre-linguistic foundations of language that were likely present in our earliest ancestors.

Each of these disciplines has its own body of data, theories, methods and terminologies. Each has one or more essential pieces of the language puzzle to contribute. Despite academics' willingness to collaborate, research within each discipline is often pursued in relative isolation, partly because of outdated educational and university structures and partly because of the intellectual challenge required to cross disciplinary boundaries. A consequence is that theories about language evolution often suffer from disciplinary dissonance: ideas proposed from one discipline, such as linguistics, invariably conflict with evidence from another, such as archaeology.[13]

The second constraint has been missing puzzle pieces; conversely, what had been thought to be important pieces did not belong at all. New research has lifted this constraint. Key discoveries have been made by archaeologists when digging in the ground, psychologists listening to children, computer scientists simulating language change, ethologists watching apes, and linguists taking language apart. The pile of new puzzle pieces from their work has been added to by geneticists decoding human genomes and neuroscientists peering inside the brain. The new evidence caused old ideas to be questioned and then discarded, notably dedicated language centres in the brain, specialised genes for language, and the notion of Universal Grammar – the idea that we are born with a ready-made and specialised mental toolbox for language acquisition. As these were removed, even older ideas acquired a new lease of life: twenty-first-century human genomics has almost caught up with Epicurean ideas about language of the fourth and third centuries BC.

A revolution in our understanding of language is underway. We now appreciate the extent of linguistic diversity throughout the world and understand how children learn the meaning of words; we are beginning to grasp how language relies on neural networks that extend throughout the brain, these constructed by complexes of interacting and multifunctional genes. Chimpanzee calls are no longer dismissed as uncontrolled outbursts; we have new insights into the material culture, behaviour and cognition of our extinct relatives and ancestors.

Biological and cultural evolution have become entirely entwined.[14] The present is now recognised as a key to the past. Just as the geologist Charles Lyell had used contemporary processes of sedimentation and erosion to explain geological strata within his *Principles of Geology* of 1830, and just as Charles Darwin had used those of inheritance, reproduction and competition to explain biological evolution in his *Origin of Species* of 1859, so too can we use linguistic change in the present to inform about that of the distant past and explore its long-term consequences.[15]

The Language Puzzle collects together all the old and new pieces of evidence and attempts to solve the puzzle. As with a jigsaw, the only way to start is by connecting pieces into a series of fragments, each a mini puzzle in itself. The edge pieces must come first to provide the overall frame of the puzzle and to hint what its middle may contain. Once the frame is complete, fragments of its interior can be assembled, ideally with each providing a satisfying picture. When all have been completed, they can be joined to reveal the bigger picture – in our case the when, the why and the how of language evolution.

Chapter 2 provides half of the jigsaw frame, with an overview of human evolution during which language evolved, introducing the species, cultures and climatic periods that feature prominently in the following chapters.[16] It begins with the last common ancestor between humans and chimpanzees that lived 6 million years ago. Humans, members of the *Homo* genus, first appeared on the African savannah 2.8 million years ago and evolved into several different species that flourished in Africa, Europe and Asia, before contracting to its sole survivor at *c.*40,000 years ago: *Homo sapiens*, the species to which we all belong. Quite why only *H. sapiens* remains is much debated. Some argue this is because we alone have language, a proposition to be tested in this book.

The jigsaw frame is completed in Chapter 3, which reviews what we need to explain: the nature of language as we know it today. This covers the nature of words and the rules by which they are combined to generate meaningful utterances, whether spoken, signed or written; how words and rules vary between languages; and the causes of such linguistic diversity.

The frame guides us to twelve further fragments of the language puzzle. The first is what the vocalisations of apes and monkeys can tell us about the foundations of language in our earliest ancestors (Chapter 4). There are two fragments that draw on the fossil evidence: what can we learn about language evolution from changes in the vocal tract (Chapter 5) and from changes in the size and shape of the brain during human evolution (Chapter 11)? Three fragments relate to past behaviour: the linguistic implications from how our ancestors made stone tools (Chapter 7), made signs and symbols (Chapter 15) and used fire (Chapter 10). Critical pieces of the puzzle come from language itself: the distinction between different types of words (Chapter 6), how language is shaped by its transmission from generation to generation (Chapter 8), how infants learn language (Chapter 9), the constant change in the meanings, roles and pronunciations of words (Chapter 13), and how language impacts on perception and thought (Chapter 14). The genetics of language contributes to our knowledge of language today and its past evolution (Chapter 12).

To assemble these fragments, I describe the work of linguists, anthropologists, philosophers and scientists of every hue who have found the puzzle pieces – the evidence. I will bring you their breakthrough moments: the experiments, discoveries and insights that have transformed our knowledge of language and how it evolved. Although the above list might suggest that I switch randomly from one subject area to another, the fragments follow each other in a logical order, as each indicates the next fragment to assemble so that the bigger picture can emerge.

With the frame and twelve interior fragments complete, the final challenge is piecing them together to solve the language puzzle. How does the evidence about the vocal tract and the brain connect to that about stone tools and the use of fire? How does our understanding of language acquisition by children influence that of language evolution by human ancestors? Was language always a tool for thought or was that a recent innovation? The concluding chapter reveals the big picture: how language evolved and its monumental impact on the lives of our ancestors and the history of the planet. It solves the puzzle of why we all love words.

2

A BRIEF HISTORY OF HUMANKIND

Our evolutionary story begins between 8 and 6 million years ago (mya) with an ape living somewhere in Africa.[1] That was the last common ancestor (LCA) for humans (*Homo*) and the chimpanzees (*Pan*), our closest living relative. We know the approximate date that our lineages diverged by the extent of difference between the human and chimpanzee genomes and the rate at which genetic mutations occurred to create that difference. Although some chronological uncertainty remains, throughout this book I will cite a date of 6.0 mya (6 million years ago) for the LCA (Figure 1).

It is commonly assumed that the LCA had strong similarities to present-day chimpanzees, some preferring to cite the long-limbed bonobo (*Pan paniscus*) that live in female-dominant societies and others the more conflict-ridden, male-dominated groups of the common chimpanzee (*Pan troglodytes*). Whether either type is a suitable model for the LCA is debated because, as with *H. sapiens*, both are products of a further 6 million years of evolution and likely possess derived features – those evolving after the time of the LCA. Unlike *H. sapiens*, however, chimpanzees have remained in the same type of closed canopy forest habitat as occupied by the LCA and maintained a similar brain size of 350–400 cubic centimetres (cm³). For these reasons, the extent of evolutionary change within the lineage leading to present-day chimpanzees appears quite limited. It is not unreasonable to suspect that the LCA had used vocalisations similar

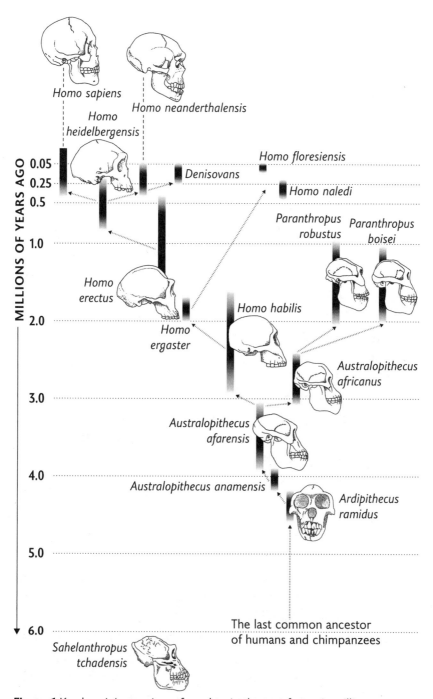

Figure 1 Key hominin species referred to in the text from six million years of human evolution

to those of chimpanzees. We will explore whether these have language-like qualities in Chapter 4.

The fossilised remains of at least four types of ape are known from Africa within the time frame of the LCA, or soon after. The oldest is *Sahelanthropus tchadensis*, coming from Chad in the north of central Africa, dating to between 7.2 and 6.8 mya, and once living in an open, savannah-like environment. The best preserved and most abundant specimens represent *Ardipithecus ramidus* from Ethiopia. These date to between 4.4 and 4.2 mya, by which time eastern Africa was thickly forested. Both species had a brain size of 300–350 cm^3, displaying anatomical similarities to the earliest *Homo* and present-day chimpanzees; they also had significant differences, removing them as viable candidates for the LCA – the fossil remains of which have yet to be discovered.

The fossil record markedly improves after 4.3 mya with as many as ten different types of ape known from eastern, central and southern Africa. These are collectively known as australopiths, some of which remain in the fossil record until 1 million years ago. They evolved during a period of increased aridity, with a shift from forested to open environments with scattered woodland.[2] The australopiths share several features with *Homo*, including bipedal locomotion, reduced facial projection and smaller teeth than those of earlier apes and present-day chimpanzees. Although the australopiths share a brain size of between 400 and 500 cm^3, there is considerable variation in body size and anatomy. That variation reflects different types of behavioural adaptation with each species having its own niche in the African landscape. The australopiths, early and all later members of the *Homo* genus are grouped together and called hominins.

Some australopiths became especially robust, with large cheekbones, facial muscles and molars reflecting an adaptation to chewing large quantities of dry and coarse plant material. These are sometimes placed into their own genus of *Paranthropus*. Other australopiths remained of a slender build, exploiting a greater diversity of foods, although still chewing tough plants. The earliest of these, dating to between 4.2 and 3.8 mya, is *A. anamensis*, which has chimpanzee-like features of a relatively narrow jaw and large canines. This species likely evolved into *A. afarensis*, known from between 3.7 and 2.9 mya, with the best-preserved specimen popularised as 'Lucy'. Although fully bipedal, Lucy's pelvis remained distinctive from that of *Homo*, and her relatively long arms, curved fingers and toes are characteristic of the much older *Ardipithecus*. Nevertheless, *A. afarensis* is regarded as the most likely direct ancestor of the earliest human.

The earliest humans

The earliest human is termed *Homo habilis*, the name coined by Louis and Mary Leakey, who found a distinctive set of fossils from Olduvai Gorge, Tanzania, in the 1950s and 60s. These were designated as representing a new species based on a larger brain, smaller molars and more human-like hand bones than the australopiths – although the diversity of that group had not been defined at the time of their discovery. Louis Leakey was undoubtedly influenced by stone artefacts from Olduvai that he believed were associated with the fossils, and hence the name 'handy man'.

Today we have fossils from Ethiopia, Kenya and South Africa that are also classified as *Homo habilis*, placing its earliest occurrence at 2.8 mya and providing this species with a brain size that ranged from 550 to 800 cm³, together with a considerable degree of post-cranial anatomical variation.[3] It seems doubtful that *H. habilis* is a species at all; some call it a 'waste bin' for an assortment of unrelated fossils. Those with a larger brain, flatter face and larger teeth are sometimes placed into a separate category of *Homo rudolfensis*.[4]

Whether the *H. habilis* remains represent one or two species is the tip of a taxonomic iceberg issue that pervades the whole of human evolution: how do we recognise a species from skeletal remains alone, especially when we are aware that males and females will differ in size, and all species exhibit a degree of variability in their morphology? An even more profound question is how a new species can be identified from skeletal remains alone.

The traditional biological view defines species as reproductively isolated from each other – members of different species are unable to produce fertile offspring. This is now known to be invalid because more than 10 per cent of primate 'species' engage in interbreeding. That has also been demonstrated for recent human 'species', with genomic evidence for interbreeding between *Homo sapiens* and *Homo neanderthalensis*, despite their considerable anatomical differences. With no resolution to these issues, fossils are grouped together on the grounds of morphological similarity and designated as 'species' without any agreed meaning for that term. Not surprisingly, anthropologists will arrange fossils into different groups, with some proposing a lot more and others far fewer species to have existed in the past.

Homo habilis / rudolfensis appears in the fossil record at broadly the same time as the first stone tools, known as the Oldowan culture. These tools were flakes removed from nodules, and the nodule remnants, which are referred to as cores. However, the earliest known stone tools pre-date the earliest known *Homo habilis* fossils and hence they may have also been made by one or more types of australopiths.[5] Whether making such tools has implications for linguistic ability will be considered in Chapter 7.

While the stone nodules, flakes and cores were likely used for a variety of tasks, including cutting plants and pounding roots, their key role was the removal of meat, fat and marrow from animal carcasses, as evident from cut marks and distinctive fractures on the bones from archaeological sites. The carcasses had most likely been scavenged from carnivore kills, with the sharp flakes being critical for quick access in the face of competing scavengers such as hyenas and vultures. Scavenging may have started by picking over carcasses after the hyenas and vultures had finished and developed into aggressive scavenging by throwing rocks and shouting to chase off those competitors before they had taken the best bits of meat and fat.

The open savannah would have been a dangerous place, requiring *H. habilis* to live and work in larger groups than its forest-dwelling ancestors to defend themselves from predators and to work cooperatively when scavenging, gathering plant foods and collecting stone nodules. The need to live in larger groups has been invoked as a selective pressure for brain enlargement: to provide the cognitive skills for negotiating the complexities of social life, including selecting mates and food sharing.[6] Such brain growth would have been fuelled by the relatively high calorific return from meat, marrow and fat, while enhanced cognition from that larger brain would have facilitated learning how to knap nodules to make the required flakes. The resulting positive feedback loop between group size, technology, diet, brain size and cognitive ability may have been critical for the incipient stages of language, an idea to be explored further in Chapter 11.

At around 1.8 mya, a new species designated as *Homo erectus* appears in the fossil record of eastern Africa, with the earliest specimens sometimes called *H. ergaster*. This is larger than earlier *Homo*, with a stature and bodily proportions approaching those of modern humans and a brain size reaching 1,250 cm³. The brain is not only larger but has some changes in shape that may relate to language, as considered in Chapter 11. An almost complete juvenile specimen, popularly known as the 'Nariokotome boy', provides an unparalleled record of post-cranial anatomy indicating a fully bipedal lifestyle. That had likely gradually evolved under several selective pressures including reaching to collect fruit, using hands to make and carry tools, reduced exposure to the sun, and needing to move swiftly across the savannah. The shoulder bones of *H. erectus* also have a modern-like appearance, suggestive of selective pressures for long-distance and accurate throwing, probably of both branches and rocks. This may have been to chase off hyenas from desirable carcasses or for hunting small game.

H. erectus fossils are widespread not only within Africa, from the far north to the south, but also beyond. An important collection comes from Dmanisi, Georgia, dating to between 1.85 and 1.77 million years old, showing a considerable degree of variation in body and brain size. *H. erectus* is securely dated in China and Java at 1.6 mya. It had spread into southern Europe by 1.5 mya but archaeological traces are sparse with the earliest European fossils coming from Gran Dolina, Atapuerca, in Spain, dating to 850,000–780,000 years ago. While some attribute these to *H. erectus*, others suggest a descendant called *H. antecessor*.

The out of Africa record is likely to derive from multiple dispersals, with *H. erectus* moving as part of the large mammal communities that responded to changing climate – travelling north during warmer and wetter periods and retreating to Africa when the climate became relatively dry and cold. Such changes arose from repeated 100,000-year-long cycles from cold (glacial) to warm (interglacial) periods within the Quaternary Ice Age that had begun at 2.6 million years ago. During the glacial periods, ice sheets expanded in high latitudes and mountainous regions, sea level fell, and low latitudes suffered drought; during the interglacial periods, the ice retreated, sea level rose, and grassland and then forest spread over what had been tundra and steppe. Within both the glacial and interglacial periods, there were further fluctuations as the climate became warmer or colder for shorter periods of time. Some of these were abrupt and intense, causing major disruption to ecosystems and human habitation. One intensely cold period happened at 1.1 mya and forced the extinction of *H. erectus* in Europe. When the climate relented, there was a new dispersal into that region at *c.*900,000 years ago.[7]

There have been eight major glacial–interglacial cycles during the last 780,000 years. The planet is currently in a warm, wet and notably stable interglacial period that began at 11,650 years ago and is named the Holocene. Some argue that the Holocene has already ended because of the intensity of human impact on the planet. They propose that a period known as the Anthropocene has started, either with the industrial revolution at *c.*1800 or the dropping of the atomic bomb in 1945. What is certain, however, is that the planet is now being artificially warmed by human action, with unknown consequence for the future of our species and all others on the planet.

Broadly contemporary with the appearance of *H. erectus* in Africa is a new stone technology called the Acheulean that involved making bifaces: large flakes or nodules that were flaked on each alternate face to create tear-shaped tools, otherwise known as handaxes. These are considerably more difficult to make than Oldowan choppers and flakes, exhibiting a deliberately imposed form that often shows marked symmetry. Whether handaxes and the out of Africa dispersals of *H. erectus* have implications for evolving language are considered in Chapter 7.

Handaxes and other bifaces with a straight edge known as cleavers are found throughout Africa, Asia and Europe for over a million years, sometimes in huge numbers at single locations. They are markedly rare from East Asia, possibly reflecting the dispersal to that region before the development of this technology and/or the use of other materials such as bamboo.[8] Handaxes and similar bifacial tools are absent in Europe before *c.*700,000 years ago. Their appearance after that date might reflect a further dispersal of *H. erectus* or a descendant species into that region.

The lifestyle of *H. erectus* appears similar to that of earlier humans with a mix of hunting, scavenging carcasses and gathering plant foods. Cooking has been proposed to reduce the effort and time of digesting raw foods, thereby releasing metabolic energy to enable an expansion of the brain, but evidence for the use of fire is sparse until c.400,000 years ago. That too may have implications for an evolving language capability, as will be explored in Chapter 10.[9]

Importantly, the anatomy of *H. erectus* had evolved in ways that likely changed the nature of social life from that of *H. habilis* times. The anatomical requirement for bipedalism required a narrow pelvis which led to a relatively short gestation period for a mammal the size of *H. erectus*. As such, offspring were born 'premature', with brain growth continuing at a foetal rate for the first year of life. This introduced a new developmental phase called childhood, one absent from the chimpanzee life course and we assume that of *H. habilis*. The role of childhood for the evolution of language is likely to be profound and its significance pervades this book, with its role in modern humans considered in Chapter 9.

The 'muddle in the middle'

This phrase refers to the most problematic period of human evolution, which occurred between 1 million and 350,000 years ago.[10] The fossil record becomes especially fragmented and diverse, defeating efforts to create coherent groups of fossils that might represent single species. While some anthropologists prefer to name just three or four species, no less than nineteen have been proposed by others. Unfortunately, this is also a critical period of human evolution because it ends with the presence of *H. neanderthalensis* in Europe and *H. sapiens* in Africa, both with evolved vocal tracts and large brains suggestive of advanced language capabilities – although not necessarily of the same type – as will be covered in Chapters 5 and 11.

The most recent African fossil attributed to *H. erectus* dates to c.780,000 years ago. Later specimens tend to have larger brains, a more rounded skull and smaller teeth than *H. erectus*, but it is difficult to draw a clear dividing line between *H. erectus* and descendant species. A sparse number of scattered and fragmentary African fossils have been designated as *H. rhodesiensis*, a name coined in 1929 but now rarely used. These and other fossils are now designated as *H. heidelbergensis*, a name derived from a 600,000-year-old jawbone discovered at Mauer near Heidelberg, Germany, in 1907. *H. heidelbergensis* has also been used for several other specimens in western Asia and Europe, implying this species had an extensive range but without providing any clarity as to where it evolved.

A marked lack of consensus about which fossils to designate as *H. heidelbergensis* suggests this 'species' might, like *H. habilis*, be a waste bin of unrelated fragments.[11] A recent proposal has been to discard the term altogether, placing the so-called African *H. rhodesiensis* and *H. heidelbergensis* fossils into a new species called *H. bodoensis* and to re-designate *H. heidelbergensis* from Europe as early *H. neanderthalensis*.[12]

A large collection of human fossils from another location at Atapuerca, Spain, called the Sima de los Huesos (Pit of Bones), represents at least twenty-eight individuals dating to *c*.450,000 years ago. These have been classified as *H. heidelbergensis*, although some wish to call these early Neanderthals. Similar taxonomic uncertainty hangs over further fossil remains from Europe, coming from Swanscombe and Boxgrove in England, Arago Cave in France, and Petralona Cave in Greece. The only region where there is broad consensus is East Asia with the designation of all fossil specimens to *H. erectus*.

The difficulties of classifying fossils dating to between 1 million and 350,000 years ago might reflect genuine taxonomic diversity arising from the ongoing climatic cycles that caused populations to fragment, become isolated and adapt to varying local conditions, or go extinct (or very nearly so). Indeed, we seem very lucky to be here because our Africa-based ancestors went through a severe contraction between 930,000 and 813,000 years ago. This is estimated to have wiped out 99 per cent of its members, leaving a breeding population of a mere 1,300 individuals – our ancestors survived by a whisker. It may have been from this calamity that the new species of *Homo heidelbergensis* emerged at around 800,000 years ago.[13]

With such changes in population numbers and distributions, it is surprising that stone tool technology remains largely consistent throughout this time, with the making of handaxes, cleavers and Oldowan-like flakes and cores in ever-changing frequencies and proportions throughout all regions.

While broadly consistent, there is a trend for handaxes to be more refined after 700,000 years ago, becoming thinner and displaying higher degrees of symmetry. By 500,000 years ago, they are found at relatively high latitudes in Europe, possibly associated with an early use of fire and the hunting of big game using spears.[14]

Homo sapiens, H. neanderthalensis and the Denisovans

After 350,000 years ago, the fossil record is better resolved. Fossils from Africa are primarily attributed to *Homo sapiens*. This species is distinguished by a suite of features including a relatively light physique, large brain (now reaching 1,100–1,700 cm³), vertical forehead, a chin, flat face and, for the more recent specimens, a relatively spherical cranium referred to as being globular, reflecting the shape of the brain inside. The processes by which the cranium and brain evolved are referred to as 'globularisation'.[15] The skulls excavated from Jebel Irhoud, Morocco, and Omo in Ethiopia, dating to 300,000 and 195,000 years ago respectively, have elongated and flat crania/brains, despite being attributed to *H. sapiens* because of their facial features and teeth.[16] A group of *H. sapiens* fossils dating to between 130,000 and 100,000 years ago, primarily from the caves of Skhul and Qafzeh, Israel, have some degree of globularity. The fully globular shape, however, is found only in fossils dating to after 100,000 years ago, most of which date to *c*.35,000 years ago and later (reflecting the sample available from the fossil record).

Globularisation appears, therefore, to have evolved gradually between 150,000 and 35,000 years ago, representing a different development pathway for the human brain compared with that of all previous types of humans. At least one other type of human was present in Africa, a diminutive species with an intriguing mix of human and australopith traits designated as *H. naledi* dated to between 330,000 and 240,000 years ago from South Africa. A similar localised evolutionary development occurred in Southeast Asia where a notably small type of human is found on Flores, Indonesia, dating to between 100,000 and 60,000 years ago. Designated as *H. floresiensis* this is likely a dwarfed form of *H. erectus*, although some claim it is derived from an early dispersal of a small-sized *Homo* or even australopith out of Africa. Either way, *H. floresiensis* and *H. naledi* are fascinating finds because they demonstrate the trend in human evolution was not always towards a larger brain.

Between 350,000 and 45,000 years ago, the fossil record in Europe is relatively abundant with all specimens attributed to *H. neanderthalensis*, other than two finds that might represent brief incursions of *H. sapiens*, dating to *c*.210,000 years ago at Apidima Cave in Greece, and *c*.54,000 years ago at Mandrin Cave in France.[17] *H. neanderthalensis* is defined by a suite of features that contrast with those of *H. sapiens*, including a relatively flat cranium and projecting face, prominent brow ridges and large nasal cavities and eye sockets. Its brain size is equivalent to that of *H. sapiens*, although it has a different shape and structure, the implications of which will be considered in Chapter 11. The Neanderthal suite of features evolved gradually, with traces present in the Sima de los Huesos collection of 450,000 years ago and becoming well defined within the later Neanderthals after 100,000 years ago. Regarding the body, the Neanderthals were shorter and more robust than *H. sapiens*, with barrel-like chests and substantially more muscle. Their bodies reflect the combined influences of a more physically demanding lifestyle and evolution in a colder climate than that of *H. sapiens*, requiring 100–350 more calories per day for fuel.[18]

Neanderthal fossils are found not only in Europe but also in western Asia and far to the east, with specimens in central Asia and Siberia.[19] While covering an extensive region, the population would have been fragmented by geographic barriers, with evidence that it fell into three main demographic clusters: western Europe, southern Europe and western Asia.[20] Neanderthals responded to their environmental conditions with a mix of big game hunting primarily using thrusting spears, plant gathering and exploiting the sea shore. They sometimes buried their dead. This should not be surprising given the need for close social ties within and between social groups, and hence inevitable grieving at the loss of a parent, child, relative or friend.

Our knowledge of *Homo sapiens* and the Neanderthals has been transformed during the last decade by palaeogenomics that extracts ancient DNA from skeletal remains. The first complete human genome was derived in 2003, and that of a Neanderthal in 2010. Comparison of their genomes has indicated that the two species shared a common ancestor between 800,000 and 600,000 years ago, usually designated as *H. heidelbergensis*. Palaeogenomics has also identified a further descendant, usually referred to as the Denisovans that diverged from the lineage leading to the Neanderthals at around 400,000 years ago.[21] The Denisovans occupied much of central and East Asia, evolving a physiology and lifestyle for cold environments, such as boreal forests and high altitudes, in contrast to the Neanderthal preference for more temperate, grassland environments.

The genomic revolution has also revealed several episodes of interbreeding, between *H. sapiens* with Neanderthals and

Denisovans, and between the Neanderthals and Denisovans.[22] Most of us today have between 2 and 4 per cent of Neanderthal DNA, and those in East Asia also have up to 5 per cent of Denisovan DNA. Chapter 12 considers the significance of such interbreeding for the evolutionary history of the three species and their linguistic capabilities.

Interbreeding arose from the mobility and interaction of populations, influenced by the ever-changing climate conditions that sometimes caused the ranges of the human types to overlap.[23] The earliest known movement of *H. sapiens* out of Africa had occurred by 180,000 years ago, documented by a specimen from Misliya Cave, Israel, and potentially by 210,000 years ago if a claimed *H. sapiens* fossil at Apidima Cave, Greece, is indeed that species.[24] A later dispersal, likely a response to a period of warmer and wetter climate that lasted between 130,000 and 115,000 years ago, resulted in *H. sapiens* in the caves of Skhul and Qafzeh in Israel at between 120,000 and 90,000 years ago. It is likely they overlapped with Neanderthals in that region, represented by remains from other caves in Israel – Tabun and Amud – dated to between 80,000 and 55,000 years ago, but with archaeological traces suggesting an earlier presence. Both *Homo sapiens* and Neanderthals used the same types of stone tools, methods of hunting and patterns of mobility. They may have interbred and exchanged cultural knowledge such as about tool making. These early dispersals of *H. sapiens* from Africa were not sustained with their lineages becoming extinct. The earliest *H. sapiens* presence in East Asia is heavily contested, with some arguing this occurred between 120,000 and 80,000 years ago, while others maintain a more conservative estimate of 65,000 years ago.

At around 350,000 years ago, humans of all species in Africa, Asia and Europe had shifted from the use of hand-held to hafted tools, notably stone points attached to shafts for use as spears. Handaxes became less prominent, being replaced by flakes and blades detached from prepared cores – carefully shaped nodules enabling flakes of a predetermined shape and size to be detached. This is referred to as Middle Palaeolithic technology in Europe, and the Middle Stone Age in Africa. Why this shift occurred has been little discussed by archaeologists. Chapter 7 will consider whether it was enabled by an evolving language capability, one that had crossed a threshold that allowed new technology to develop.

The use of fire became habitual after 400,000 years ago, with the first appearance of managed hearths. This was followed by the first body adornments and decorated objects, both appearing after 200,000 years ago. Neanderthals in Europe collected red ochre, used minerals that produced black pigment, made body adornments from birds' feathers and talons, and, in rare circumstances, made incisions into stone and pieces of bone. *H. sapiens* in Africa and in western Asia were similar, although their body adornments were made from shell beads and they made much greater use of red ochre, this becoming intense after 100,000 years ago when the first engravings were also made on stone. Chapters 10 and 15 consider the implications of fire and the new interest in signs and symbols for language capabilities. The extent of these developments in southern Africa after 100,000 years ago has led *H. sapiens* from after that date to be designated as 'modern humans'.

Modern humans and their global diaspora

After 70,000 years ago, modern humans dispersed out of Africa, as documented by the fossil, archaeological and genomic records (Figure 2). Unlike earlier migrations, their journeys were swift, sustained and extensive, implying goal-directed exploration rather than a mere response to environmental change. One route out of Africa was northwards, via the Rift Valley into Southwest Asia – present-day Occupied Palestinian Territories,

Figure 2 Early and later dispersals of *Homo sapiens* from Africa

Israel, Jordan and Syria. Here they encountered Neanderthals, with whom they shared the same landscape for several thousand years, sometimes at a distance and sometimes so close that there was interbreeding. It was in this region that a new technology emerged soon after 50,000 years ago involving the production of long flint blades, which provided the basis for Upper Palaeolithic technology that would be taken into Europe after 45,000 years ago.[25]

Another route out of Africa was by crossing the Bab el-Mandeb Strait from eastern Africa into Arabia. From there a coastal route was followed into south and southeastern Asia, where interbreeding with Denisovans occurred and the earliest known figurative art was made at *c.*40,000 years ago: a hand stencil and the painting of a pig-like animal on a cave wall in Indonesia.[26] Boats were constructed that took modern humans into Australia by 60,000 years ago.[27] The modern humans spread throughout Asia, with a confirmed presence in China at 45,000 years ago and contested claims for an even earlier date.[28] They reached the far northeast, crossed the Bering Strait into North America, and swiftly spread south, colonising a diverse range of environments including the Amazon rainforest to reach Tierra del Fuego by at least 10,000 years ago.

The modern human colonisation of Europe has been documented in considerable detail. It is possible that there was at least one incursion before 50,000 years ago, represented by claimed modern human remains and artefacts at Mandrin Cave in France.[29] It was not until 41,000 years ago, however, that modern humans established themselves throughout Europe, their presence denoted by a material culture quite different from that of the resident Neanderthals: the new Europeans used tools from long blades, made extensive use of bone and ivory, wore beads and pendants, carved animal and human-like figurines from ivory and stone, and made flutes from hollow bird bones.

In my 1996 book *The Prehistory of the Mind*, I characterised this new material culture as reflecting cognitive fluidity: the ability to blend knowledge and ways of thinking about different entities of the world to devise new types of tools, personal ornaments and art objects. Beads and pendants became highly variable in their raw materials, shapes and colours, suggesting that they were intended to send specific social messages to specific types of people. The design of tools for hunting now integrated knowledge of raw materials with an understanding of the physiology and behaviour of the prey being hunted to create a series of specialised weapons. This was often expressed by carving animals into the tools themselves, such as an ibex depicted on the end of a spear thrower from the Ice Age site of Mas d'Azil, located in the Pyrenees where ibex was the targeted prey. Human and animal forms were sometimes blended into a single carved figurine or image painted on a cave wall, such as the 'lion man', a figure carved from mammoth ivory between 40,000 and 35,000 years ago, with the head of a lion and the body of a man. This all reflects a new way of thinking, one that enabled a degree of creativity and innovation never previously witnessed in humankind (Figure 3).

Africa also experienced a wholesale technological change after 40,000 years ago. As in Europe, blade technology became prominent, with small blades being chipped to form microliths, which were set into wooden or bone handles. New tool types emerged including arrow heads, fishing equipment and polished bone points. Beads made from marine shell and ostrich eggshell become widespread, along with engraved decoration on bone and wood.

The Neanderthals, Densiovans, *H. erectus* and *H. floresiensis* became extinct by or soon after 40,000 years ago, leaving *H. sapiens* as the sole remaining member of the *Homo* genus. When seeking to explain why that is the case, academics from many disciplines have suggested *H. sapiens* had an enhanced capability for language over that possessed by those species that went extinct – although without specifying what form that may have

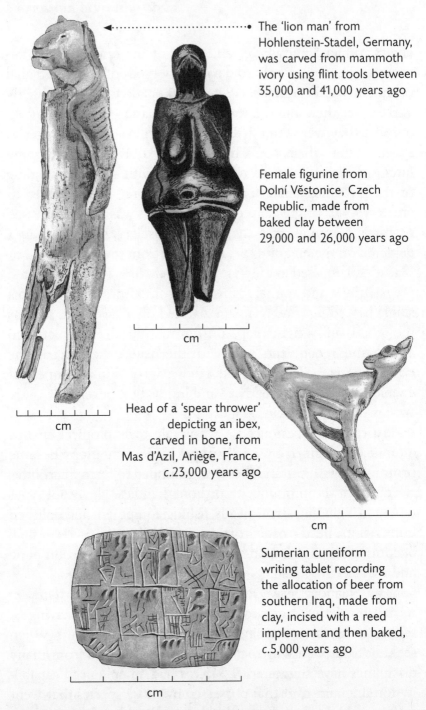

The 'lion man' from Hohlenstein-Stadel, Germany, was carved from mammoth ivory using flint tools between 35,000 and 41,000 years ago

Female figurine from Dolní Věstonice, Czech Republic, made from baked clay between 29,000 and 26,000 years ago

cm

Head of a 'spear thrower' depicting an ibex, carved in bone, from Mas d'Azil, Ariège, France, c.23,000 years ago

cm

cm

Sumerian cuneiform writing tablet recording the allocation of beer from southern Iraq, made from clay, incised with a reed implement and then baked, c.5,000 years ago

cm

Figure 3 Early art and writing by modern humans

taken. At this stage, we can confidently agree that by 40,000 years ago *H. sapiens* had language of the type we possess today – which I will call the 'fully modern language' capability. It is simply inconceivable that they could have painted caves, built boats and colonised the world without fully modern language, the nature of which will be dissected in Chapter 3. We cannot, however, yet deny the same fully modern language to the Neanderthals, Denisovans and others, or attribute them with any other type of language, until a detailed consideration of their anatomy, behaviour and culture has been undertaken in the following chapters.

From the height of the last glaciation to the end of the Stone Age

By 40,000 years ago, the global climate was heading towards the height of the last glaciation, which arrived at 20,000 years ago. Ice sheets expanded across high latitudes, causing sea levels to fall and so expose extensive coastal shelves. Low latitudes suffered aridity, causing forest and woodland to retreat. Human communities responded by relocating, adapting their technology, adjusting their diets and social lives, and most likely suffering considerable demographic decline. The most striking response was in Europe. New technology and hunting methods enabled the mass slaughter of migrating reindeer herds while investment in ritual, evident from the painting of cave walls, enabled resilience to the harsh, glacial conditions by intensifying social bonds within and between communities. In central and eastern Europe large dwellings were constructed from mammoth bones and tusks; symbols that bound far-flung communities together into social networks took the form of female figurines, either carved in ivory or bone, or baked in clay (Figure 3).[30]

Similar innovations were happening throughout the world, creating a level of cultural diversity never witnessed before in the history of humankind. Equally, humans were having an unprecedented environmental impact: throughout Europe, Asia, Australia and the Americas, megafauna such as mammoth and giant sloths became extinct. Climate change was a major factor but human activity, either from the hunting of such animals or by influencing habitat change, likely tipped the balance from population decline into extinction. Megafauna has survived only in tropical Asia and more notably in Africa where we can still see elephants, hippopotamus, rhinoceros and giraffe.

Although the modern human response to the most severe ice age conditions at 20,000 years ago and their immediate aftermath displayed a new degree of innovation, this paled in comparison with the cultural revolution that was to come. Following a period of marked climatic fluctuations, dramatic global warming occurred at 11,650 years ago. Temperatures rose by around 4°C in a matter of decades while atmospheric carbon dioxide increased by 50 per cent. Ice sheets melted, sea level rose, and landscapes were transformed as woodland spread and animal communities changed to those of warmer-adapted species. The Holocene began, a period of warmer, wetter and more stable climate within which the modern humans would flourish.

Human communities recolonised landscapes that had been lost to ice and extended into new regions, now entering the High Arctic and travelling to Pacific islands. They did so through a constant stream of innovation and culture change. The manufacture of small blades and microlithic tools became prevalent in many regions, these providing the most efficient use of stone. New technology was devised to collect and process the newly abundant plant foods, ranging from pottery vessels in eastern Asia to flint sickles and stone mortars in the west. Marine and coastal foods became prominent in the diet with a new range of fishing technology and the accumulation of huge shell middens in coastal regions throughout the world.

The invention of farming was of most significance for

human history. This first occurred in Southwest Asia where the intensive exploitation of wild cereals led to the evolution of domesticated strains that were as dependent on human harvesting as humans were on their regular supply of grain. Similarly, the hunting of wild goats was intensified, leading to the management of herds and the emergence of domesticated strains.

By 10,000 years ago, hunter-gatherers in Southwest Asia were living in permanent villages; they were soon reliant on domesticated plants and animals, becoming the first farmers. That lifestyle entailed a host of other innovations: new architecture made from stone, mud-brick and plaster; new technology; new social organisation for sedentary lifestyles; and new ideology, art and ritual. Populations began to grow and had to disperse into new lands, taking the farming lifestyle into Europe, North Africa and central Asia (Figure 4).

Much the same occurred in China, where rice and millet were domesticated by 10,000 years ago, leading to farming communities that spread throughout the east and into South Asia. Within a few thousand years domesticated plants and animals emerged in other regions of the world: beans, maize and peppers in Mesoamerica; taro and bananas in Highland New Guinea; quinoa, llamas and potatoes in South America. Hunting-and-gathering lifestyles soon became restricted to environments where farming could not be sustained, notably those of high aridity and within the thick forests of the Amazon, West Africa and Southeast Asia.

The earliest farming communities are designated as Neolithic – the New Stone Age. Other than pottery, they remained reliant on the same raw materials that humans had always used, notably stone and wood, even if they were now able to manipulate and transform these in entirely new ways. But the emergence of farming foreshadowed the inevitable end of the Stone Age. Villages soon became towns and then urban communities connected by networks of trade. The means to smelt

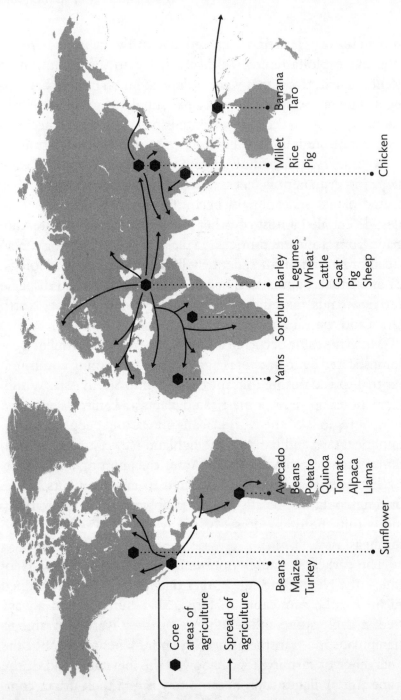

Figure 4 The origin of key domesticates and the spread of farming

copper was discovered, rapidly leading to bronze and then iron to provide the tools for work and warfare. Social hierarchies emerged with a constant thirst for prestige items and new forms of wealth, both supplied by ornaments of silver and gold. Population growth, technological innovation, economic change and social competition coalesced into the early civilisations of Mesopotamia, China and Mesoamerica. Within these a further step in the evolution of language occurred: the invention of writing.

The earliest writing took the form of marks imprinted onto clay tablets known as the cuneiform script of the Mesopotamian civilisation (Figure 3). The marks began at $c.5,500$ years ago as iconic signs known as phonograms and gradually became more abstract to represent the sounds of speech. That is the first definitive proof for the presence of a language capacity equivalent to that found in the modern world. Writing was independently invented in China and Mesoamerica, indicating the linguistic capacity was a feature of *Homo sapiens* throughout the world.

Six million years of language evolution

We have swiftly moved through 6 million years of human evolution, from the time when our ancestors used vocalisations comparable to those of a chimpanzee today to the use of fully modern language by 40,000 years ago, and potentially much earlier. Throughout those 6 million years there were changes in anatomy, brain size, life course, technology, diet, behaviour and geographical distribution. It would be perverse to think that vocal and cognitive capabilities did not also change and hence we should expect a gradual evolution of the present-day language capacity. Whether that was at a steady or intermittent pace of change, whether words and the rules evolved together or consecutively, and when we might wish to designate vocal communication as having crossed a threshold of complexity to become language, of a fully modern type or otherwise, cannot yet be specified.

To answer those questions, we need to find and assemble more fragments of the language puzzle. In this chapter we have noted the particular need to consider the linguistic implications of primate vocalisation, the evolution of the human vocal tract and brain size, stone tool technology, the control of fire, and the appearance of visual symbols (which will be covered in Chapters 4, 5, 7, 10, 11 and 15). Before assembling those fragments, we need to complete the jigsaw frame by defining and dissecting what we mean by 'fully modern language'.

3

WORDS AND LANGUAGE

I have used the term 'fully modern language' for the type of language we use today and which we can be confident was spoken by the modern humans who were colonising the globe by 40,000 years ago. Fully modern language may have been present long before, used by Neanderthals, *Homo erectus* and other ancestors. We will find that out only when all the fragments of the language puzzle have been assembled and joined together. Having assembled the first of these to provide the timescale, species, cultures and climatic framework for human evolution, the next fragment must define what we are trying to explain: what is 'fully modern language'?

Fully modern language is the composition of words into meaningful utterances by using rules to modify and arrange them into a particular order. The utterances can be either spoken, signed or written as sentences. Because the meaning of an utterance depends on both the meaning of the individual words and how they are arranged, fully modern language is described as having compositionality. It is this which delivers the versatility and power of language, the ability to express an infinite number of meanings from a finite number of words. Any form of linguistic expression requires a combination of motor actions and mental processes to embed meaning into either the sounds, signs or marks that others will see or hear. This definition begs two questions: what are words, and what are the rules? Adequate answers cannot rely on those from a single language but must encompass the linguistic diversity found within the *c.*7,000 languages still spoken in the world today.

Words

Words are the sounds we make that have meanings agreed by those who use the same language – our language community.[1] Spoken words can consist of a single sound, referred to as a phoneme, or sequences of different phonemes, some of which are combined to make larger units of sounds known as syllables, which are the building blocks of words.[2] Phonemes can be divided into vowels and consonants, these referring to the different ways in which the sounds are made by the passage of air through the vocal tract.[3]

The shape of the vocal tract and the ability to control that passage of air define the range of sounds that can be made. Those of the chimpanzee, and by implication of the last common ancestor, impose severe constraints, limiting not only the range but also the consistency of sounds. As such, the evolutionary history of the vocal tract through the *Homo* genus, from *H. habilis* to *H. sapiens*, provides an essential fragment of the language puzzle, one to be assembled in Chapter 5.

Words can be divided into two types: lexical and grammatical. Lexical words are those that have meanings, for which there are four classes in English: nouns, verbs, adjectives and adverbs. There are large numbers of these words, and more can be introduced easily as need arises. Moreover, any one of them can have two or more meanings. By 'meanings' I mean they are associated with concepts stored in the brain. The relationship between words, concepts and thought is a further fragment of the language puzzle, one that will be assembled in Chapter 14.

Grammatical words, sometimes called empty words, are articles, pronouns, prepositions and conjunctions – the 'a's, 'the's, 'she's, 'it's, 'in's, 'at's, 'but's and 'and's of language. These are fewer in number than lexical words and constitute a closed class because new members cannot be introduced easily. They have no identifiable meanings by themselves but help define the meanings of lexical words and of sequences of words in a spoken, signed or written utterance. There is a fuzzy boundary between lexical and grammatical words, with some words falling between the two categories. Auxiliary verbs, for instance, are those which provide additional meaning to a main verb, as in *You <u>are</u> reading this book* and *I <u>have</u> excavated a Neanderthal fossil.*

Lexical words can be ranked on a continuum from those that are more concrete to those that are more abstract.[4] Concrete words refer to things or actions that can be experienced directly through one or more of the five senses. These can be nouns (e.g. *banana, zebra* and *foot*), verbs (*jump, fly*) or adjectives (*red, coarse*). We can learn about the meaning of such words and demonstrate our knowledge about them from entities in the world, by pointing to the picture of a zebra or experiencing the colour of a banana. One important distinction within nouns is between what we call common names for general categories of things, such as *women* and *men*, or *mummy* and *daddy*, and proper names for specific things and often individuals, such as Lauren and Nick (who might also be a mummy and daddy).

Abstract words are those that refer to entities that cannot be experienced directly – their meanings are only defined by other words. The words *freedom, justice* and *democracy*, for instance, relate to concepts that have no sensory manifestation – they can be neither seen, heard, smelt, touched nor tasted. Such words can only be defined, and we can only demonstrate our understanding of them, by using other words.[5] To define *fun*, for example, one needs to invoke words such as *wonderful, enjoyable* and *laughing* or describe funny events. Perhaps surprisingly, most words, whether nouns, verbs or adjectives, are towards the abstract end of the spectrum.

Words have single or multiple parts, referred to as morphemes. *Walk*, for instance, is a 'root' morpheme, otherwise known as a free lexical morpheme, because it represents a complete word. Its meaning can be adjusted by adding a grammatical morpheme, referred to as an inflection or an affix, such as *-ed* or *-ing*. Free grammatical morphemes are grammatical words such as *of*, *to* or *the*. Bound lexical morphemes are affixes that have been used to create another word, such as '-ment' in the word *amazement*. A single word can have several morphemes, each tweaking its meaning, and combined according to rules of the language to which the word belongs, referred to as its morphology. The word *nationalised*, for instance, consists of a free lexical morpheme, *national*, two bound lexical morphemes, *al* and *ise*, and a bound grammatical morpheme, *ed*. Languages vary: in English we change *walk* into *walking* by adding *ing*, whereas in French two changes are required to turn the equivalent *marche* into *en marchant*.

Walking means moving at a regular pace by lifting and setting down each foot in turn, never having both feet off the ground at once. The closest word in Welsh is *cerdded*, in German *gehen* and in Icelandic *gangadi*. I could go on, but this is sufficient to illustrate one of the key features of words – most of them have arbitrary meanings. There is nothing inherent to the combination of phonemes that constitute *walking* or *cerdded* that associates those words with that act of bipedal locomotion. The speakers of English and Welsh have simply 'agreed' to use these words as a matter of convention.

The arbitrariness of words has been described as one of the key design features of language, distinguishing human communication from that of other animals.[6] There are exceptions, most notably onomatopoeias, such as *plop*, *honk* and *hiss*.[7] Onomatopoeias are a type of 'iconic' or 'sound-symbolic' word. These types of words have phonemes that mimic either the sound, size, weight, movement or some other perceived quality of the object being referred to. Words for small things, for instance, are often made by turning the mouth into a small cavity, such as *bee* and *flea* in English, and vice versa, such as *hippopotamus* and *enormous*. Many words combine iconic and arbitrary sound segments and can be referred to as hybrid words. Iconic words have long been recognised by linguists but they were attributed with little significance until the last two decades. They are now identified as a key fragment of the language puzzle, as considered in Chapter 6.

Before moving on, we must note a class of word-like entities that fall outside the categories we have considered so far. These are expressions such as *tsk-tsk* to express disapproval, *oh-oh* when something goes wrong, *shh!* to ask for silence, and *oops* when a mistake happens.[8] They are used as stand-alone exclamations and have a wide range of variation in how they sound and no one knows how they should be spelt. They do, however, have an agreed meaning that must be learned and can be used for communication, although may not require a response. A further class of vocal expressions is even more distant from words because they are instinctive and we have little if any control over them: sobs, screams, chuckles and laughs. As far as we know, these are universally found within all human communities, whatever language they speak.

Rules . . . or guidelines

Although words can be used by themselves, such as when we shout *Stop!* to a child who is about to encounter a fierce dog, they are more often combined into sequences with other words, such as *the fierce dog bit the child*. The meaning of this sentence derives from the meanings of its component words and the order in which they come – the principle of compositionality. The rules by which words are ordered are referred to as syntax, which together with the rules of word morphology constitute the grammar of a language. The same words can be placed into a different order to create a different meaning, such as *the fierce child bit the dog*. Just as languages have different words and rules of morphology, they also have different rules of syntax. In English we always place the adjective before the noun – *fierce dog* – whereas in French it can be the other way round – *chien féroce*. Such variance in the rules of word order, along with those of word morphology, add to the challenge of learning a new language as an adult.

Word order is the simplest form of syntax. The next level of complexity is the use of short strings of words as single phrases, which are combined to create what linguists refer to as a hierarchical phrase structure. This often involves combining what is known as a noun phrase (NP) with a verb phrase (VP). For instance, the sentence/utterance *The dog loved the girl* consists of the NP *The dog* and the VP *loved the girl*. These phrases can then be manipulated within an utterance as if they were a single word.[9] They can be embedded within another phrase as a single unit, such as *By watching them play, I saw that the dog loved the girl who gave him biscuits*. This itself can be embedded within a phrase, creating another level of a hierarchy: *Yesterday, by watching them play, I saw that the dog loved the girl who gave him biscuits, and that made me happy*. Languages may include a special type of hierarchical phrase structure known as recursion. This involves embedding a phrase within another phrase of the same type. For instance, *The dog with the fluffy tail, that had a red collar, and that liked biscuits, loved the girl*.

In principle, an infinite number of phrases could be embedded into a single sentence/utterance. They are easier to follow in a written sentence than in a spoken utterance because that avoids relying on memory alone to keep track of the meaning. Indeed, the rules of language, whether those of morphology or syntax, are more easily and frequently followed in writing than in speaking, for which they sometimes appear little more than guidelines. It is usually easier to understand the meaning of an ungrammatical spoken utterance than of an ungrammatically written sentence, especially if we are within sight of the person speaking and it is someone who we know. That is because we can draw on the context in which the spoken utterance is made, such as what the speaker is looking at or holding, or what we know about the speaker's interests and feelings. With plenty of situational information, strings of words that fail to follow any of the rules of language can be full of meaning. The way words are spoken in terms of their intonation can also be indicative of their meaning. This is known as prosody.

Prosody

Prosody is sometimes called the musicality of language because it refers to the use of intonation, stress, rhythm, pauses and tempo to influence the meaning of an utterance. This is achieved by changing the pitch, length or loudness of either a syllable, a word or the whole utterance. Simply by adding short pauses, a speaker enables a listener to process and sometimes disentangle what has been said and to grasp its meaning before the next flow of words begins.[10]

In English, an utterance with a rising pitch asks a question and one with a falling pitch makes a statement, even though the words used and their order are exactly the same. For instance, the phrase *Neanderthals had fully modern language* can be spoken as either a question or a statement depending on the intonation; at this stage in the book please read it with a rising pitch in your voice. One might also utter this phrase using a mix of rising and falling pitch to express incredulity at the idea that Neanderthals had fully modern language.

The meaning of an utterance can be changed by shifting the stress from one word to another. If I were to say *The Neanderthals didn't have language* with a stress on the word *Neanderthals*, such as by making it higher pitched, louder or longer, that would imply that another human species did have language. Alternatively, if I stress the word *language* that would imply the Neanderthals had something else other than language.

Prosody can not only adjust the meaning of an utterance, but also completely reverse it. When I say *clever Neanderthals* in my normal voice, I mean they had high levels of intelligence. If I were to say *clever Neanderthals* with a lengthening of the words and speaking more loudly than usual, I would be implying they were in fact rather stupid by using irony. This example illustrates an important aspect of prosody: to draw on its use in someone's speech, one needs to know their usual range of intonation when speaking so that any difference can be appreciated. Prosody can also be expressed through gesture – the use of hands, arms, facial expression and body posture. These can also help convey, influence or radically change the meaning of a spoken utterance.

Language acquisition: Universal Grammar and its demise

As evident from my brief summary, English is complicated: thousands of words with different meanings, coming in several cross-cutting categories and used for different purposes, with rules for how words are modified and ordered to generate meaningful utterances, which must be interpreted by attending to context, prosody and gesture. All the *c.*7,000 known languages have a similar level of complexity. Not surprisingly, one of the most enduring and puzzling questions for linguists has been how children acquire the words and rules of the language community they happen to be born within. They do this from scratch, but so quickly and effortlessly that they become chatterboxes by the age of three. Put the same newborn baby in Chelmsford (England), Chonqing (China) or Chennai (India) and by three years old the toddler will likely be speaking English, Mandarin or Tamil respectively (recognising that other languages might also be acquired in these cities, especially in Chennai).[11] How is that possible?

The most prominent idea was proposed by Noam Chomsky in the 1950s and developed throughout his long academic career, which at the age of ninety-four is continuing today (in 2023).[12] Chomsky argued that children could not possibly learn language by simply listening to speech because that provides them with insufficient information in the available time. He referred to this as the 'poverty of the stimulus', reflecting the view expressed by the nineteenth-century psychologist William James who had described the infants' environment as 'blooming, buzzing confusion'.[13] With no more than chaotic noise as input for language, Chomsky argued that infants must be aided by a genetically endowed capacity which he termed 'Universal Grammar'. This relies on the idea that all languages are so similar that children simply need a set of specialised mental tools to extract the specific grammatical rules of their own language from the utterances they hear, such as whether to place the adjective before or after the noun.

Chomsky's idea of Universal Grammar (UG) was compelling. It seemed to resolve the otherwise impossible feat of language acquisition and came to dominate research in linguistics and influence many areas of cognitive science. UG became prominent in proposals for how language evolved because it lends itself to the idea of a language capacity being constructed by natural selection, shaped by a sequence of genetic mutations. As such, UG has long been viewed as a critical fragment of the language puzzle.

Having had some critics from the start, concerns about UG have gradually increased, although it still has ardent supporters. The influential linguist Steven Pinker remains a strong advocate and describes UG as a language-learning toolkit.[14] Unfortunately, linguists are unable to agree about the specific tools within the kit, where they come from and how they operate. Not surprisingly, they have had even more difficulty finding evidence for UG's existence.[15]

A language-learning toolkit implies one or more dedicated areas in the brain for language, supported by language-specific genes. Despite the incredible advances in neuroscience and genetics during the last few decades, neither of these have been found. The capacity for language is now known to be widely distributed throughout the brain and reliant on a multitude of genes, each contributing to several and probably a multitude of cognitive and physiological processes. These findings challenge the idea of Universal Grammar, while providing new pieces for the language puzzle, to be described in Chapters 11 and 12.

A further challenge to Chomsky, Pinker and Universal Grammar has come from our new understanding about how children acquire language. Rather than being poverty stricken in terms of information about language as Chomsky had argued, the blooming and buzzing environments of babies, toddlers and children are now known to provide an abundance of clues about the possible meanings of words and the rules by which they are combined to create meaningful utterances. Not only that, but even babies' brains can extract those meanings and rules by using general-purpose learning processes, a feat that was unimaginable to Chomsky working in the 1950s. This new understanding provides another key fragment of the language puzzle, one for Chapter 9. Moreover, by studying how language is passed down from generation to generation, linguists have discovered that syntax can emerge spontaneously – another new fragment of the puzzle, one to be assembled in Chapter 8.

As our knowledge of language acquisition progressed during the last two decades, so too did that of linguistic diversity – the myriad and fundamental ways in which languages differ from each other. That also challenged Universal Grammar because Chomsky and his advocates had drawn primarily on English, French and other European languages of Proto-European descent. When the quite different types of languages coming from Asia, Africa and Australasia are added to the mix, the range of variation becomes far too great for a single language-learning toolkit to cope with. The linguists Nicholas Evans and Stephen Levinson brought the conflict between Universal Grammar and linguistic diversity to a climax in their seminal 2009 publication 'The myth of language universals'.[16]

The extent of linguistic diversity in the world today contributes a major part of the frame of the language puzzle. When this is combined with our new knowledge about language transmission between generations, how children learn language, the brain and the genetics of language, there is neither room nor need for the idea of Universal Grammar. It is unlikely to exist.

Languages and their loss

There are around 7,000 languages in the world.[17] No one knows for sure the exact number – it is difficult to measure because a language to one linguist can be a dialect to another. The most widely spoken are English and Mandarin, each with over a billion speakers, followed by Spanish and Hindi with over half a billion, and then French, Arabic, Bengali, Russian and Portuguese.[18] The majority of the world's population are either bilingual or multilingual. As the linguist Viorica Marian explains in her 2023 book *The Power of Language*, in many countries of the world children grow up with two languages and acquire additional languages as adults. People living in countries whose national language does not extend beyond its borders, such as Norwegians and Estonians, are frequently bilingual or multilingual. Almost two thirds of the entire population of the EU speak at least two languages. Over one fifth of people in the United States speak a language other than English at home, with this approaching half the number of households in the larger cities.

Languages are disappearing at an alarming rate. Currently, 96 per cent of the global population speak a mere 4 per cent of the

7,000 languages – that is about 7.7 billion people speaking about 280 languages. A quarter of existing languages, about 1,750 of them, have fewer than 1,000 speakers. Some have no more than a dozen speakers and others may have a lone survivor. That was the reported situation in 2020 for Resígaro, a language spoken in the Amazon by the Arawak people of Peru.[19] Present trends suggest that 90 per cent of the existing languages will have disappeared by the end of the century.[20] This is serious because only around 500 of our known languages have been fully documented; we risk losing a huge swathe of linguistic and hence cultural diversity without even knowing what it contains. Much of this has already been lost: half a million languages may have emerged, flourished and become extinct since people began to talk.[21]

These figures leave the 500 documented languages as a tiny sample on which to base any understanding of 'fully modern language'. Even within this, however, the range of variation is remarkable, described as a 'linguistic jungle' by Nicholas Evans and Stephen Levinson. The extent of linguistic diversity must be fully appreciated when just one language, English, has dominated the study of language evolution. When not English, other European languages are most prominent, notably French, Spanish, Dutch and German. The study of language can sometimes appear as if European languages, notably English, are the evolutionary pinnacle of language development to which Neanderthals and other human ancestors were striving to attain and against which their linguistic capabilities should be measured. We need to rid ourselves of that view, just as we have done with the nineteenth-century belief that European culture was the pinnacle of civilisation, with people of other regions and cultures dismissed as barbarians and savages.

Linguistic diversity

While all languages have words and rules, and the same function of communicating thoughts and information, they are tremendously diverse in their range of sounds, numbers of words and ways of arranging them together.

To start, we can simply note that languages have different numbers of words for the same type of things: people like to talk about the same things in different ways and with differing levels of detail. Take colour as an example. Some languages have many colour words, and others just a few. English has eleven: black, white, red, green, yellow, blue, pink, grey, brown, orange and purple – and then a myriad of infrequently used words such as vermillion, ochre and azure. The Papua New Guinean language Berinmo has only five colour terms, and the Amazonian language Tsimané only three, corresponding to black, white and red. The Pirahã language categorises colours only as 'light' and 'dark'.[22]

The same applies for number words. In English we have words for any quantity of objects – we can count as high as we like. This is because we have a recursive system of numbering, enabling us to generate labels for precise quantities of whatever amounts we wish. Many other languages are restricted. The most extreme are the Amazonian languages of Pirahã (which lacks numbers beyond 1 and 2, and possibly even for those amounts) and Mundurukú (which has numbers up to 4, although only uses the first three consistently).[23] Other languages rely on words that express approximate or inexact numbers, equivalent to 'few' or 'many', sometimes combining these with words for exact numbers of small quantities. Some languages have object-specific counting systems. The Austronesian language of Takuu uses different words when counting humans, fish, canoes and coconuts.

The same exercise can be used for all other entities in the world: languages vary in the number and types of words they have for, among other things, body parts, kin relations, time, emotions, common actions and basic geometric shapes.[24]

Languages use different ranges and types of sounds. Some have no sounds at all – the sign languages used by the deaf. There are around 300 documented sign languages and dialects in the world. They have different sets of signs to form their words and grammar – some have more and some have fewer signs, just as spoken languages vary in their numbers of words and sounds.[25]

The number of sounds (phonemes) within spoken languages varies from around a dozen to almost 150 – the precise number depending on how sounds are distinguished from each other.[26] English has forty-four, twenty of which are vowels and twenty-four consonants. The Taa language, spoken by people in Botswana, has 144 sounds, involving a range of clicks, choking sounds and different tones; the Rotokas language spoken on the island of Bougainville, Papua New Guinea, makes do with a mere eleven.[27] An analysis of 317 languages found a total of 558 different consonants and 210 vowels. Only one type of consonant was common to them all – the stop consonant. This is made by closure of the vocal tract, using either the lips as in /b/, front of the tongue as in /d/, or back of the tongue as in /g/ – but none of these variants are universal. Similarly, no single vowel occurs in all languages.[28]

Languages with fewer sounds tend to have longer words. Hawaiian, for instance, has between thirteen and thirty-three sounds (depending how they are counted), and long words, such as *aloha kakahiaka* for good morning, pronounced *a-lo-ha kah-kah-hee-yah-kah* and *aloha ʻauinalā* for good afternoon, *a-loh-ha ah-wee-na-lah*. When many sounds are available, there is a greater opportunity to use short words to refer to the objects, actions and ideas that need talking about; with few sounds, one soon runs out of sufficiently distinctive short words, resulting in long, multisyllabic words as found in Hawaiian.[29]

The extent and character of prosody varies between

languages, with Japanese and Mandarin being notably 'musical'.[30] Prosody takes an exaggerated form in tonal languages such as Vietnamese and Thai from Southeast Asia and Yoruba from West Africa. Within such languages, the same word can change its meaning according to the tone with which it is pronounced. In Mandarin Chinese, for instance, the sound *ma* can mean either mother, hemp, horse or scold, depending on the tone (or tones) in which it is spoken. Different languages also make use of different gestures, some conventionalised with specific meanings, others generalised but often iconic, such as wide eyes to emphasise surprise and a clenched fist for anger.[31]

Diversity in rules of morphology, word classes and syntax

The rules of word morphology also greatly vary between languages. The words of some languages, such as Vietnamese, are primarily restricted to a single morpheme. These are called isolating languages. When their speakers wish to make a plural or change the tense, rather than adding another morpheme as an affix, as we do in English, they either add an entirely separate word or simply rely on context. English has a moderate use of affixes – its words are never very long – and it is called a synthetic language. A third category of languages is referred to as polysynthetic. These convey information through inflectional morphology by adding a succession of affixes, sometimes described as packing whole English sentences into a single word.[32] In Cayuga, for instance, a language spoken by Northern Iroquois Native Americans, the word *Ęskakhelona'tayęthwahs* means 'I will plant potatoes for them again', while in the Australian language of Bininj Gun-wok the single word *abanyawoihwarrgahmarneganjginjeng* means 'I cooked the wrong meat for them again.'

Polysynthetic languages are sometimes described as being over-specified because they seem to provide excessive information.[33] Yagua, a language of Peru, has inflections that differentiate five levels of remoteness. A verb used in the past tense must take a suffix that denotes whether the event happened a few hours ago, on the previous day, a week, a month, or even longer ago. In English we have the option of not referring to how long ago in the past an event occurred; if we wish to, we add additional words to our utterance rather than adding an affix to the verb.

Word classes offer a similar level of diversity. It was once assumed that all languages would have the four major word classes of nouns, verbs, adjectives and adverbs. Many of them do, but they vary in their numbers – some languages have thousands of verbs and others no more than thirty. Others lack one or more of these four word classes.[34] Galela, a Papuan language spoken in Indonesia, has verbs and nouns but no adjectives, whereas Wambon, from the same region, has adjectives but no adverbs. Ngiti, a language from central Africa, combines adverbs and adjectives into a single word class. Neither nouns nor verbs are sacrosanct. In Samoan, as occasionally in English, the same word can be a noun, a verb or an adjective depending on context. Ngiyambaa, an Australian language, has verbs but all other words can be used in different ways.

Word classes exist that are not found in English.[35] Some languages have 'ideophones', described by Nicholas Evans and Stephen Levinson as words that 'encode cross-modal perceptual properties . . . [that] depict the sight, sound, smell, or feeling of situations in which the event and its participants are all rolled together into an undissected gestalt'. They give the lovely example of *ribuy-tibuy* from the Mundari language of India which means 'the sound, sight or motion of a fat person's buttocks rubbing together as they walk'. Ideophones are complex forms of iconic words that will be considered further in Chapter 6. Positionals are another type of word class, one found within Mayan languages but neither in English nor in most other languages. These words describe both the position and form of persons and objects, such as the word *latz'al* which means 'of flat items, arranged in vertical stack'.

While word order is frequently critical to the meaning of an utterance, some languages have no such constraint. In Jiwarli and other Australian languages, words are tagged with suffixes to indicate how they relate to other words in the utterance, such as being the object of a verb. Words can be spoken in any order to produce utterances with the same meaning. When interviewed, speakers of Australian languages such as Warlpiri, seemed unable to repeat a sentence using the same order of words. Such free-word-order languages do not embed one phrase within another to make complex statements, the device known as hierarchical phrase structure. Nevertheless, such languages can express complex propositions by other means, such as by placing greater emphasis on context – who is speaking, how the words are expressed, attention to body language and gesture.

Languages, dialects and idiolects

The approximate figure of 7,000 languages fails to convey the extent of linguistic diversity. There are, for instance, 150 dialects of English around the world. More than 19,000 languages and dialects are spoken in India alone.[36] Within any dialect cluster, one will have emerged as the standard and is often used to denote the language. That is normally the dialect that had been spoken at the centre of political and economic power and may have been deliberately propagated to create national unity. As such, the notion of a language is often closely associated with the idea of a nation state: it has been said that a language is a dialect with an army and navy.[37] Had it not been for political borders, Spanish and Portuguese might be considered as mere dialects of a single language, as might Norwegian, Swedish and Danish. Conversely, some languages, such as Mandarin, contain dialects that could easily be considered as separate languages.

Why stop at dialects? Even in my small circle of English-speaking friends – all of whom would be classed as speaking the same dialect of 'standard English' – there are significant differences in their vocabulary, pronunciations and use of grammar. An individual's unique use of language is referred to as their 'idiolect'. With a global population of 8 billion people, that makes many more than a mere 8 billion languages, because we each have several idiolects, adjusting the way we speak – the words we use and how we pronounce them – according to who we are speaking with. Moreover, many people are bilingual and some multilingual. I suspect both conditions were more prevalent in the past than they are today. In fact, we have only touched the tip of a linguistic-diversity iceberg because languages, dialects and idiolects are continuously changing through time. Indeed, it is said that we reinvent language every time we speak.[38]

The causes and constraints on linguistic diversity

Why does 'fully modern language' show such variation in so many ways? Why do we find similarities between languages that are spoken in distant parts of the world? Why are languages not even more diverse in the types of sounds, words and syntax they use?

One reason is because any language is constrained by the anatomical and cognitive features that everyone shares by virtue of being a member of the same species, *Homo sapiens*. There are only so many sounds that can be generated by the vocal tract and distinguished by the auditory system we possess: the 144 sounds of the Taa language in Botswana are likely to be close to the upper limit; languages with fewer than ten sounds would struggle to communicate effectively and may be unfeasible. While stops were the only type of consonant found in each of 317 analysed languages, other types of consonants referred to as nasals, fricatives, liquids and glides were found in over 90 per cent of those languages. Similarly, although there is no universal vowel, over 80 per cent of those 317 languages used /i/, /u/ and /a/.

Similarly, the size, shape and flexibility of our hands, face and other physical attributes constrain the number of discrete signs and facial expressions that can be made for sign language. Our brains also impose constraints: there are limits on the number of words, clauses and especially embedded clauses that we can keep track of within a single spoken utterance, whether speaking it or hearing it from someone else.[39]

That constraint is relaxed when language is expressed by writing. Sentences can be written with a large and potentially infinite number of embedded clauses. These can be understood by others because when reading we do not need to remember the whole sequence and order in which the clauses come, as we do when hearing a spoken utterance. This is just one way in which writing differs from speech.[40] Unfortunately, some linguists have confused the two by assuming the capacity for spoken language also needs to encompass those of writing and reading, such as the ability to form sentences with an infinite number of embedded clauses. That is like arguing that to explain the evolution of counting we also need to account for our ability to undertake differential calculus.[41]

Noting that spoken language is constrained by anatomical structures and cognitive capabilities does not negate the variability within these and their impact on linguistic diversity. We all differ in our physical and mental properties, and this may cause us to speak and hear in slightly different ways. Among the many genes we possess that influence brain growth and development are *ASPM* and *MCPH*. These come in variants because of random mutations that have occurred at various times in the past and which have spread to some but not all populations in the world. Communities that speak tonal languages tend to have relatively low frequencies of two variants (or alleles) of these genes, known as *ASPM-D* and *MCPH-D*.[42]

Individuals and populations also vary in the morphology of their vocal tracts. Minor differences can lead to variation in the manner by which sounds such as /r/ are formed, which makes some people more receptive to sound change than others.[43] People who speak the Khoisan languages of eastern and southern Africa tend to have a relatively diminutive bony ridge just behind the teeth on the upper and lower jaw compared with other people, this being known as the alveolar ridge. Khoisan languages are the only languages in the world that use clicks as a consonant – made by a sucking motion to produce either a sharp popping or smacking sound between the tongue and the roof of the mouth or a sucking sound between the lips (the kiss click) or teeth or at the side of the mouth. A small alveolar ridge reduces the effort in making clicks.[44]

The evolution of languages

The diversity of languages that are encompassed within the category of 'fully modern language' is normally accounted for by their evolutionary history, one that can be partially identified via historical records. The standard account is that there was once a single language – although when, why and how that arose is not addressed. As people conversed, errors arose: words were mispronounced, misheard and misunderstood. This introduced random variation into the language, just as genetic mutations introduce random variation into our DNA. Some of the language mistakes caught on because, by chance, they improved the efficiency of communication – they either reduced the effort required in speaking or conveyed information in a more effective manner. Such mistakes were adopted and spread through the population to become an accepted way of speaking, in an equivalent manner to how advantageous genetic mutations are spread by natural selection. Other mistakes may have been sustained not because they were useful but simply by chance, the linguistic equivalent of genetic drift. As the population increased, dispersed and fragmented into separate speech communities, each generated their own suite of linguistic mistakes, some of which became fixed. Languages began to drift apart in the sounds and words they used, how the words were combined and what they meant; dialects developed; some became so different from each other that they became mutually unintelligible – separate languages.[45]

This process for how languages evolved accounts for many of the similarities and differences in the languages of the world today. Similarities are found when languages have descended from a common ancestral language; the differences are from relentless change as people kept on talking and listening to each other, combined with population dispersals and ultimately the isolation of speech communities.

Although this is a standard account for linguistic diversity, it needs refining. For a start, the analogy between linguistic mistakes and random genetic mutations is problematic. Unlike genetic mutations some linguistic mistakes are more likely to arise than others because of the way our vocal and auditory tracts are designed. We are, for instance, more likely to mistakenly say the vowel /i/ for /ee/ than for /o/, because both /i/ and /ee/ are made when the tongue is high and at the front of the mouth, whereas the tongue is low and towards the back when saying /o/. People with a relatively small alveolar ridge are more likely to make a click sound by chance when speaking, which might then become adopted as part of a word. Similarly, those without the *ASPM-D* and *MSCPH-D* genetic variants are more likely to be sensitive to tonal variation in speech and may 'mistakenly' associate a meaning with the specific tone with which a word is spoken.

Moreover, some linguistic novelties can be intentionally devised, making them quite unlike genetic mutations. People deliberately invent new words and may consciously change their way of speaking to forge their social and cultural identity, even if this involves more rather than less effort and makes their utterances more difficult to understand.

Despite these concerns with the standard account, it largely explains how languages have changed and diversified through time, enabling us to reconstruct their evolutionary histories. Just as we trace the evolution of living things by grouping them into families and attributing common ancestors, we can do the same with languages. When two languages have similar words and grammatical constructions, they are likely to have derived from the same ancestral language.

More than 400 languages in the world today are termed 'Indo-European' because they all derive from a single and now extinct language estimated to have been spoken 6,500 years ago in the Bronze Age (4500 BC, Figure 5). We call this extinct language Proto-Indo-European and it is thought to have been spoken in the region now covered by Ukraine and southern Russia.[46] The Bronze Age people dispersed both east and west. As they did so and as they formed separate speech communities, their language evolved by the process of 'mistakes' and selection, while also being influenced by the languages of the people they came into contact with. Today the Indo-European language family includes most European languages, as well as some found in northern India and the Iranian plateau, such as Bengali, Farsi and Kurdish.

Uralic is another Eurasian language family, although much smaller than Indo-European with just thirty-eight languages. These include Hungarian, Finnish, Estonian and several languages with relatively few speakers such as Mari, Sami and Vepsian, all spoken in northern Eurasia. The largest language families are Austronesian, with an estimated 1,300 languages that are spoken in communities as far apart as Madagascar and Taiwan, and the Niger-Congo family with about 1,500 languages spoken over the majority of sub-Saharan Africa.

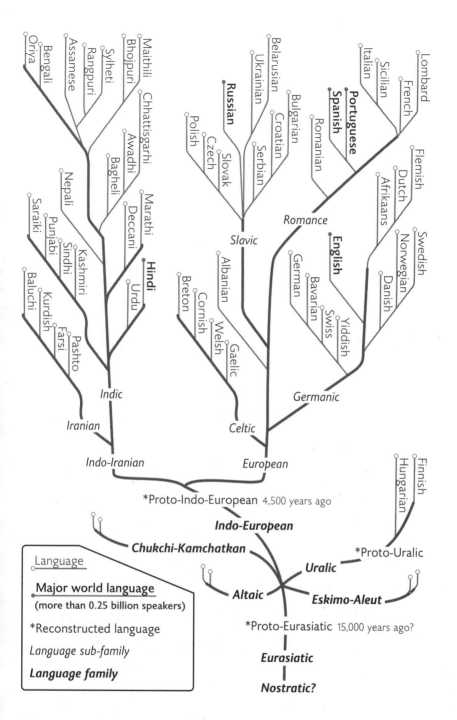

Figure 5 The evolution of Indo-European languages

Indo-European and Uralic have been grouped with three other Eurasian language families into a super-family, known as Eurasiatic. This is based on common words which are described as ultra-conserved because they have changed so little over time – words that are expressed as *mother, bark, ashes* and *worm* in spoken English today – although this is a heavily contested issue.[47] Proto-Eurasiatic, the ancestral language to all those languages within the Eurasiatic super-family, is claimed to have been spoken about 15,000 years ago, towards the end of the last ice age. Eurasiatic, in turn, has been grouped with other super-families to create a mega-super-family called Nostratic.[48] Although there is consensus about the principle of reconstructing language families and we can be confident about reconstructing some aspects of ancestral languages spoken up to 8,000 years ago, many are sceptical that we can go further back in time, questioning the proposals for Eurasiatic and Nostratic.

Why do some words and rules survive and others disappear?

The standard account of linguistic diversity by descent with modification has paid considerable attention to the generation of variability but given insufficient concern to why some variants become adopted and spread within a speech community while others fade away. This has led to a view that linguistic diversity is essentially random. As recently as 1989 a distinguished linguist stated: 'There is no correlation whatever between any aspect of linguistic structure and the environment. Studying the structure of a language reveals absolutely nothing about either the people who speak it or the physical environment in which they live.' This was reiterated in 2003 by another linguist: 'There is no ecological regularity in how the major linguistic types are distributed around the world.'[49] These views are incorrect: there are patterned regularities between the characteristics of languages and the social, economic and physical environments in which they are found.

In 1911 Franz Boas, the so-called father of anthropology, made the reasonable suggestion that the words people use 'must to a certain extent depend upon the chief interests of a people'.[50] The most discussed example is the word – or rather, words – for *snow*. Boas had noted that the Inuit (then known as Eskimo) have more words for snow than are found in English. Some anthropologists, committed to the idea of language as a free variable, not only questioned Boas' observation but also exaggerated his proposal to the point of ridicule leading to claims of the 'great Eskimo vocabulary hoax'.[51] Fortunately, common sense returned and a formal academic study confirmed Boas' view: the Inuit do indeed have words to make much finer distinctions between types of snow than are found in English.[52] This was further developed by a study that found that languages from warm, low-latitude regions often have a single word to refer to both ice and snow, while in colder climates these have their own words. Moreover, by examining messages from different climatic regions posted on Twitter between 2009 and 2013, it was demonstrated that people in colder climates do indeed talk about snow more often than those in warmer regions.[53] The same principle explains why the rice-farming Hanunóo people of the Philippines have many words for rice and the Scots likewise for rain.[54] How could it possibly be otherwise?

Climate can influence the frequencies of sounds within a language, irrespective of lifestyle: languages spoken in warmer climates have relatively more vowels and those in colder climates have relatively more consonants.[55] This has been explained by a higher frequency of outdoor communication in warmer climates where speakers and hearers are more often separated by greater distances than in the more frequent indoor settings of colder climates. Vowels have higher levels of sonority than consonants – they are louder and can be heard over greater distances.

Speakers in warm climates also use a higher frequency of sonorous consonants such as /r/, /l/ and /n/, whereas the most frequent consonants in cold climates are the relatively quiet fricatives and stops such as /s/, /z/ and /t/.[56]

By invoking the principle of least effort and assuming that sonorous sounds involve greater effort to make than non-sonorous sounds, the former will have less likelihood of arising and becoming selected into languages when conversation takes place indoors with speakers and hearers near each other. This is not unreasonably claimed to be a feature of cold-climate communities although we need further studies to verify that such environments do involve a greater frequency of indoor speaking. Moreover, other environmental factors need to be considered such as topography, vegetation cover and humidity, all of which have been claimed to influence the costs and benefits of talking with sonorous vowels and consonants.[57] Further factors also need to be explored. There are physiological differences between people whose families have lived for long periods in cold and warm climates: the former leads to larger body size to conserve heat (known as Bergmann's rule).[58] Might that influence the vocal tract and sounds that are made? My own personal experience is that my posture becomes more drawn into itself in cold weather than in warmth; does that change the sonority of the vowels and consonants I make?

Further lifestyle influences on vocabulary

As I noted when describing linguistic diversity, some languages have many colour terms and others just a few. If there are only two words for colours, they will be black and white (or a dark and light equivalent). The third term is always red, while blue is often missing or combined with green. Whatever the number of colour words people know, they are most adept at identifying and talking about warm colours – reds, oranges and yellows.

This is simply because most objects in the world are of these colours, while backgrounds tend to be the cool colours of blues and greens; language has adapted to people's needs. The number and range of colour terms is significantly greater in the languages of modern-day western cultures than among traditional hunter-gatherers. This is because craft activities associated with farming communities and industrial manufacture within urban societies have created objects with many more colours than are naturally found, such as by applying pigments and using dyes. The languages of such communities have adapted accordingly by increasing their number of colour terms, whether by deliberate word invention or via the unconscious creep of language change.[59]

Similarly with words for smell, although in this case the number of words has reduced with farming and urban lifestyles. Most people have greater difficulty finding words to describe odours than for colours. This is normally explained because vision became the more important sensory mechanism during human evolution after bipedalism elevated our noses away from the ground. While there is genetic evidence that this is partly the case – 60 per cent of olfactory receptor genes are functionally inactive in humans – lifestyle is also important.[60] A study compared the ability of Semaq Beri hunter-gatherers and Semelai swidden horticulturalists, both living in the Malaysian rainforest and speaking closely related languages, to name odours.[61] The hunter-gatherers were as adept at naming odours as they were with colours, but the horticulturalists struggled to come up with any suitable words for the odours. The finding has been repeated with other hunter-gatherers who have been found to be as proficient in naming odours as they are colours, in contrast to those of us in western industrialised cultures.[62]

Here (and everywhere else) we must be cautious to avoid imposing our own western categories such as odour and colour onto communities for whom these may not be entirely appropriate. The anthropologist Diana Young spent time with the Anangu hunter-gatherers of the Western Desert of Australia and found they have a strong correspondence between the colour green and the smell that is released when the first rains hit the desert floor.[63] Noting the deficiencies of her own words for odours, she described this smell as 'eucalyptus with a top note of dust and shit – perhaps dog, camel or human'. The Anangu use the words *panti wiru* to mean 'a good smell'. This smell is associated with the colour green, that of the sprouting plants that follow the rain and which attract game to hunt. One of the most favoured of those green plants is wild tobacco, *mingkulpa*. Its leaves are chewed to absorb the smell of rain and to keep the body moist. The leaves are described as *wanka*, a word that has multiple meanings of green, healthy, raw and alive. Young summarised her findings by arguing that the colour green and a particular odour are culturally constructed as inseparable – the same word can invoke both sensations. This is what we call a synaesthetic experience, one in which perception evokes more than one sense at the same time. As we will see in Chapter 6, such experiences may have played a key role in the invention of words.

The contrast between a hunting-and-gathering and a farming way of life also influences the frequency of sounds within a language. Labiodental sounds are made by positioning the lower lip against the upper teeth, including the fricatives /f/ and /v/ and affricatives /pf/ and /bv/. A study of languages throughout the world found that those of hunter-gatherers have about 27 per cent the number of labiodental sounds as found within the languages of food-producing communities.[64] This is explained by the consumption of harder foods by hunter-gatherers than by farmers and all of us who rely on cultivated foods. Staples such as wheat and rice are boiled and softened before consumption, whereas hunter-gatherers tend to chew and grind up less processed foods. This wears their teeth so that by adolescence the typical hunter-gatherer bite configuration has changed:

the vertical and horizontal overlap of teeth that we all develop when young, known as overbite and overjet, becomes replaced by an edge-to-edge bite. That makes labiodentals harder to say, requiring 30 per cent more muscular effort than with overbite and overjet. As such labiodentals are both more likely to arise by chance and become adopted within speech communities reliant on cultivated foods.

Social influences on vocabulary

Joining environment and lifestyle as an influence on language are social aspects of the speech community: its size, its contact with communities speaking other languages, and its extent of internal variability – whether it has craft specialists and a range of social classes. Two significant correlations have been found between population size and language characteristics. First, larger populations tend to have a greater number of sounds within their languages – a larger phoneme inventory.[65] This correlates with having relatively shorter words. Smaller populations tend to have fewer sounds resulting in long, multisyllabic words, as we saw for Hawaiian when considering linguistic diversity. Note that a correlation does not necessarily mean there is a causal relationship: many factors are likely to influence the phoneme inventory, and disentangling which of these are a cause and which a consequence is non-trivial.

Second, languages spoken by fewer people tend to have greater morphological complexity than those spoken by larger populations.[66] That means they add affixes to words to denote their case, tense, negation and so forth, rather than using separate grammatical words to do these tasks. This also adds to the length of words, which can become extremely long, complex and difficult for another adult to learn.

In general, languages spoken by small and relatively isolated populations tend to be more complex and have unusual

sounds and difficult sound combinations; they have highly specific lexical items and lack consistent rules for word order. By
having few rules, such languages are especially difficult to learn
by adults – although infants have no problem acquiring them as
their first language. Papua New Guinea provides many examples of such languages. Its current population of 10.3 million
people speak more than 830 different languages, some by no
more than a few dozen individuals. Almost every village has its
own distinctive language, each of which is quite different from
any other and can exhibit baffling complexities. Similarly for languages found within the Amazon rainforest.

Why are there such differences between languages spoken
in relatively small and large communities? Several factors have
been proposed. One is the extent of economic self-sufficiency.
In today's world, such languages tend to be spoken in tropical environments with little seasonal variability. That allows
for continuous food production throughout the year, enabling
small communities to be self-sufficient, requiring limited trade
and contact with each other.[67] As such, their languages evolve in
their own unique ways.

Another factor is the number of adult language learners within the speech community.[68] Infants and children learn
language in a fundamentally different way to adults who are
learning a second language. Children's brains are primed to learn
language without any conscious effort during what is termed a
critical period, one that ends around adolescence. Adult learners
need to teach themselves or be explicitly taught the meaning of
words and the rules used to put them together. This is especially
difficult when the new language is from a different family: an
English speaker will find it harder to learn Mandarin as a second
language than German. Larger speech communities are more
likely to have more second-language learners than smaller ones
are, often with a mix of indigenous ethnic groups each with their
own first language, immigrants, traders and travellers. The types

of second-language learners have changed throughout history, once including enslaved populations and now likely to include higher frequencies of tourists. This presence of second-language learners creates a tendency for the dominant spoken language(s) to develop features that facilitate adult language learning: shorter words which easily map onto objects and actions in the world, with regular rules for how such words are combined.

Another factor is the extent of labour division and social differentiation within a population.[69] When groups of people within a single speech community develop specialised skills, such as making pottery or working metal, or have differences in wealth and status, they will develop their own vocabulary and ways of speaking. This may either be as a deliberate strategy to forge their identity or simply out of need: potters need to talk about different types of clay and temper, whereas metalworkers discuss hammers and moulds. As part of a single speech community, however, these groups still need to converse. This will have the same impact as having second-language learners: a gradual shift to a more systematic language, one with shorter words and more rules.

The number of adult language learners and the number of specialised skills present, along with simply a greater number of people, each of whom will have their own idiosyncrasies of speaking and hearing, create a wider range of sources for infants to use when learning language.[70] Although parents and siblings are the primary source, children are constantly listening to, learning from and often copying anyone they can hear. In the diverse linguistic environment of a large population, infants will encounter a higher frequency of speech variants, some of which they will adopt, sometimes contrary to how their parents speak. This multi-sourced process of language learning enhances the speed of language change towards more efficient and effective communication – one with more sounds, shorter words and a greater consistency for how words can be combined to create additional levels of meaning.

The converse will happen in small communities where

infants have a limited number of sources from which to learn language. Moreover, language evolution in such communities may be prone to random change, analogous to genetic drift in biological evolution.[71] The chance death of members might reduce the overall number of sounds being used or words known within the speech community. When most conversation is with closely related people, complex and inefficient grammatical structures will accumulate, just as deleterious genes accumulate within inbred populations.

The linguists Alison Wray and George Grace characterised the differences between languages in small and large communities as being for *esoteric* and *exoteric* communication respectively.[72] By the former, they mean talking to members of one's own community among whom much information can be taken for granted, and hence does not need to be communicated by words. Such communities tend to have limited differences between its members regarding skills, knowledge and social status, and have limited contact with other speech communities. They will have few members who were not born into the community and did not acquire its language during infancy. Within such communities, languages will evolve to be highly distinctive and not easily learned by outsiders.

Exoteric communication is the converse: it is outward facing and facilitates contact with members of other communities for whom there will be limited overlaps in knowledge, people whom Wray and Grace describe as 'strangers'. To facilitate communication, information is encoded into words that are relatively short and combined using a consistent set of rules. This enables swift language learning by an adult outsider and makes the meaning of spoken utterances more transparent – guesswork can be effective to deduce the meaning of a word.

'Fully modern language'

The task of this chapter was to join the remaining edge pieces of the jigsaw to complete the frame of the language puzzle by defining and dissecting 'fully modern language'. We have found that words come in several different categories, the frequencies of which vary between languages. The most basic division is between lexical and grammatical words, those which have meanings and those used to combine words together into meaningful utterances. Within lexical words, we have noted the distinctions between concrete and abstract words, and between arbitrary and iconic words. As we considered words, three fragments of the language puzzle were identified that will need assembling: the influence of the vocal tract and auditory tracts on speaking and hearing; the distinction between arbitrary and iconic words; and the relationships between words, concepts and thought (to be addressed in Chapters 5, 6 and 14).

The long-held notion of Universal Grammar was questioned and found to be wanting. Its now vacant place in the language puzzle can be filled by four closely connected fragments: those concerning language transmission between generations; how children learn rather than automatically acquire language; how language is distributed in the brain; and its genetic basis (Chapters 8, 9, 11 and 12).

A further fragment identified in this chapter is perhaps the most significant: the ongoing process of language change, with its causes and constraints (Chapter 13). As we have seen, linguistic diversity exists because all known languages are continually changing by the process of cultural transmission within and between generations. Variation is constantly arising within any spoken language by verbal mistakes and accidents. The need for efficient and effective communication selects which of the variants will survive – influenced by the environment, lifestyle and population size of the speech community. The extent of linguistic diversity that can evolve, however, is constrained by the common anatomical and cognitive features of *H. sapiens*, notably the anatomy of the vocal tract and auditory pathway. The same processes of linguistic change have been documented since the time of the earliest recorded language, the written texts of ancient Mesopotamia. The same processes must have been going on long before, ever since the first words were spoken.

We have now completed the frame of the jigsaw puzzle of language. This and the previous chapter have identified the twelve fragments that we must assemble to fill its interior and reveal how language evolved. Where shall we start? The obvious place is at the beginning, or at least as near to the beginning as we can get. That is with the communication – vocal and other-wise – of chimpanzees, our closest living relative and our model for the last common ancestor of 6 million years ago.

4

MONKEYS AND APES

What can we learn about the evolution of language from monkeys and apes? Probably a great deal because we are so closely related. We shared a common ancestor with the two chimpanzee species, the common chimpanzee (*Pan troglodytes*) and the bonobo (*Pan paniscus*) between 8 and 6 million years ago (mya). Before that we shared ancestors with the gorilla (10 mya), orangutan (13 mya), gibbons (20 mya) and monkeys (40–30 mya). These may sound to be long periods of time ago but evolution works slowly, especially for something as complex as language.

Language draws on a wide array of anatomical features and cognitive processes that most likely evolved at different times and for different reasons before being pieced together to provide our capacity for words and grammar. We can gain some estimate when they appeared by the extent to which they are shared with other living species. The anatomy of the inner ear, for instance, is essentially the same in all mammals, which places its origin way back in time to around 200 million years ago. If we find similarities between chimpanzee calls and human words, such as the intentional sending of information, we can assume they were also present in our last common ancestor of 6 million years ago. If we also find it within monkeys, we can push the date of its appearance back to 40 million years ago.

We refer to such features of language as being ancestral because we inherited them from our ancestors. Features of language that have no trace within apes and monkeys must have evolved within the *Homo* lineage and are known as 'derived'. To date their appearance, we need to find evidence from the fossil and archaeological record.

A dilemma we face is that apes and monkeys have been evolving over the same length of time as *Homo sapiens*. If we find similarities between chimpanzee calls and human language, these may have arisen from the convergent evolution of independently derived features – evolution finding the same solution to a common problem of communication in both lineages long after the time of the last common ancestor. Equally, it is conceivable that the common ancestor had linguistic-like capabilities that became lost in the lineage leading to the chimpanzee but maintained in the *Homo* lineage, leading us to think mistakenly that they are derived rather than inherited from an ancestor. There is, therefore, a constant challenge in identifying which are ancestral and which are derived features of language and its cognitive prerequisites.

Being aware of these challenges, I will argue that some pieces of the language puzzle do indeed reside within the minds, calls and actions of our present-day primate relatives because they were also present in our common ancestors. These puzzle pieces have been found by psychologists undertaking experiments with those in captivity, meticulous observations of those in the wild, and by making subtle manipulation of natural settings to see how primates respond. This chapter assembles those pieces into a fragment of the language puzzle, one that provides a baseline for interpreting the evidence from the fossil and archaeological records in the following chapters.

Poor, quiet Viki

Research on chimpanzees did not get off to a good start. Between the 1950s and 1970s psychologists tried to teach chimpanzees to speak with words. From today's perspective, it is difficult to imagine how any scientist could have imagined this was worthwhile, let alone ethical. It tells us how not to undertake research on language evolution and confirms what the scientists must have already known: chimpanzees cannot speak with human words.

The most telling study was undertaken in the 1950s by the psychologists Keith and Catherine Hayes.[1] They were aware that caged chimpanzees, devoid of social contact, would never show signs of speech and so they adopted a baby chimpanzee a few days after her birth. They named her Viki and she grew up in their household, being treated as similarly as possible to their own child of a similar age.

In 1951, when Viki was three years old, the Hayes reported that her development had closely paralleled that of a normal human child. She had similar interests and abilities: playing (although being rather more athletic than a human child), seeking the company and attention of people, and exploring objects. Viki was, however, less vocal than children of a comparable age. She had babbled a little during her first year, persuading the Hayes that she had the vocal mechanisms for producing an approximation of human speech, but then became effectively silent. In contrast, the Hayes' own child had produced an almost continuous stream of babble and then early words.

With no evidence for spontaneous speech, the Hayes began a speech training programme when Viki reached five months old – one that would now fail any ethical evaluation. They were surprised that it took five months to teach Viki to vocalise on command. That produced an unnatural whispery 'hoarse staccato grunt'. The Hayes then manipulated Viki's lips while she vocalised so she said 'mama' and they taught her to say what were described as 'satisfactory approximations' to 'papa' and 'cup' by persuading Viki to copy their own spoken words. They explained that Viki did not use her three words meaningfully at first, but 'when we required her to employ them appropriately' (a phrase that makes me shudder) she learned to use them for some of the time, although often got the words and their meanings muddled up.

Poor Viki. She never learned any more words and died at the age of seven from viral meningitis. Whatever she was doing when she learned to say 'mama', 'papa' and 'cup', she was neither learning language nor demonstrating a capacity for speech.

Can apes understand and use visual symbols?

Following the failure of experiments to teach chimpanzees to speak, attention switched to their minds: could chimpanzees understand words even if they lacked the ability for vocal expression? From the 1960s, attempts were made to teach them 'words' using American Sign Language and via symbols depicted on plastic tokens and computerised keyboards. The outcomes remain heavily contested as to what they tell us about the mental capacities of chimpanzees and whether they reveal anything about the cognitive foundations of language.

Just like human children learning language or any other task, chimpanzees varied in their interest and capability to learn sign language and symbols. In general, they needed 300 or more trials to learn their first symbol, after which their learning become more rapid but ultimately hit a low threshold, usually well below fifty symbols.[2] A chimpanzee named Washoe (1965–2007) was one of the higher achievers. Between the ages of two and five years old she was raised by R. Allen Gardner and Beatrix Gardner in their house just as one would raise a child. She was taught American Sign Language and is said to have acquired 350 signs and was able to put different signs together into simple combinations, such as *Gimmie sweet* and *You me go out hurry*.[3]

The star achiever was a bonobo called Kanzi.[4] In 1980 he was a six-month-old infant sitting close to his mother, Matata, while the psychologists Sue Savage-Rumbaugh and Duane Rumbaugh were trying to teach Matata how symbols on a keyboard represent objects, actions and ideas. At the age of two, Kanzi began to use the symbol keyboard himself, having had no direct instruction. By the age of eight, he is said to have learned the use of several hundred symbols, to know up to 3,000, and to make novel three-symbol combinations. Kanzi is also said to have acquired an impressive comprehension of spoken English, being able to parse a variety of grammatical constructions, including those with subordinate clauses. Some of these were anomalous sentences, requesting actions that he could never previously have heard or seen performed, such as *Take the lettuce out of the microwave* and *Put the soap on the apple*. Overall, Kanzi reached the level of a proficient two-year-old child.

What is the significance of Washoe and Kanzi's achievements? Some believe these indicate the cognitive foundations for language exist within the chimpanzee mind, needing no more than a particular type of cultural environment to become expressed. Others dismiss Washoe and Kanzi as little different from trained animal acts, whose gestures were over-interpreted and often arose from inadvertent cues provided by their trainers.[5]

Their contrast with human children is unquestionably vast. After reaching two years of age, children swiftly learn hundreds and then thousands of words and begin to order them in novel sequences with simple but correct grammatical rules to create new meanings. Kanzi, still active in his forties, has never progressed beyond the language abilities of a two-year-old. Unlike Viki, he seems likely to have begun learning symbols by using the same mental processes as a human infant but remained with these whereas children harness additional learning mechanisms for language acquisition – as will be described in Chapter 9. It seems doubtful that Kanzi ever grasped the concept of words as building blocks for meaningful utterances.

Are gestures the ape equivalent of words?

With no clear outcome from the sign and symbol learning experiments other than a low threshold of achievement, attention turned to the role of gestures by apes. Might these be the ape equivalent of words and the precursor to language? To be more accurate, attention *returned* to gesture because the idea of a gestural origin for language had first arisen during the Enlightenment, notably by the French philosopher Étienne Bonnot de Condillac (1714–80). He argued for a continuity between animal communication and human language, suggesting that human speech had evolved from the 'language of action' found in animals, notably their gestures.[6] An interest in gesture remained in the margins of academia throughout the modern period before reaching greater prominence during the last two decades.[7]

Both captive apes and those in the wild use a wide range of gestures, involving the face, limbs and hands.[8] Jane Goodall described around a dozen used by the chimpanzees in Gombe National Park, Tanzania, with similar numbers noted for captive chimpanzees and up to thirty for orangutans and gorillas. Typical gestures involve offering a particular body part to another, tearing strips from leaves with the teeth, giving another ape a gentle nudge, and exaggerated scratching on the body. Some of these are used to initiate a social interaction, such as when an infant touches its mother's back, causing her to lower it so the infant can climb on. Others are about getting the attention of another, such as by slapping the ground or throwing an object.[9]

When compared with their calls, ape gestures appear to be less tied to emotion and more evidently used for communication. Some gestures are directed to specific individuals and modified according to their audience, implying intentionality. As with words among humans, some gestures are group-specific, while individuals vary in their repertoires; captive apes invent gestures and these spread within their community by social learning. Moreover, gestures using the arms are either absent or more obscure among monkeys. This suggests such gestures evolved after 30 mya, making them chronologically closer than vocalisations to the origins of words and language.

These observations lead some academics to propose that gestures rather than vocalisations provide the most likely foundation for human language.[10] They claim the cognitive capacities for language evolved to enhance the meanings and provide the syntax for gestural communication. At some later date in human evolution, those capacities found expression via the oral-auditory pathway to create spoken language. Why and how that happened remains unaddressed. To many, including me, this is implausible. A key problem is that the highly specialised nature of human anatomy and neurology for speaking and listening implies a long and gradual modification of a pre-existing system for vocal expression. What could have caused those changes if language was evolving in the gestural mode?

A further reason to doubt a gestural origin of language is the absence of iconic gestures within the repertoire of apes. We make frequent use of these such as beckoning to someone, blowing a kiss, and mimicking how to move, use a tool or work a machine. Infants and children use iconic gestures, notably when they point and ask *What's that?* with their outstretched arm or finger mimicking the direction of gaze they wish in their caregiver. Despite the prevalence of gestures within chimpanzees and other apes, there have been no unambiguous observations of iconic gestures in either captivity or the wild.[11] Neither have chimpanzees been able to use iconic gestures made by humans in laboratory-based experiments.[12]

Recognising the problems with a gestural origin for language, while also being sympathetic to the role of gesture, some argue for a multimodal (vocal and gestural) origin.[13] I remain unconvinced: gesture supports but does not constitute language unless it is heavily formalised into sign language. We must, therefore, return to the vocalisations of apes and monkeys and ask whether these have any connections to words as used in human language.

Word-like features of monkey calls

Our understanding of monkey vocalisations was transformed during the 1980s by the work of Dorothy Cheney and Robert Seyfarth, who studied vervet monkeys living wild in Amboseli, Kenya.[14] Some word-like features are evident in their calls.

Vervet monkeys are found in eastern and southern Africa. They have black faces and grey bodies, are quite small reaching up to 50 cm in size, with males being larger than females. Vervets are herbivores, living in groups of up to seventy individuals, with dominance hierarchies within both the females and males. They have a range of social interactions ranging from mutual grooming to aggressive encounters. Males move to neighbouring groups when they reach sexual maturity. We shared an ancestor with them 30 million years ago.

Cheney and Seyfarth began working in Amboseli in the 1970s and were intrigued by reports that vervet monkeys give different types of alarm calls to different types of predators, with each call eliciting a different response from nearby vervets. A loud barking call is given to leopards, causing vervets to run into trees; a short double-syllable cough to eagles, causing vervets to look up or run into bushes; a *chutter* is made on seeing a python, to which other vervets stand on their back legs and peer into the grass around them. When the snake is spotted, they group together and mob it while repeatedly giving the snake alarm call.[15]

Cheney and Seyfarth undertook a long and rigorous investigation of these calls using playback experiments – employing loudspeakers to broadcast each of the different calls in different contexts and with different acoustic properties. They also placed stuffed leopards into trees to watch the vervets' response. These interventions introduced the rigour of a laboratory-controlled experiment into the messy natural world.

They concluded that vervet alarm calls have several properties of human words. They not only relate to a specific type of predator but also provide information about the identity, level of fear and anxiety of the caller, with the listeners responding to both strands of information. Moreover, a degree of learning was involved. While infants were evidently born predisposed to the different types of calls, those younger than six months old gave eagle alarm calls to harmless birds or even falling leaves, and leopard calls to small mammals. The vervets also had voluntary control over their alarm calls. Sometimes they kept quiet and at other times made their calls more intense, notably when they have close kin nearby.

There were also key differences from words. Although false alarms were occasionally given in the absence of a predator, Cheney and Seyfarth concluded that the intention behind an alarm call was to modify behaviour rather than to influence minds: the purpose of a leopard alarm call is to make vervets run into the trees rather than to think there is a leopard nearby. It seems unlikely that vervets understand that others have thoughts and knowledge, let alone beliefs and desires.

Cheney and Seyfarth were unable to resolve what the calls had meant to those hearing them. Did the *chutter*, for instance, mean *snake*, or *Stand up and investigate the grass*, or *Snake! Let's mob it*? The calls were not onomatopoeias – they had no acoustic similarity to the sounds of their relevant predators. As such, Cheney and Seyfarth designated the calls as being arbitrary, which, if correct, would provide another similarity to human words.[16] An added complexity comes from the observation that the same *chutter* is used during aggressive encounters between groups. Similarly, the barks produced by males in response to leopards are also used during aggressive interactions and may function as a sexual display to advertise male fitness.[17] Meaning, therefore, appears highly dependent on context, allowing the limited range of sounds that vervets can make to be used in different ways.

Playback experiments were also used to learn about the grunts that vervets routinely make in social interactions. These are used when an individual approaches either a dominant or a subordinate member of the group, when they see members of another social group, and when initiating a group movement across the open plain. Unlike alarm calls, grunts are made in relatively relaxed situations and rarely elicit a response; they sound like humans acknowledging each other rather than having a chat. Cheney and Seyfarth's experiments showed that although all the grunts sound the same to the human ear, the vervets hear them differently and, like words, subtle acoustic variations carry information about the caller. Moreover, also like the words used by human hunter-gatherers, the grunts play an important role in mediating social and foraging behaviour.

In other ways, however, the grunts are quite unlike words. First, they are restricted in use to a limited number of circumstances. Cheney and Seyfarth could think of many situations when vervets remain silent although grunting would have been beneficial, situations in which a human would spontaneously introduce a word. Vervet mothers, for instance, often leave their infants in vulnerable situations by simply moving off without a grunt to say *Follow me*. Second, there is no evidence for grunt-combinations that could be a forerunner of syntax. Whenever an individual made two or three grunts in sequence, it was always the same grunt being repeated. Similarly, whenever a grunt was given in response, it was always a repeat of what the first vervet had said.

Chimpanzee calls have similarities to words

By using the playback and fake predator methodology pioneered by Cheney and Seyfarth, our understanding of chimpanzee vocalisations has been similarly transformed, notably through studies undertaken by Klaus Zuberbühler and his colleagues. We now recognise that chimpanzee calls have numerous word-like qualities. This is a sea change from no more than two decades ago, when most academics drew an absolute distinction between animal cries and human words, agreeing with the Enlightenment and Classical thinkers whose views we will explore in Chapter 6.[18]

Consider chimpanzees' 'rough grunts', an acoustically variable call made when they encounter food. To test whether such calls carry specific meanings that can be understood by other chimpanzees, a playback experiment was conducted at Edinburgh Zoo.[19] The chimpanzees living there prefer eating bread to apples. Recordings were made of their rough grunts made in response to both food types. Artificial trees were constructed in the chimpanzees' outdoor enclosure and the chimpanzees were shown how one tree would drop bread and the other apples; they learned there was a 'bread tree' and an 'apple tree'. When the chimpanzees were inside their indoor enclosure, a mix of rough grunts was broadcast from close to the trees to mimic the discovery of food. When the first chimpanzee emerged, either the apple rough grunt or the bread rough grunt was broadcast and chimpanzees' search for food was observed. Because no food had been dropped from the trees, the only information available was the sound of rough grunts.

To minimise interventions in the daily routine of the chimpanzees, only the behaviour of one individual, a five-year-old male known as LB, could be sufficiently monitored for this task. He was almost always the first to emerge from the indoor enclosure once the recordings were broadcast. In the first four of six initial experiments LB rushed to the tree that corresponded to the playback, to find either the apples or the bread that was evidently being conveyed by the type of rough grunt emitted by the recording. For the fifth and sixth trial, however, LB always went straight to the bread tree. He had learned that whichever rough grunt he heard, there was never any food on the ground and hence he might as well look for his favourite type. While this experiment demonstrated that chimpanzee rough grunts carry specific meanings about food types, it could not determine whether the calls are made to intentionally inform others and whether they mean 'bread' and 'apples' or are more generic, such as 'good food' and 'bad food'.

Further insights into rough grunts were attained by an observational study of the Sonso chimpanzee community living wild in the Budongo Forest in Uganda.[20] The chimpanzees search for food daily, often travelling in small groups while looking for trees with suitable fruit and leaves. Rough grunts are sometimes made when a food source is found; these are relatively quiet and directed towards the foraging group, in contrast to the loud pant-hoots that are sometimes made to announce the discovery of food to distant individuals.

Nine male chimpanzees were closely monitored. Records were kept of whether they made a rough grunt when finding food, who they were with, the quantity and quality of the food. Rough grunts were made on just over half of the 367 observed feeding events. They were more likely when the food-finder was in the company of his allies – the other males with whom he regularly grooms to build mutually supportive relationships. The quantity of food was also significant: larger amounts also increased the likelihood of a rough grunt. When a food source was found that could be easily monopolised, the likelihood of a rough grunt reduced, especially by the sub-adult males. The overall size of the group and the presence of a female in season had no influence on rough-grunting. This study showed that the male chimpanzees had voluntary control over when to make their rough grunts and used these to strengthen the social bonds with their existing allies.

Intentionality, meaning and displacement in predator alarm calls

Another study with the Sonso community demonstrated that chimpanzees are aware of what other chimpanzees know and adjust their calls accordingly.[21] This involved the use of an artificial snake, a viper. Although vipers are not predators of chimpanzees, they can respond with a deadly bite if trodden on. When seeing a viper, chimpanzees typically make a quiet type of call known as an 'alert hoo'. These calls do not scare the viper away but cause the chimpanzees who hear the call to move cautiously and in silence. Making a loud call, as happens when the chimpanzees see a leopard or python, would be dangerous because it might attract either a predator or a hostile chimpanzee.

Detailed observations were made on chimpanzee groups who went on foraging trips together. An artificial snake was placed on their well-used trails and records were made about which chimpanzee saw the snake first and whether it made one or more alert hoos. Records were also made about which other chimpanzees saw the snake, whether they also made a call, and which chimpanzees made a call without seeing the snake but in response to other calls. Further records were kept for the number of times each chimpanzee had previously seen the snake to monitor their level of habituation.

When the resulting mass of data was analysed, it became clear that the likelihood of a chimpanzee making an alert-hoo call in response to a viper was primarily dependent on whether other members of the group had also seen the snake or been alerted to its presence. The perceived level of risk from the viper, the composition of the group and the extent of habituation to the artificial snake had no influence. In place of these, the chimpanzees were keeping track of each other's knowledge and informing those who were ignorant about the snake's presence.

A similar experimental study conducted with wild orangutans in Sumatra found that the predator alarm calls of this ape – one slightly more evolutionarily distant from humans than chimpanzees – had a feature that is often described as unique to human language: displacement.[22] This refers to the ability to transmit information that is not immediately present in time or space. This experiment was undertaken by Adriano Lameira and Josep Call, whose predator model was a human moving on all fours below the forest canopy while draped in a tiger skin.

This human-tiger-predator prowled for two minutes in the vicinity of a known female orangutan, with the experiment repeated for the seven females that formed the resident population with their offspring; male orangutans are solitary and rove through the forest. When the females saw the 'tiger', they would always halt their activity and exhibit signs of distress such as urinating and monitoring the floor to see where the predator was heading, while climbing higher into the canopy. Despite such anguish, they made alarm calls on only half of the 'tiger' sightings, and these were almost always when the 'tiger' was no longer visible. The average delay was seven minutes, with this being longer for the older females; in one case an orangutan waited for seventeen minutes before giving her call.

Lameira and Call dismissed the possibility that lack or delay in calling reflected a state of petrified fear – if that were the case, the orangutans would not have immediately begun climbing into the canopy. They concluded the delay was an intentional response to reduce the risk of attracting attention and causing an assault on either themselves or another member of their group. They interpreted the longer delays by the older females as reflecting this concern for others because they were more likely to have offspring in the vicinity who would have also seen the predator – orangutan infants remain with their mothers for nine years. If this is correct, the question that Lameira and Call had to ask was: why, then, did the females alarm call at all? They suggest it was to enable their infants to learn about the risk of predation. If their calls had been fully suppressed the infant would have been unable to learn from safety that encounters with tigers were dangerous. To achieve such learning, the mother had to displace her alarm call, while the infant had to connect the call when it occurred to its previous experience of seeing the predator.

Whether chimpanzees also delay their alarm calls for the same reason is unknown. But they certainly make these when their favoured companions are facing danger. This was demonstrated by another experiment using the Sonso community in which chimpanzees were presented with an artificial python, a lethal predator.[23] The python was hidden in leaves along a trail. When a targeted chimpanzee approached, a line attached to the snake was pulled to make it move which caught the chimpanzee's attention. This generated three different types of alarm call, known as a 'soft hus' (SH), 'alarm hus' (AH) and a 'waa barks' (WB). While the SH call appeared to have been an involuntary response to the python (yikes!), the AH and WB calls were voluntary and goal directed. By following the eye gaze of the chimpanzees, the scientists could see that the calls were primarily directed at individuals within the group who were either more dominant individuals or allies – those individuals with whom close social relationships are maintained by grooming and proximity when resting. As such, when making the AH and/or WB calls, the callers monitored those individuals whom they most cared about and only stopped calling when they all appeared safe.

The sociality of ape calls

The fake predator experiments demonstrated that chimpanzee and orangutan vocalisations are closely woven into their social relationships, just as Cheney and Seyfarth had found with the grunts of vervet monkeys and is the case with human utterances. This is also evident from another call made by the Sonso community, known as a 'travel hoo'. The chimpanzees make this before engaging in travel, such as to find food, a nesting site or to interact with a neighbouring community.[24] Detailed records were kept about which chimpanzees made these calls, when they made them, who they were directed towards and what responses arose. The data indicated that travel hoos were also intentional and goal directed. They were made by both sexes and were most common when their allies were in the vicinity. Travel hoos were rarely made before travelling alone or with dependent offspring. They tended to immediately recruit the targeted individuals; if they didn't, the caller waited, while monitoring his or her allies and making further calls.

The social role of chimpanzee calls was further revealed by an analysis of the 'pant grunt'.[25] This call is produced only when greeting a higher-ranking group member and has traditionally been interpreted as a rigid, formulaic greeting – the chimpanzee equivalent of bowing one's head in deference and saying, *Good day, sir*.

The pant-grunts of nine adult females within the Sonso community were monitored regarding to whom they pant-grunted and in what social contexts. This found that pant-grunts were preferentially given to the alpha male. All other males received pant-grunts but with no correlation between the number they received and their position in the social hierarchy. Females were more likely to pant-grunt to males when the alpha male was absent, when there were fewer bystanders (especially males) who could watch the interaction, and when the alpha female was absent. The study concluded the pant grunt is 'a reflection of a caller's wish to interact with a socially relevant group member, a way of providing a recipient's mood, and motivations, and a vocal tool to make one's presence known and to convey respect'.[26] That sounds rather like someone smiling brightly and saying, *Hello, how are you today?*

Do chimpanzees learn their calls?

The voluntary, meaningful and goal-directed nature of chimpanzee rough grunts, alert hoos, travel hoos and pant grunts are features shared with words. Do wild chimpanzees also learn their calls in the manner that children learn words from their caregivers and wider speech community?[27] We have already seen that some learning is involved by vervets when acquiring their alarm calls, although they are predisposed to these calls at birth and the learning is primarily about context of use.

Learning to blow raspberries has been observed within captive communities of chimpanzees. It has been independently invented within several separate groups, presumably by one individual from which it spread to other members of the group. A study at Yerkes Research Centre in the United States found that the raspberry and another invented call known as an extended grunt were predominantly used to capture the attention of human carers.[28] The significance of such vocal innovations and social learning is problematic because of the unnatural captive contexts of the chimpanzees. Of far greater interest is when these occur within wild populations. The required research is inherently difficult to undertake because of the practical and ethical issues involved in the close monitoring of infants with their mothers and other group members within wild communities.

An indication that some learning is involved has come from a comparative study of the pant-hoots made by male chimpanzees in one distant and three neighbouring communities in the Taï National Park of West Africa.[29] Pant-hoots are long-distance calls made by chimpanzees when either feeding or travelling. They consist of four distinct units, an introduction, a build-up, the climax and a let-down. An acoustic analysis of pant-hoots found that those made in the three neighbouring communities were significantly different from each other. As such, when a pant-hoot is heard, the community that the caller belongs to can be identified. Moreover, each male had distinctive features to their pant-hoots, making such calls individually specific as well as indicative of a particular community.[30] The pant-hoots made in the distant fourth community, however, had no community-specific features, overlapping acoustically with the pant-hoots from all three of the neighbouring communities.

These findings suggest that male chimpanzees within each of the neighbouring communities had modified their pant-hoots via social learning to become community specific. Because the pant-hoots of the fourth community were never heard by members of the other three communities, there was no need for members of that distant community to develop their own distinctive call. DNA analysis of the chimpanzees via their faeces excluded the possibility of genetic influences on the similarities and differences in the pant-hoots between individuals within communities and those between communities, confirming the role of innovation and social learning.

The process of acoustic convergence within a chimpanzee community has been documented for chimpanzee food calls at Edinburgh Zoo.[31] As previously noted, these chimpanzees make distinctive rough grunts in response to apples. In 2010, the Edinburgh chimpanzees (ED) were joined by a group from Beekse Bergen Safari Park in the Netherlands (BB) to make a single community. The BB chimpanzees had a stronger preference for apples over other food types than those in Edinburgh, and an acoustically different apple rough grunt. Over a period of three years, the BB apple rough grunt converged to become the same as the ED apple rough grunt (although the BB chimpanzees maintained their preference for apples). This rough-grunt convergence happened over the same time frame that full social integration occurred, indicating it was closely related to the development of strong social bonds between the ED and BB chimpanzees. Convergence of the apple rough grunt onto that of the ED rather than the BB community most likely reflected a need for the incomers to conform to the host community.

Both the Taï National Park and Edinburgh Zoo studies emphasise the role of sociality in vocal learning. The role of group size has been demonstrated in a study of orangutan calls from wild-living populations in Borneo and Sumatra. Orangutan communities are organised in loose female groups with roving adult males. Although this type of organisation prevents a measure of group size, which is a key aspect of sociality, population density in the forest provides an adequate proxy.

The study measured the variability in the orangutans' primary predator alarm call, known as a kiss-squeak, within and between six communities that had different population densities.[32] Individuals living in higher densities were found to be more original and unpredictable in their calls than those at lower densities whose calls were both more repetitive and similar to each other. The calls within low-density populations were also more acoustically complex. Such variability was attributed to the extent and diversity of social influences when infant orangutans learned their kiss-squeak alarm calls, both being greater within higher-density populations. This finding resonates with the influence of human population size on language. As described in Chapter 3, languages spoken by larger populations tend to have a wider range of sounds, with shorter and less morphologically complex words, than those found within smaller communities.[33]

Multifunctional calls

The multifunctional role of calls within the same group may be one of the most important features of chimpanzee calls, because this is a feature shared with human words. Our use of words is independent of context – I can say what I like where I like, requiring listeners to attend to context when seeking to understand what I mean. I can, for instance, say *I feel excited* when I see an attractive plate of food, before going on holiday, after taking a stimulant drug, or even when I feel the precise opposite. Context can also change the meaning of the same word, as in *Pass me the bat* when said during a game of cricket or visiting a wildlife centre, and this is prominent when words are used as a metaphor, as in literal and metaphorical *throwing the book at him*. While most monkey and ape calls appear tied to a single context and function, a glimpse of such flexibility is seen within the vervet calls, for whom *chutters* can be used either in response to snakes or in aggressive situations. More than a glimpse has come from a study of bonobo peep calls.

The peep is a high-frequency call made by bonobos with closed mouths.[34] Kanzi, the star performer of language-learning experiments, often used peeps to communicate with his caregivers and they are widespread in wild populations. A study of their use by members of the Bompusa bonobo community of West Africa (Democratic Republic of the Congo) found that peeps were used in numerous contexts including feeding, resting, travelling, grooming, nesting and aggressive encounters. When used in positive or neutral contexts (e.g. feeding or nesting) they were acoustically identical, which suggests that their meaning was dependent on the listeners considering contextual information, whereas peeps used in negative contexts (such as aggressive encounters) were distinct. This study concluded by suggesting that bonobo peep production might represent an intermediate stage between functionally fixed primate calls and those with flexible meanings dependent on context that characterises human words.

Syntax-like sequences in chimpanzee calls

The experiments and observations I have described indicate that chimpanzee calls are not words but have word-like qualities. Do chimpanzees use syntax-like rules to combine their calls into sequences that have a secondary level of meaning to the individual calls themselves? Probably not, but the way in which calls are combined provides the opportunity for them to do so should a minor cognitive shift occur.[35]

Before chimpanzees, consider the combinations used by two types of small West African monkeys. Male Campbell monkeys make two alarm calls, krak and hok.[36] These indicate that danger is imminent. An -oo sound is sometimes attached to these calls. This dilutes their meanings to become less urgent, such as referring to the sight of a distant predator rather than one that is about to attack. Experiments using edited krak and krak-oo alarm calls indicate that the -oo lacks any meaning in itself but modifies the meaning of krak, just as a suffix does to a spoken word.

Putty-nosed monkeys also have two types of alarm calls, pyows and hacks.[37] These are often assembled into sequences. Pure pyow sequences tend to be given to terrestrial dangers whereas pure hack sequences are given to the sight of a crowned eagle, a key predator. Males sometimes produce mixed sequences of pyows and hacks. Playback experiments have shown these calls trigger group movements, a 'meaning' that cannot be derived from the meaning of its pyow and hack components. As such, despite involving a combination of otherwise independent calls, these monkey calls remain as holistic utterances quite different to the compositional utterances used in human language.[38]

Shifting our concern to chimpanzees, we can consider their multi-component pant-hoots. These are their long-distance calls that consist of four components: introduction, build-up, climax and let-down. Acoustic analysis of those made by male chimpanzees of the Sonso community of the Budongo Forest in Uganda established that each component has a different primary meaning, although there is some degree of overlap.[39] The introduction mainly encodes the identity of the caller; the build-up indicates age; the climax encodes social status; and the let-down encodes the behavioural context of the call, such as whether it is made while travelling or feeding. Each of the four components can also be given on its own, but the stand-alone meanings are unknown. Without knowing those meanings and with only a limited understanding of the overall meaning of the pant-hoot, we cannot yet consider this as an example of chimpanzee syntax.

Bonobos also combine acoustically separate calls into sequences.[40] When responding to food they have five acoustically distinct calls: grunts (G), barks (B), peeps (P), peep-yelps (PY) and yelps (Y). A study undertaken at Twycross Zoo found that peeps and barks are more frequently given to kiwis, a preferred food, while grunts and barks tend to be given to apples, a less preferred food. There are, however, no exclusive associations and the bonobos frequently combine their calls into long sequences, such as P-P-PY-P given to kiwis and as PY-B-B-PY given to apples.

Playback experiments were used to detect whether pre-recorded call sequences from an individual who had found either kiwis or apples could direct other bonobos to the most likely source of food in their compound, to either the kiwi or apple feeding location.

That was the case. Those sequences which had higher frequencies of Bs and Ps resulted in the bonobos spending more time foraging in the kiwi field. This indicates that the bonobos were inferring the meanings of the calls they heard by integrating information from their individual components, which is similar to how we infer the meaning of a spoken utterance from its individual words. Syntax, however, relies on the specific ordering of words and in this bonobo case there was no evidence that the order of the Ps, Bs, Ys, and PYs had any relevance – the bonobos were only paying attention to their relative frequencies.

The most intriguing study of call combinations concerns those made by chimpanzees in the Taï National Park.[41] This showed an unprecedented degree of call combination for a non-human primate, with features that are otherwise only known in human language. The chimpanzee calls were categorised into twelve distinct types. These included grunts (GR), pants (PN), hoos (HO), barks (BK), screams (SC), panted-scream (PS), pant-ed-hoos (PH), panted-barks (PB) and panted-grunts (PG). Within a database of almost 5,000 utterances made by forty-six chimpanzees, two-thirds of these used a single call, almost a fifth were a two-call combination known as bigrams, and the remainder were either three-call combinations (trigrams) or combinations of between four and ten calls. Overall, the twelve calls resulted in 390 different sequences, including fifty-eight bigrams.

Analysis of the bigrams showed that some call types, such as HOs, GRs and PHs, were more frequently used in the first position whereas others, such as SCs, BKs and PGs, were preferentially used in the second position. Fourteen of the bigrams were produced at a higher level than by chance (i.e. from a random coupling of the calls made at their known frequencies). Of these, seven were produced by at least ten individuals, including GR-PN, HO-PH and PH-PS. Within these seven, the first call never constrained which would come next, but the second was often bound to what came before. A PS, for instance, nearly always followed a PH.

Forty-nine of the bigrams were also emitted with a third call attached to make a trigram. Four of these trigrams were found at above chance level and were emitted by at least ten individuals. Three of these had a consistent pattern for the bigram element to be positioned at the start of the trigram. The bigram GR-PG tended to be followed by a GR, while PH-PS was predominantly preceded by an HO.

What do these patterns all signify? First, they demonstrate a high degree of flexibility in vocal sequence output, an ability to combine twelve discrete types of calls into almost 400 different combinations ranging from two to ten units. Second, there is a bias as to the order in which two call types are combined into a bigram, with some calls tending to come first and others to come second. This also applies to how bigrams are used within trigrams. Third, some call sequences have a recurrent use by more than ten chimpanzees.

This all appears suspiciously similar to how human language combines words with individual meanings to make utterances that have a secondary level of meaning, with some word sequences having a more frequent occurrence than others. We do not know if this similarity is significant because we have little idea of what the individual chimpanzee calls mean – their grunts, hoos, barks and so forth – and absolutely no idea about the meaning of their bigrams, trigrams and longer call sequences. These might be like the pyow-hack sequences produced by putty-nosed monkeys, the meaning of which is entirely unrelated to the meanings of the individual pyow and hack elements. Alternatively, we might be witnessing a combinatorial system that amounts to syntax, one that uses call order to generate utterances with secondary levels of meanings – almost 400 different phrases from a mere twelve units. Even if that is the case (which I strongly suspect it isn't), the chimpanzee syntax would be restricted to the ordering of calls and missing what some believe to be the most critical element of human syntax, its hierarchical structure that involves the embedding of one phrase within another.

Until we find a way to extract the meaning from the chimpanzee calls, bigrams and trigrams, it would be premature to conclude that their call combinations are evidence for syntax. But just as chimpanzee calls have word-like qualities, so do their combinations have syntax-like qualities.

Two long journeys

Our understanding of vocalisations by monkeys and apes has come a long way since these were claimed to have absolutely no continuity with human language. If language does remain as the bastion of human uniqueness, it is now on shaky ground and looks set to topple.

Most progress has been made by introducing the control of laboratory experiments into the real-world setting of monkeys and apes living in the wild.

The data acquired by Dorothy Cheney, Robert Seyfarth, Klaus Zuberbühler and their colleagues will always remain open to alternative interpretations. Mine is that although monkeys and apes entirely lack words, their calls sometimes have word-like qualities, especially those made by chimpanzees. I sense that only a small cognitive shift would transform their vocal communications into something more recognisable to us as language.

Are those word-like qualities ancestral or derived? Are they found within monkeys and apes because they were present in the common ancestors that we shared with them or have they evolved following the divergence of our lineages? If the former, we can confidently attribute them to the descendants of our common ancestors, such as the australopiths and *Homo habilis* at 2.8 mya, the earliest member of our genus – although further steps in the evolution of language may have already occurred by that time. If the word-like qualities of chimpanzee vocalisations are derived, it is possible that early ancestors such as *Ardipithecus ramidus* had evolved the same vocal adaptations by convergent evolution. They also lived in social groups within forested environments, had a brain size of *c.*300 cm^3 and would have faced similar selective pressures for effective communication.

Either way, whether ancestral or derived, we now have a robust fragment of the language puzzle to place within the interior of the frame: chimpanzee word-like calls provide a starting point for the evolution of fully modern language. Those calls, however, had a long evolutionary journey to make before they would become recognisable as words, a journey that can be tracked only through the fossil and archaeological record for the human past. We can start that journey by assembling the next fragment of the language puzzle, the evolution of the vocal tract. How does the modern human version differ from that of the chimpanzee, when did it evolve, and what implications does it have for language?

5

SPEAKING AND HEARING

Chimpanzees' vocalisations may have word-like qualities, but they are not words. One reason is that the chimpanzee brain lacks the capabilities to attribute arbitrary meanings to different call types, meanings that are learned and shared within the calling community. We will consider the neurological basis for language and evolution of the brain in Chapters 11 and 12, but here we focus on a second reason why chimpanzees do not have words: their vocal tracts. These are unable to generate call types with a sufficient consistency, diversity and frequency on each breath to cross the evolutionary threshold into the world of words. The human vocal tract is finely attuned for those tasks, as is the auditory pathway for hearing the words spoken by others. This indicates a long period of moulding by natural selection for speaking and hearing, which must have gone hand in hand with the evolution of a linguistic capacity in the brain.

The key to deciphering how the vocal tract evolved to make linguistic utterances is to assemble evidence from several disciplines, all of which contribute to this fragment of the language puzzle: from human physiology for how the modern vocal tract operates; by comparing the vocal tract anatomy of modern humans to that of the chimpanzee; and by a careful scrutiny of the fossil record for how the vocal tract changed in shape during the course of human evolution.

Beginning a word

The vocal tract is the anatomical passage from the vocal folds, located in the windpipe, to the lips and nose, averaging about 17 cm in length for an adult. Its anatomy originally evolved to be – and continues to be – used for breathing and ingesting food but was further shaped by natural selection to generate a variety of sounds that we refer to as phonemes and which provide the basis for syllables and words (Chapter 3). The vocal tract has also been influenced by selection for other tasks we undertake, notably walking upright on two legs while carrying a large and heavy brain. Untangling the multiple influences on our anatomy is challenging.

For a modern human, speaking a word begins with a stimulus – something we see, hear or think. That leads to the firing of neurons in the brain, releasing chemicals known as neurotransmitters. These generate electrical signals in neighbouring neurons which propagate in a wave-like fashion to thousands of other neurons that activate concepts, ideas and memories in the brain. From these the motivation to speak and the selection of a word may emerge. We are rarely aware that concepts are being activated in the brain because our words just tumble out.[1] Quite why, when and how brain waves generate conscious awareness about the process of speaking is unclear; when it occurs, we have the opportunity to carefully select the most appropriate word from within our mental lexicon.

Whether carefully planned or otherwise, we start to speak with the first phoneme of the first word – which may be its only phoneme.[2] Doing so involves activating a large complex of muscles throughout and beyond the vocal tract. These manoeuvre multiple anatomical elements in a coordinated manner to impose an acoustic wave onto an exhaled breath of air that carries the phoneme to the ears of the speaker and those of anyone else within hearing distance.

We begin by expelling air from our lungs in a controlled

manner, maintaining a higher pressure in the lungs than in the mouth. The air passes along the windpipe (the trachea) and then through the larynx. That is a valve consisting of an opening, the glottis, covered by two slivers of muscle, known as the vocal folds – sometimes called the vocal cords. When we speak, the two folds vibrate in response to the passage of air, which is known as 'voicing'. This generates ripples of air at a frequency that defines the pitch of the speaker's voice, combined with harmonic overtones that are multiples of that frequency.

The air continues its journey into the space between the larynx, the nose and the mouth known as the supralaryngeal vocal tract. This space acts as a resonator. It contains four moveable structures: the lips, mandible (lower jaw), tongue and soft palate. By bringing these into a particular articulation with three immoveable structures, the upper teeth, hard palate and rear wall of the pharynx (the throat above the larynx), frequencies within the incoming air stream are amplified into formants (bands of frequencies that have particular prominence) within the outgoing air stream. These create the phonetic quality of the sound produced, while also imposing the specific qualities of the speaker's voice. The lips, mandible and tongue can also be used to constrict the end of the vocal tract such as by narrowing the gap between our teeth or pursing our lips. This modifies the sound by adding a so-called fricative element, such as /f/ and /sh/, as the airwave leaves the speaker's body and enters the outside world.

Vowels and consonants

Sounds made when we restrict the vocal tract in this manner are referred to as consonants; those with an unrestricted air flow are vowels. Consonants and vowels are normally combined to make syllables which are the building blocks of words. They alternate with each other; consonants being made when the mouth is relatively closed and vowels when it is relatively open. Syllables typically have a vowel at their centre, bracketed on either side by a consonant.

Consonants and vowels are not simply two varieties of sounds, but different types of sound. They are processed independently in the brain, such that a cognitive defect can impair the ability to hear and say vowels while leaving consonants unaffected, and vice versa – a phenomenon known as a double disassociation.[3] Consonants and vowels also play different roles in language acquisition: consonants are more helpful for identifying words, while vowels make a greater contribution to learning the rules by which words are put together to make meaningful utterances.[4]

In spoken English we use seven short vowels, five long vowels and twelve vowels made from two short vowels joined together, known as diphthongs. Vowel sounds are written by using symbols from the International Phonetic Alphabet (IPA). For instance, /ɪ/, as in *pit*; /e/, as in *pet*; /æ/, as in *pat*; /ʊ/, as in *put*; and /ɒ/ as in *pot*. Long vowels are denoted either by the addition of two tiny triangles after the IPA symbol or by the combination of two symbols: /iː/ as in *week*; /ɑː/ as in *hard*; /aʊ/ as in *mouse*; and /ɪə/ as in *clear*.

Vowels sounds are primarily defined by the position of the tongue in the mouth, this influencing the formants. High vowels, such as /iː/, are made with the tongue elevated in the mouth, and low vowels, such as /ʊ/, by the converse, holding the tongue low down. Backness refers to the horizontal position of the tongue. In a forward vowel the tongue is positioned at the front of the mouth; it is held towards the back, against the soft palate, to make a backward vowel. Roundedness refers to whether the lips are rounded.

Some examples using the IPA symbols are:

/uː/, as in *food*, is a high, back, rounded long vowel
/ɑː/, as in *palm*, is a low, back, unrounded long vowel

/i:/, as in *seat* or *cheap*, is a high, front vowel

/e/, as in *bed*, is a mid-front vowel, made by placing the tongue at the front of the mouth and in a mid-vertical position

/æ/, as in *bad*, is a low, front vowel.

There are three main classes of consonants within spoken English: plosives, fricatives and nasal.[5] The first arise when the air flow from the larynx is interrupted by a complete closure of the mouth, and hence plosives are sometimes called stops. The air flow builds up behind the closure; when released the sound seems to explode out, as in /p/, /t/, /k/, /b/, /d/ and /g/. Note that unlike vowels, the IPA symbols for consonants frequently correspond to alphabetic letters. Exceptions include /θ/ which represents *th* when used in words such as *think*. Similarly for digraphs, such as /ʃ/ which represents *sh* when used in *ship*. Fricatives are made by allowing the air to pass through a restricted space, such as by making a small gap between the teeth or lips. This causes turbulence and a hissing sound, creating /f/, /s/, /v/ and /z/. Nasal consonants arise when the air is stopped from going through the mouth by a lowering of the soft palate, redirecting it to the nose. This is used for /n/ and /m/ in words such as *seen* and *seem*.

Variation in the sound of plosives, fricatives and nasals is caused by the relative positioning of the tongue, hard and soft palates, the teeth and use of the lips when the sound is made. Rather than trying to explain how this is done, it is easier to find out yourself by saying /d/, /b/, /p/ and /k/ and reflecting on what is happening in your mouth as you do so. Variation also arises by the extent of voicing – the vibration of the vocal folds when the sound is made. This causes the difference between /s/ and /z/, and between /p/ and /b/, the second of these pairs being the 'voiced' consonant. Consonants can also influence the sound of vowels. If the consonant is nasal, a preceding vowel may also take on a nasal quality, as in the word *man*. If the consonant is voiceless, the preceding vowel becomes shorter in length.

One further source of variation for both vowels and consonants comes from the individual: we all have different lung capacities and windpipe lengths; our mouths, teeth, tongues and lips have subtly different shapes, and we can manipulate them in slightly different ways. Consequently, there is idiosyncratic variation in how the same vowel or consonant sound will be made and how these are joined together. Moreover, there is variation in how we hear and interpret the sounds others make, and our abilities to replicate them.

Finishing the word and making an utterance

The first phoneme may be both the start and end of a word, as in the personal pronoun 'I', but it is usually followed by other phonemes to make a syllable with a consonant-vowel-consonant sequence. This will be followed by further syllables required for the word. We continue without a pause to say the next word, and then the next to complete an utterance. This involves the repetitive opening and closing of the mouth to produce the appropriate string of vowels and consonants for the syllables and words. Across all languages, this opening and closing exhibits a rhythm of between 3 and 8 Hz (cycles per second), generating an average of five vowels and five consonants per second. That speed appears to be constrained by both our vocal tract and auditory apparatus; any disruption to this speed decreases intelligibility.[6]

The whole utterance is usually made within one exhalation of air lasting several seconds while maintaining a constant lung pressure. We then then take a quick in-breath of air, either before making our next utterance or when pausing in the expectation that someone will reply – turn-taking is an essential facet of speech.[7]

As we open and close our mouths during that single exhalation, we continuously vary the vibrations of our vocal folds, the positions of our tongue, palette and teeth, to make the required sequence of vowels and consonants that constitute an utterance. By so doing, we produce not only a variety of phonemes, syllables and words but can put these into a multitude of combinations, each with its own unique meaning and some never created before. Moreover, by varying the amount of energy we use for each syllable, we vary how loud they sound and how long they last. In this manner, we add prosody to our utterance that can further influence its meaning.

There are two points about the process of speaking that are critical to note for our evolutionary discussion. First, the whole process of pushing air from the lungs through the larynx, and then shaping the resulting acoustic wave by manoeuvring our lips, teeth and lower jaw, usually occurs without any conscious awareness. The phonemes, syllables and words simply flow. We can, however, impose such control if we wish: we can deliberately change the pitch of any vowel or consonant; we can make them nasal or plosive, louder and longer. We can change how we pronounce either a single word or a whole utterance, even remaining silent when expected to speak in response to a question or event.

Second, while the vocal tract can generate over 300 distinct phonemes, only a fraction of these are required for a language. As we noted in Chapter 3 when exploring linguistic diversity, the number of sounds (phonemes) within spoken languages varies from around a dozen to almost 150. There is nothing to say that a language with more phonemes is better or more advanced than one with fewer. English has forty-four phonemes, lacking a range of clicks and guttural choking sounds that are prominent in other languages. Linguists disagree whether any vowel occurs in all languages; some argue this is the case for /i/, /u/ and /a/, but Peter MacNeilage, author of the most authoritative book on speech, denies this, simply placing them at high frequencies: 91.5 per cent, 88 per cent and 83.9 per cent respectively.

Monkey and ape vocal tracts

Turning now to our model for the non-linguistic common ancestor of 6 million years ago, the vocal tracts of monkeys and apes are not dissimilar in their anatomy to those of modern humans, having also evolved for breathing and ingesting food. Their vocal tracts operate in the same manner: pushing air from the lungs through the larynx and into the space behind the nose and mouth (the supralaryngeal vocal tract), where it is shaped to have a range for formants that deliver specific sounds. There are, however, subtle differences in the vocal tracts between monkey and ape species, and more substantial differences when they are both compared with that of modern humans. These differences influence the type, range and rate of sounds that can be produced.

The majority of monkey and ape calls are 'voiced', generated by oscillations in vocal folds just as vowels are made in human speech. As described in the previous chapter, chimpanzee calls are described using terms such as barks, grunts, hoos and pants. Apes also make voiceless calls. These are produced by articulations of the tongue, lips and jaws to make clicks, smacks, raspberries, kiss-squeaks and whistles, and can have similarities to the consonants in human speech. It is not unreasonable to assume they had been present within the vocalisations made by the last common ancestor of 6 million years ago. If so, that suggests early human ancestors were using both vowel-like and consonant-like sounds within their call repertoires.

A key difference from human speech is that monkeys and apes vocalise when both breathing in and out. A second is that they can make only a single sound on an individual movement of air. This is quite different to the human pattern in which a long sequence of different sounds can be made during one exhalation.

Assuming the vocalisations of the 6-million-year-old common ancestor were similar to those of monkeys and apes today, two evolutionary processes were required to transform them into speech. One is the evolution of a brain that stores and processes words and the rules to combine them into meaningful utterances; we will consider that in Chapters 11 and 12. The other is the evolution of a vocal tract that can be controlled to generate a sufficient diversity of sounds in the required order and speed to express those words and utterances in a manner such that their intended meaning can be acquired by another individual.

For such a vocal tract to evolve, three constraints on the calls of monkeys and apes today had to be overcome: on their number, on their clarity and on their control. We will consider these in turn with the evidence for when they may have been overcome during the human past.

How many call types are needed to make a language?

The limited range of sounds made by monkeys and apes is the most widely cited reason for why they lack the capability for human-like speech. This is also the most problematic because a viable human language needs no more than a dozen phonemes and that is certainly within the range of monkeys and apes today.

Counting the number of call types used by a monkey or an ape is difficult because they often blend into each other, as evident by the use of vague descriptive terms such as hoos, barks and grunts. The most systematic count for chimpanzees reached twelve, while counts for monkeys range from six to fifteen – all sharing the same problem about where to draw a line between one call and another.[8] Overall, twenty seems to be the maximum limit, which may be rather generous. That number contrasts with the capability of the human vocal tract to generate over 300 sounds.

The number and production of chimpanzee calls is

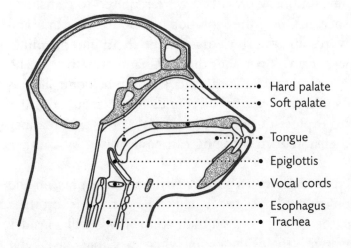

- Hard palate
- Soft palate
- Tongue
- Epiglottis
- Vocal cords
- Esophagus
- Trachea

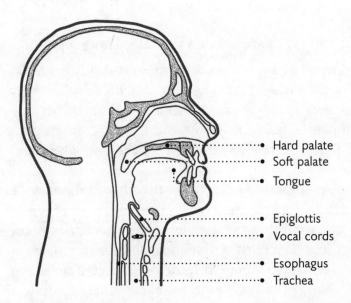

- Hard palate
- Soft palate
- Tongue
- Epiglottis
- Vocal cords
- Esophagus
- Trachea

Figure 6 Chimpanzee and human vocal tracts

constrained by the size and shape of its supralaryngeal vocal tract (SVT) (Figure 6).[9] The SVTs of most mammals, including humans, has two components: one is horizontal and stretches from the lips to the start of the throat (pharynx), and the other is vertical, extending from the pharynx to the larynx. In humans, the horizontal (oral) and vertical (throat) components are the same length, but in all other mammals, including chimpanzees, the vertical component is considerably shorter.

The short length of the vertical component of the chimpanzee SVT causes its epiglottis – a flap of cartilage that prevents food from entering the larynx – to touch its soft palate (velum), making a partial barrier between the vertical SVT and the back of the tongue. As such, the vertical SVT makes a limited contribution to the role of the whole SVT as a resonator of the vibrating air being received from the larynx. Hence for chimpanzees, the types of sounds they can make are dependent on the horizontal component of the SVT alone. In contrast, by having a long vertical SVT, the epiglottis in modern humans is removed from the soft palate enabling the vertical and horizontal SVT components to combine into a single, large resonating chamber.

The capability of that resonating chamber is influenced by the shape of the tongue. This is a key organ of speech, having an internal musculature that allows it to be manipulated into various shapes. Because its overall shape matches the SVT configuration, the human tongue has a relatively large vertical component and is more globular than the chimpanzee's, which is largely restricted to the horizontal component of the SVT and is almost entirely flat. By manipulating the tongue, modern humans can change the shape of both their vertical and horizontal SVT components, either at the same time or independently of each other, thereby generating a diverse range of sounds. The tongue's globular nature helps to make the /i/, /u/ and /a/ sounds that are pervasive, if not universal, within languages of the world. Rapid tongue movements enable equally rapid changes to the resonant properties of the whole SVT and hence allow a sequence of different phonemes to be produced within a single short exhalation of breath. In contrast, the movements of the chimpanzee's tongue have minimal impact on the resonant properties of its vertical SVT, reducing the range of sounds and the rate at which these can be made compared with modern humans.

It had long been thought that the relatively long vertical SVT component in humans arises from a unique developmental pathway during childhood referred to as the 'descent of the larynx'. That should, in fact, be called the descent of the tongue, which takes the larynx with it so that the larynx becomes situated between the fourth and sixth vertebrae, creating the long SVT component. This descent enables the formation of a swallowing mechanism in early childhood – its evolution was not driven by requirements of speech. Recent research has shown that chimpanzees also experience a descent of the tongue/larynx during infancy, although this is of a shorter distance than in humans, leaving the larynx at the level of the third vertebra.

Chimpanzee infants undergo another developmental change that does not occur in humans: they grow a long, protruding face. This creates the flat tongue and extends the horizontal SVT component so that it loses almost all connection with that of the vertical SVT component. In contrast, the human face remains short and flat, that of an adult being little different from a child. This facial configuration appears to have evolved gradually under the influence of several factors including changes in diet, social factors and – potentially – selective pressures for speech.[10] The evolved short, flat face keeps the tongue globular and the horizontal and vertical components of the SVT at an equivalent length, enabling them to function as a single resonating chamber. In essence, the human capability to make a varied and rapidly changing range of sounds primarily derives from the maintenance of a juvenile-like flat face rather than the descent of the tongue/larynx. Infants are able to mimic adult speech, generating vowel-like sounds before the descent of their larynx.[11]

The extent to which chimpanzee calls are restricted by their almost partitioned SVT and flat tongue remains to be fully established. Some academics argue that this entirely removes the ability to make the three key vowel sounds of /i/, /u/ and /a/.[12] I am sceptical, because comparable sounds can be generated in the absence of a modern human vocal tract. A recent acoustic analysis and anatomical study of a baboon found that it produced five distinct types of voiced calls, each being comparable to a human vowel.[13] Also, animals can be effective mimics of human speech. Tilda is an adult orangutan who has been in captivity since the age of two. She was trained for human entertainment before becoming housed in Cologne Zoo where she developed 'faux-speech' by imitating her keepers. Hoover (1971–85) was a harbour seal who developed a remarkable ability to imitate some aspects of human speech.[14]

Although fixated on the descent of the larynx and its liberating effect on making vowels, anthropologists have attempted to discover when the modern human configuration of the SVT arose. Being composed of soft tissue, neither the larynx nor tongue survive in the fossil record. Consequently, anthropologists are reliant on searching for traces on fossilised skulls and mandibles that indicate where the larynx might once have been located: had it been relatively high as in the chimpanzee or low as in modern humans?

A great deal of attention has been given to the degree of curvature of the base of the skull.[15] The degree of basicranial flexion has been used by some researchers to argue that Neanderthals had a descended larynx, and by others to claim they didn't. As an indicator of the vocal tract, basicranial flexion has become mired in controversy. There are so many influences on the base of the skull that its curvature is unlikely to provide a reliable indicator for the position of the larynx.

We can have more confidence in the rare discoveries of a small bone known as the hyoid. This is attached to the base of the tongue and supports the larynx, being held in a cradle of muscle. The hyoid has a U shape; that of the chimpanzee has longer and more parallel horns than that of modern humans, along with a larger and more globular base.[16] Being small and fragile, only five hyoid bones have been discovered in the fossil record. The best preserved is the oldest, coming from a specimen of *Australopithecus afarensis* known as DIK-I-I dating to 3.3 million years ago.[17] This hyoid is virtually identical to that of the chimpanzee. The next oldest are fragments from skeletal remains attributed to *Homo heidelbergensis* at Sima de los Huesos, Spain, dating to 450,000 years ago. These have the modern human shape, as do two Neanderthal samples, one from Kebara, Israel, and one from El Sidrón, Spain, dating to 60,000 and 43,000 years ago respectively.[18]

Further evidence for a modern form of human tract from at least the time of *Homo heidelbergensis* comes from a small groove found on the mandible, known as the mylohyoid groove.[19] Muscles attached to this groove support the hyoid bone and larynx. As such, the groove's angle has been claimed to indicate where the larynx is located. Those on the mandibles of Neanderthals and *Homo sapiens* from the fossil record are identical in position and angle to that of modern humans; unfortunately, those from earlier fossils such as *H. erectus* and *H. habilis* have not been reported.

Although further evidence is required, that currently available suggests that a vocal tract not significantly different to that of modern humans had evolved by the time of *Homo heidelbergensis*. There would, however, have been some differences in the types of sounds that could be made. The greater extent to which the face of *H. heidelbergensis* and later Neanderthals protruded to the front, rather than being flat like that of modern humans, would have changed the shape of the SVT and the vowel sounds they produced. Similarly, the larger nasal cavities of the Neanderthals influenced the SVT, suggesting that a larger number of vowels and consonants may have had a nasal quality. Whether the sounds generated from the vocal tracts of *H. heidelbergensis* and *H. neanderthalensis* were combined into syllables and words, and whether those words were ordered using sets of rules, remains to be considered.

Acquiring clarity

Although the larynxes of monkeys and apes operate in the same manner to that of modern humans, they have two minor anatomical differences that have major impacts on the types of sounds they generate – vocal membranes and air sacs. The membranes are small ribbon-like extensions of the vocal folds; air sacs are cavities connected to the larynx that fill with air exhaled from the lungs. Because membranes and air sacs are present in all great apes, the common ancestor is likely to have had them too. Their loss during the last 6 million years of human evolution has helped deliver the capacity for speech because both membranes and air sacs impede the clarity of sounds coming from the vocal tract.

A characteristic of human speech is that its phonemes have stable frequencies enabling specific sounds to be consistently matched with the intended syllables, words and their meanings. The calls of monkeys and apes have less clarity because both vocal membranes and air sacs interfere with the airwaves produced by the larynx.[20] The membranes vibrate to a greater extent than the vocal folds, creating sounds with fluctuating frequencies. Although these are effective for loud screams and barks, they lack the consistency required for speech.

Air sacs have a similar impact. They vary between species in their size, shape, position, and likely function.[21] For some, their key role is likely to amplify calls by acting as a resonance chamber; for smaller primates they may serve to increase the duration and alter formant frequencies to make the animal sound larger than it really is. Another hypothesis is that air sacs enable primates to call faster and for longer periods without hyperventilating because they can re-breathe the exhaled air from their air sacs rather than having continually to inhale and exhale. Whatever their role, a general consequence is that air sacs interfere with the sound emanating from the vocal folds, reducing the ability to produce a set of consistent and distinctive sounds of the type required for speech.

The loss of both vocal membranes and air sacs during human evolution played a key role in providing the stable vocal qualities required for speech. This is an interesting twist from most evolutionary stories in which it is the addition not loss of attributes that enables cognitive and behavioural complexity to emerge: the addition of a bigger brain, a larger body and longer lifespan. In the case of the vocal tract, however, modernity has come from the loss of features: less anatomy has meant more vocal clarity.

Quite when that loss occurred is not known. Vocal membranes and air sacs leave no trace in the fossil record. The only hint of their presence comes from the well-preserved hyoid of the 3.3 mya *A. afarensis* specimen. Chimpanzees have their air sacs attached to the globular base of their hyoid bones, and hence we might suppose that they were also attached to a near-identical globular base found on this fossilised hyoid.

Taking control of vowels . . .

Monkeys and apes have limited voluntary control of their vocalisations. The majority of their calls are instinctive responses to stimuli: they are unable to help themselves from either making or not making a call; nor can they select the type of call to make or how to moderate the way it sounds, such as its intensity and duration. Note the word 'majority'. A decade ago, this qualifier would have been unnecessary but, as described in the previous chapter, recent research has identified a degree of intentionality and control behind the calls of several species, these being modified according to the circumstances and who is around to hear.

Even if monkeys and apes have some voluntary control over their calls, they lack the breathing and muscular control for human speech-like sounds. Speech requires a rapid succession of different sounds generated during a single exhalation of air, involving the rhythmic opening and closing of the mouth at between 3 and 8 Hz (cycles per second). A single breath containing an utterance normally ranges between two and six seconds but can last for more than twelve seconds. When modern humans generate multi-sound utterances, we maintain a constant lung pressure while air is passed along the windpipe, through the larynx and out into the world. That requires control of our abdominal and intercostal muscles (those surrounding the ribcage) with as much subtlety as we use for controlling the small muscles of the hand when manipulating an object. Such muscular control also enables us to stress one or more of the syllables we generate during the single exhalation of breath, thereby adding prosody to our words.

Monkeys and apes lack such muscular control. While they have the same set of abdominal and intercostal muscles as in humans, they lack the nerve endings coming from the spinal cord for their manipulation.[22] As a consequence, they only have time to make one sound on each inhalation or exhalation of air. The reason for their limited muscular control is evident from the cross-sectional areas of their vertebral canals, the passage in the backbone containing the spinal cord with its mass of nerves.

When controlled for body size, this area is always substantially smaller than that of modern humans, indicating smaller quantities of nerves and hence limited ability for muscle control.

This was discovered by the anthropologists Ann MacLarnon and Gwen Hewitt, who went on to measure the vertebral canals from a sample of fossil specimens to establish when the requisite degree of breathing control may have evolved. They found that the vertebral canal dimensions of australopiths and of *Homo erectus / ergaster* dating to 1.6 mya fall into the range of non-human primates, suggesting that their vocalisations were constrained to the same short durations and single-sound events that we find among chimpanzees today.[23]

Unfortunately, specimens dating to between 1.6 mya and 100,000 years ago either have not been found or remain unmeasured and hence we cannot determine when modern breathing control evolved to enable a more rapid sequence of changing sounds on a single exhalation. The vertebral canals of Neanderthals and *Homo sapiens* dating to after 100,000 years ago were significantly larger than those of *H. erectus* and equivalent to that of modern humans today, indicating the same level of control over vowel production.

That does not necessarily mean that the vowels produced by Neanderthals and *H. sapiens* were the same. We have already noted how differences in facial anatomy will have influenced the SVT and hence the formants of generated vowels. We can also note that the Neanderthals had greater muscularity, reflecting more active lives. Their lung capacities have been estimated to have been 20 per cent greater than those of modern humans.[24] This suggests their vowels could have been louder and held for longer. Moreover, because human utterances are made on a single exhalation of breath, those of Neanderthals had the potential to be 20 per cent longer than those of modern humans. The anatomical differences between Neanderthals and *H. sapiens* would also have influenced the sounds of consonants.

Greater lung pressure may have made more powerful and explosive stops, while larger nasal cavities may have imposed a greater nasal quality onto consonants.

. . . and consonants

Regarding consonants, monkeys and apes have greater control over the articulation of their tongue, lips and jaws than their voiced calls. The most detailed studies have been undertaken on the orangutan, with whom we shared a common ancestor 13 million years ago. One of its voiceless calls is the kiss-squeak, a sharp intake of air through pursed, trumpet-like lips, produced to express alarm, either at other orangutans or potential predators. An analysis of almost 4,500 kiss-squeaks made by forty-eight different orangutans living within four wild populations demonstrated considerable acoustic variability, especially regarding their frequencies and duration.[25] The kiss-squeaks were found to contain information about the identity of the caller, its population, body size and the context for the call.

Further orangutan control over voiceless calls is evident from how some populations produce raspberries while nest building, whereas others produce smacks or remain completely silent. These are social traditions sustained by vocal learning. Captive orangutans have imitated voiceless calls by humans, notably whistles, and captive chimpanzees have innovated blowing raspberries to gain human attention.[26]

One form of facial muscle control exhibited by monkeys and apes is lip-smacking, a rapid opening and closing of lips. Rather than to produce sound, this is a visual signal that coordinates and can prolong grooming between individuals.[27] The sound that lip-smacking makes is less significant than its speed, which occurs at an average rate of 4.5 Hz.[28] That is within the range that humans use when opening and closing their mouths to generate syllables when talking, and is considerably faster than the speed at which humans, apes and monkeys move their jaws when chewing food. The same speech-like rhythm has been found in the song-like calls of gibbons, and in the voiceless (consonant-like) clicks and voiced (vowel-like) 'faux-speech' sounds made by Tilda, the fifty-year-old captive orangutan.[29] She has acquired numerous gestures and calls that are otherwise absent in wild orangutans, including the mimicking of human speech. To do so, Tilda synchronises her lip movements at a rate of 4.5–6.0 Hz, within the range of human speech, and makes consonant-like clicks with her tongue at the same rate.

Lip-smacking, Tilda's calls and her clicks suggest that some aspects of the facial muscular control required for human-like speech were present within the common ancestor of 6 million years ago. Evolution often works by repurposing existing structures, and that used for lip-smacking may have been recruited for the rapid mouth movements required for speech. Similarly, the ability of apes to produce and control a variety of voiceless calls provides an antecedent for human consonants, giving another insight into the potential abilities of the common ancestor.

There would have been little point in our ancestors evolving the capacity for rapid and rhythmic multi-sound utterances if the mechanisms for hearing those sounds had not kept pace. Consequently, another source of evidence for the evolution of both speech and language is the fossil record for the anatomy and function of the auditory tract constituted by the outer, middle and inner ear.

Hearing

A spoken word is represented by nothing more than vibrating particles of air. That wave enters the outer ears of all those in hearing distance, including those of the speaker. By having two ears, our brains can detect where the sound is coming from – we can identify who is speaking. The acoustic wave travels through

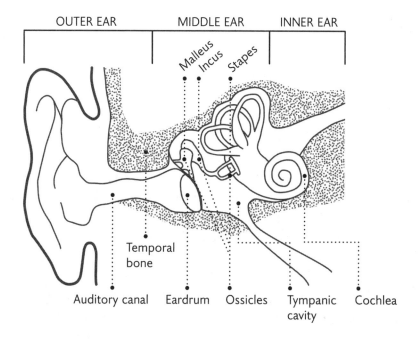

OUTER EAR MIDDLE EAR INNER EAR

Malleus
Incus
Stapes

Temporal
bone

Auditory canal   Eardrum   Ossicles   Tympanic   Cochlea
cavity

Figure 7 The human auditory pathway

narrow passages called the external auditory canals to the ear-drums, more formally known as the tympanic membranes. They vibrate and each ear sends their vibrations to three tiny bones of each middle ear, the malleus, incus and stapes, collectively known as the ossicles and contained within the tympanic cavity. These amplify the vibrations and send them to the cochlea, a snail-shaped cavity within the bony labyrinth of the temporal bone (Figure 7).

The cochleae are filled with fluid and contain an elastic partition known as the basilar membrane which is covered by tiny hair-like cells. Those near the wide end of the snail-shaped cochlea detect high-pitched sounds, such as the yell from a child when bitten by a dog, and those close to their centre detect low-pitched sounds, such as the dog's bark. The vibrations coming from the ossicles cause the fluid within the cochlea to ripple, creating a wave along their basilar membranes. This in turn causes their hair cells to move up and down, to bump and then bend against their overlying structures. When they bend, pore-like channels at their tips open up and chemicals rush in to create electrical impulses that deconstruct the acoustic wave into its basic frequencies. These frequencies are mapped onto dedicated slots in the primary auditory cortex, at which point the acoustic wave is replaced by brain waves and we have the sensation of sound – that of the first phoneme of the first syllable of the word. It will be followed by that of the second phoneme, because having generated one acoustic wave, the speaker will have immediately sent another pulse of air, and then maybe a third and more to complete the word.

The next minor miracle is that the stream of sound is 'heard' as a single word. We draw on our memorised mental lexicon to attach a meaning to the sound, usually one that is shared by members of our language community.[30] When hearing the phonetic sound stream of 'dog' we think dog – more specifically we activate the concept of dog in our mind – which, unless we misheard, is the same concept that the original speaker had in their mind when generating the sound stream. We may also think dog when we hear a sound stream of a dog barking. Other sound streams will not trigger any concepts in the mind that are necessarily shared with other listeners. Music, for instance, is not associated with shared meanings because we impose our own interpretations onto its stream of sound, or more frequently just let it flow into our brains and bodies without any thoughts at all.

This astonishing process of conveying thoughts from one brain to another via brain and acoustic waves is even more remarkable because it often occurs in the context of other sound streams – that of traffic when we are chatting with someone while walking down a street, or background noise when at a party or with the radio on. Our brains direct our attention to what we wish to listen to, while unconsciously monitoring the other sounds. When chatting on the street, for instance, we will react to the siren of a police car; while at a party we will hear and respond to our name when mentioned in another conversation despite not 'hearing' anything else being said.

Chimpanzee and human auditory anatomy compared

The anatomy of chimpanzee ears and how they operate is essentially the same as ours.[31] Indeed, this anatomy is one of the most conserved (i.e. unchanged) features in the whole of mammalian evolution, with the appearance of the three tiny ossicle bones, the malleus, incus and stapes, being a defining feature of the earliest mammals in the fossil record at more than 200 million years ago. Such conservation is reflected in development: human ear ossicles are fully formed and have reached their adult size at birth; unlike most other bones in the body, they do not remodel during the first two years of life.[32]

While chimpanzee auditory anatomy is essentially the same, there are subtle differences in the shapes of the external auditory canals, ossicle bones, tympanic cavities and cochleae that influence the degree of sensitivity to different frequencies of sound. The malleus bone is about 75 mm long and looks like a tadpole, with a chunky head and a slender tail known as the manubrium. The main difference between the chimpanzee and human forms is that the manubrium is shorter in humans. The incus is of a similar size; it also has a chunky head, which is at right angles to its slender body. The human incus is larger than the chimpanzee's and has a more open angle between the head and tail. The stapes bone is an elongated semicircle shape, with a small knob at the top called its head and a flat base called its footplate. Again, the difference from that of the chimpanzee is principally size – that of the human is larger in all dimensions. The size of ossicles alone might be less important than their ratios. One crucial measurement is known as the malleus–incus lever ratio, the extent to which these are unequal in length. This has a low value in chimpanzees and a high value in humans.

Turning to the inner ear, there are differences in the shape of the human and chimpanzee cochlea.[33] Both have the same overall design – spiral-shaped cavities in the bony labyrinth. The human cochlea has a larger volume, a slower gradient of curvature, fewer turns, a larger and more oval cross-section of the first turn of the spiral, and is thicker than that of the chimpanzee.

The combined impact of differences in the ossicles and cochlea is that humans and chimpanzees have different auditory capabilities.[34] These are measured by comparing their audiograms – charts that record how well sounds of different frequencies can be heard. Those for chimpanzees indicate two peaks of maximum sensitivity near 1 kHz and 8 kHz, and a loss of sensitivity to mid-range frequencies at around 4 kHz. This type of audiogram is found in many non-human primates and most likely represents the ancestral state. Modern human audiograms differ by having a cut-off at around 6 kHz for high frequencies, a broad region of heightened sensitivity in the mid-range of 1–6 kHz, with the greatest sensitivity at 2–4 kHz.

This contrast in audiograms has been interpreted as reflecting the reliance of modern humans on complex, short-range vocal communications for which sensitivity to mid-range frequencies would be advantageous.[35] In particular, consonants such as /t/, /k/, /f/ and /s/ have a considerable amount of energy concentrated within the frequency range of 3–5 kHz. As noted above, vowels and consonants are processed differently in the brain and may have contrasting histories regarding the evolution of speech. Vowels are more prominent than consonants in chimpanzee vocalisations as reflected in their audiograms. If those of the common ancestor of 6 million years ago were similar, then greater sensitivity to the frequencies for consonants may have been a derived feature in human evolution, indicating that consonants had a later introduction than vowels into the evolutionary history of human speech. Might we be able to test that by exploring change in the ossicles and cochlea of our human ancestors?

Auditory anatomy and capabilities, from australopiths to Neanderthals

Because ossicles are such tiny bones, they are rarely preserved in the fossil record. Nevertheless, a small and valuable collection has slowly accumulated, coming from australopiths, *Homo heidelbergensis*, *H. neanderthalensis* and *H. sapiens*.[36] They have been meticulously measured and compared with each other, and with those from chimpanzees and modern humans of today. As valuable as they are rare are those that form a complete ossicular chain – the malleus, incus and stapes from the same ear of a single individual.[37] Only three of these have been recovered, two from Neanderthal infants, from La Ferrassie and Le Moustier in France, and one from a 30,000-year-old *Homo sapiens* from Darra-i-Kur, Afghanistan. Usually, just one ossicle specimen is known from a single individual. These include a 3.5-million-year-old stapes of an *Australopithecus africanus* from South Africa, an incus of a 60,000-year-old Neanderthal from Amud in Israel, and seven ossicles of *H. sapiens* dating to *c.*100,000 years ago from Qafzeh, Israel, all from different individuals.

Moving beyond measuring ossicles, the advent of computerised tomography has enabled three-dimensional images to be made of the hollow structures within bony labyrinths, allowing the size and shape of ear canals, tympanic cavities and cochleae to be derived.[38] Notable studies are those of the cochlea of a 130,000-year-old Neanderthal from Krapina, Croatia, and of ten 450,000-year-old *Homo heidelbergensis* individuals from Sima de los Huesos, Spain, comprising six adults and four adolescents.

A further technical advance has provided estimates for the

audiograms of extinct species. This involves constructing a virtual 3D 'sound power transmission' (SPT) model for the outer and middle ear by drawing on the fossil finds, computerised tomography (CT) scans, and making reasonable assumptions for the soft tissue that no longer survives.[39] SPT models explore how sound is filtered as it moves through the various shaped cavities and materials of the outer and middle ear to create a particular set of frequencies at the entrance to the cochlea, which then shapes the resultant audiogram. This enables an estimation of the occupied bandwidth (OBW), defined as the frequencies that the species were most attuned to. A wider OBW allows for a larger number of easily distinguishable acoustic signals to be used in the oral communication – modern humans have the largest OBW of all known primates. Armed with these tools, anthropologists have begun to gain insights into the evolution of hearing, providing further puzzle pieces for the evolution of language.

The australopiths and *Paranthropus* provide a mix of chimpanzee-like and human-like features.[40] They have a human-like malleus, but a chimpanzee-like incus and stapes, providing them with a low malleus–incus lever ratio, comparable to that in chimpanzees. Similarly, they have a shorter cochlea and smaller tympanic membranes than found in modern humans. When their SPT models are explored, estimated audiograms indicate a similar sensitivity for sounds up to 1 kHz as found in chimpanzees, enhanced sensitivity over both chimpanzees and modern humans for sounds between 1.5 and 3.0 kHz, and a fall-off to minimum sensitivity levels between 4.5 and 5.0 kHz, also similar to chimpanzees.

The enhanced mid-range sensitivity between 1.5 and 3.0 kHz has been explained as an auditory adaptation to the more open habitats that these early hominins were occupying, compared with the wooded environments of the 6.0 mya common ancestor (assumed to have a chimpanzee-like audiogram). Open habitats would have favoured short-range communication which requires a sensitivity to mid-range frequencies.

From 3 million years ago, we must jump to a mere 450,000 years ago before finding the next insight into ancestral hearing. This comes from the specimens of *Homo heidelbergensis*, the common ancestor to modern humans and Neanderthals, preserved at Sima de los Huesos, Spain, described in Chapter 2.[41] Although only three ossicle bones are known, meticulous CT scans of temporal bones provide a unique insight into the auditory anatomy of an extinct species.

The Sima de los Huesos *Homo heidelbergensis* specimens have similarities to those of the early hominins by showing a mixture of human-like derived and chimpanzee-like ancestral features. The outer and middle ear – the ear canal, ossicles and tympanic cavity – are very similar to those of modern humans, while the volume, size, shape and other details of the cochlea resemble those of chimpanzees. SPT models and OBW estimates indicate that *Homo heidelbergensis* had similar hearing abilities to modern humans, but a narrower band of maximum sensitivity and an elevated upper limit for the audible frequency range. This mix of features arises from the outer and middle ear anatomy evolving at a faster rate than that of the inner ear.

From 130,000 years ago in Croatia to 50,000 years ago in France, Neanderthal ossicle specimens show marked similarities to those of modern humans, while maintaining subtle differences: they had a larger head to the malleus and a smaller footplate to the stapes, as well as minor differences in the shape of the tympanic cavity.[42] Apart from these minor differences, the Neanderthal cochlea is near identical to that of modern humans.[43] Accordingly, SPT models and OBW estimates indicate a wider range of maximum sensitivity and a higher cut-off than found in *H. heidelbergensis*, matching the 2–4 kHz range and 6 kHz cut-off found in modern humans.

Even following the emergence of *Homo sapiens* as a species

at *c.*300,000 years ago, the modern human auditory tract continued to evolve.[44] Ossicles from 100,000-year-old fossil specimens from Qafzeh, Israel, are relatively small compared with those of present-day humans. Interestingly, the extent of variability in *H. sapiens* ossicle shape and size coming from Qafzeh alone is greater than that found within the entire Neanderthal sample from the whole of Europe and western Asia. This likely reflects the much lower degree of genetic diversity within the Neanderthals than *Homo sapiens*.

An evolutionary mosaic and early developments

Although the fossil evidence is sparse and fragmentary, it suggests that the evolutionary route to modern human vocal and auditory capacities was a mosaic rather than a straight line, with different elements evolving at different rates. The human vocal tract is a consequence of multiple interacting factors, only one of which was selective pressures for vocal communication. It involved the loss of anatomical features (vocal membranes and air sacs), the gain of others (voluntary muscular control) and probable repurposing of primate lip-smacking for rapid speech.

We can be confident that by 450,000 *Homo heidelbergensis*, the common ancestor to *H. neanderthalensis* and *H. sapiens*, had a vocal tract and auditory pathway not significantly different to those of modern humans, suggesting an ability to generate and hear a similar range of sounds. Anatomical differences between *H. neanderthalensis* and *H. sapiens* in facial anatomy, dentition and lung capacity would have created differences in their speech sounds and the length of single utterances they could make.

These are profound conclusions to be drawn so early in the piecing together of the language puzzle. They suggest that major developments in human spoken communication had occurred by 500,000 years ago, long before modern humans had evolved. With so few fragments of the puzzle interior assembled, however, it remains too early to designate vocal communication by *Homo heidelbergensis* as a form of language, let alone fully modern language – although this cannot be rejected. To cross that evolutionary threshold, vocal sounds need to have become words with learned meanings that are shared within the speech community. How might words have arisen from the grunts and barks likely used by the last common ancestor of 6 million years ago with its chimpanzee-like vocal tract? To answer that, we need to assemble the next fragment of the language puzzle.

6

ICONIC AND ARBITRARY WORDS

A long-standing principle of linguistics is that the sound of a word is arbitrarily related to its meaning. The word *dog* neither looks nor sounds like a dog. *Chien, hund, kuri* or *pies* could equally be used, as they are in French, German, Māori and Polish. None of these words for dog are intuitive, with meanings that can be guessed: they need to be learned and are shared within a speech community by consent. Had *Homo heidelbergensis* with its modern-like vocal tract already begun using, learning and sharing words with arbitrary meanings? If so, how had that radical change from chimpanzee-like grunts and barks come about?

Another type of word has long been recognised but traditionally given minor attention within linguistics: iconic words, also known as sound-symbolic words. Rather than being arbitrary, these words mimic one or more of their referents' properties – the way they sound, their size, shape, movement or texture.[1] Both names can be confusing: *iconic* can mean famous as well as the way a sound or gesture mimics something else; *sound-symbolic* is misleading because symbolism can be a synonym for arbitrary and the essence of these words is that they are non-arbitrary. A third name for such words is imagistic.

Onomatopoeias are the most obvious example, as in *bang, quack* and *gargle*, but other words sound the way their referent looks, feels or moves. Having been neglected by linguists throughout the last century, they were still being dismissed as no more than a 'quaint curiosity' by Steven Pinker his 1994 book

The Language Instinct.[2] During the last two decades, however, iconic words have risen to prominence concerning the role they play in the acquisition of language during childhood and potentially in the evolution of language by our human ancestors.

Recent research suggests that iconic words may have provided the bridge between the grunts and barks of the 6-million-year-old last common ancestor and the first use of arbitrary words – whenever that occurred. That, however, is far from a new idea, being no more than a rediscovery of the significance that iconic words had attained in the views of Classical and Enlightenment thinkers about the invention of words and origin of language. We have, therefore, an especially fascinating fragment of the language puzzle to assemble, one that brings together the most ancient and the most recent ideas about the evolution of words.

The role of iconic words had been prominent in the work of Jean-Jacques Rousseau (1712–78), Moses Mendelssohn (1729–86), Johann Gottfried Herder (1744–1803) and other Enlightenment thinkers about how human language could have emerged from animal calls. They worked with knowledge of neither human evolution nor linguistic diversity, without audio recordings and big data. Their ideas have been dismissed as no more than 'bow-wow', 'pooh-pooh' and 'ye-hey-ho' theories for language origins. Such disrespectful terms fail to recognise the insights, subtlety and intellectual challenge of their arguments.[3] We must start, however, by going even further back in time, by making a visit to Classical Athens.

Ancient wisdom about words

Cratylus sits on the step of an Athenian temple and talks with his friend Hermogenes.[4] The date is 420 BC and they debate about how things should be named. Cratylus argues there must be a correct name for things regardless of what they might be currently called; Hermogenes counters that any name can be correct because names are arbitrary, agreed by convention within the speech community. Socrates arrives at the temple and joins the conversation. He persuades Hermogenes that his view implies that any new name given to a thing must be just as correct as any previous name. That causes Hermogenes to worry about the consequences of multiple names for the same thing, all with equal validity.

After a lengthy explanation about the etymology of names, Socrates suggests there had once been primary names, from which all others were derived. The primary name would have imitated the thing being named by capturing its essence via the movements of the tongue or mouth when the letters that make up the word were pronounced. The letter /r/ would have been used to signify something that moved because the tongue is agitated when making this sound; words for things that are sleek and smooth would have used the sound /l/ because this requires the tongue to glide across the mouth. He continues with further examples of how sounds capture the essence of things, citing instances of what we would today call iconic words.

Having appeared to support Cratylus' view that there must be correct names for things, Socrates gradually shifts his position. Eventually, he challenges Cratylus to explain how there can be a correct name for every word. Number words, for instance, such as 'two' and 'thirty-seven', can never be imitated by a sound. Socrates concludes by agreeing with Hermogenes that whereas names should be as much like things as possible, they are frequently arbitrary and agreed on by convention.

This scenario draws on the *Cratylus* dialogue written by Plato (424–348 BC) in 383 BC, or thereabouts.[5] In it, Plato suggests there had once been skilled and knowledgeable name-givers that 'baptised' items in the world with correct (i.e. iconic) names, thereby creating a perfect language. The passage of time introduced convention and arbitrariness, resulting in the messiness of language as he knew it. Plato's 'correct' names would now be described as iconic words.

Aristotle (384–322 BC) disagreed with Plato's idea that names had once been 'correct' and were later replaced by those agreed by convention. In *On Interpretation* (c.350 BC), he argued that names had always been arbitrary. Epicurus (341–270 BC) and his followers, the Epicureans, disagreed with both Plato and Aristotle – rejecting the idea of name-givers while accepting that arbitrary words had evolved from sense impressions. Their view of the natural world was fundamentally different from the well-ordered, timeless structure that Plato and Aristotle believed; the Epicureans perceived the world as chaotic and believed that thoughts emerged from bodily sensations rather than from an Aristotelian-type mind containing well-formed concepts.

In Epicurus' *Letter to Herodotus*, likely written sometime in the early third century BC, he proposed that words arose in two stages. The first was when:

> men's own natures underwent feelings and received impressions which varied particularly from tribe to tribe, and each of the individual feelings and impressions caused them to exhale breath peculiarly, according also to the racial differences from place to place. Later particular coinings were made by consensus within the individual races, so as to make designations less ambiguous and more concisely expressed.[6]

Quite what is meant by 'impressions' and 'exhale breath peculiarly' can be debated. My understanding is that 'impressions' refer to the sensory perception of nature, while 'exhale breath peculiarly' implies an attempt to capture that perception, whether by an onomatopoeia or some other means. Elsewhere in the *Letter*, Epicurus refers to how the sound of a falling tree would have been different in a desert, next to a waterfall, surrounded by sheep or in the midst of a thunderstorm, with this variation leading to different sounds made by different tribes.[7]

Epicurean ideas about language, and the nature of being in general, were marginalised throughout the medieval period as scholastic theology found its intellectual base in an Aristotelian view of the world. Epicureanism was rediscovered in the late sixteenth century with its critique of Aristotle playing a key role in the emergence of the Enlightenment.[8] Gottfried Leibniz (1646–1716) developed Epicurean ideas about the origin of language by arguing that arbitrary meanings for words had not arisen by deliberate naming-acts of individuals but through a long-term social process that gradually transformed initial sound impressions. This evolutionary view, however, was soon contested by virtually all Enlightenment thinkers and writers.

Enlightenment views on the origin of language

Jean-Jacques Rousseau led the critique within his *Discours sur l'inégalité* of 1754. He argued the gap between animal-like natural cries and arbitrary words was too great to have been overcome by a gradual process of change. More fundamentally, Rousseau and other Enlightenment thinkers were committed to the idea that human language and rational thought were inseparable. How then could arbitrary words have ever been devised? Without such words, the required thought to invent them was absent. Moreover, how would any agreement be made about which arbitrary words should be used if people lacked the capacity to converse, for which arbitrary words, with their meaning already agreed by convention, were required?

Moses Mendelssohn, a German-Jewish philosopher and theologian, translated Rousseau's discourse into German and appended an essay in reply. He suggested that arbitrary words might have arisen by an association of ideas which had then become disconnected. Mendelssohn's example used sheep and flowers. When still reliant on imitating natural sounds, people might have encountered a sheep surrounded by flowers in a meadow. While for most people, the sound of bleating would be associated with the sheep, for some it might summon the entire scene: sheep, flowers and meadow. Gradually, over a large span of time, elements of this scene might be lost, leaving the bleating connected to the flowers alone – providing a word that sounded like a sheep's bleat but that meant flower. Mendelssohn, however, did not address Rousseau's other concern: how this new arbitrary word for flowers could have spread and become adopted throughout the speech community. Other members of that community, going through a similar process, might have concluded that the bleat sound meant sheep, grass, green or any other single element of the original scene.

By the mid-eighteenth century, the Academy of Sciences in Berlin, the capital of Prussia, had become a court for discussing intellectual ideas, aspiring to attain the prominence of London and Paris. To further promote its status, the Berlin Academy began a series of prize essay competitions. The essay set for 1759 concerned the relationship between language and thought, beginning with the question: *To what extent do the opinions of a people influence its language, and alternatively how does language influence the opinions?*[9] It went on to ask how language would emerge in a child that lived in isolation from all other human beings.

The essay competition was won by Johann David Michaelis (1717–91), a German-born biblical scholar. His doctoral dissertation had been on Hebrew vowels (or rather their absence in written texts) and he promoted expeditions to Arabia to search for the roots of western civilisation. His essay addressed the emergence of language by suggesting that the first words had expressed the physical aspects of objects that struck speakers most vividly – implying some form of iconic sounds. Over a long period of time and in the context of social life, the original sounds were changed, making the words seem completely arbitrary.

The next prize essay relating to language was in 1771 and continued the theme addressed by Michaelis in 1759. It asked: *Abandoned to their natural capacities, would human beings be in a condition to invent language? And by what means could they achieve this invention on their own? We require a hypothesis explaining the matter clearly and satisfying all the difficulties.*

Johann Gottfried Herder, a German-born clergyman, poet and literary critic, rose to the challenge and won the competition – despite admitting he had no answer to the question.[10] He began his essay with the bold and striking statement that 'while still an animal, man already has language'.[11] By 'still an animal', Herder is referring to an absence of self-awareness, rational thought and intentionality. As with Rousseau, Herder believed the divide between the language of man while still an animal and the arbitrary nature of human language was so vast that it ruled out any natural transition between the two, whatever length of time and number of sheep in flowering meadows had been available. Equally, he was unprepared to accept the notion of divine intervention for the origin of language – arguing that it was worthier of God to let his creation, the human mind, develop language on its own.

Our interest is with Herder's conception of the natural language possessed by man while still an animal – who we might equate to our earliest ancestors as now known from the fossil record. Herder challenged the idea that such natural language had merely been the imitation of natural sounds by referring to multimodal sensations. By doing so, he pre-empted the science of synaesthesia – the phenomenon by which activation of one sense stimulates that of another, such as when someone automatically experiences a colour when they hear a number. Two hundred and thirty years after Herder's essay, synaesthesia was directly linked to iconic words and the evolution of language. Herder is worth quoting at length:

> At the basis of all the senses there is sensation, and this established for the most varied forms of sensation so intimate, so strong, so ineffable a bond that from this interconnection the strangest phenomena arise. I know of more than one instance where an individual – naturally and possibly under the impression of something retained from childhood – could not but associate, by a direct and rapid impulse, this particular sound with that particular colour, this particular phenomenon with that particular dark and quite different feeling, where a comparison through slow reason could detect no relationship whatever. For who can compare sound and colour of phenomenon and feeling? We are full of such interconnections of the most different senses. . . . The philosopher must abandon one thread of feeling as he pursues another. But in nature all threads are one single tissue. The darker the senses, the more they commingle; the greater a man's lack of experience, that is, the less he is trained to use one without another, to use it with skill and clearly, the greater the darkness! – Let us apply this to the beginning of language. . . . Most visible things move. Many sound while moving. If not, they lie close to the eye in its early state, directly on it as it were and can be felt. The sense of feeling is close to that of hearing. Its epithets – such as hard, rough, soft, woolly, velvety, hairy, rigid, smooth, prickly etc. which all concern surfaces and do not penetrate – all sound as though one could feel them. The soul, caught in the throng of such converging sensations and needing to create a word, reached out and grasped possibly the word of an adjacent sense whose feeling flowed together with the first.[12]

Iconic words become marginalised, but not forgotten

Such holistic views of language became lost as Romanticism took hold of European thought during the late eighteenth century, and even more so during the early nineteenth and into the twentieth century with the rise of positivism and modern science. It was within that intellectual context that the Swiss linguist Ferdinand de Saussure (1857–1913) promoted 'l'arbitraire du signe' – the idea that words have arbitrary meanings agreed by convention.[13]

Saussure was a remarkable linguist, having made a break-through in the reconstruction of Proto-Indo-European at the age of twenty-one. Between 1907 and 1911 he delivered a series of lectures at the University of Geneva that were posthumously published in 1916 as *Cours de linguistique générale*, a work that established his reputation as the 'father of linguistics'.[14] L'arbitraire du signe soon became adopted as a fundamental principle of language. It was promoted as one of the seven design features of language by Charles Hockett (1916–2000) in his influential *A Course in Modern Linguistics* of 1958. It has been an axiom in every textbook on linguistics I have ever opened.[15]

Despite the overwhelming acceptance of l'arbitraire du signe, it soon became challenged, although the challenge made little impression on the discipline. Otto Jespersen, a professor of English in Copenhagen, devoted a short chapter in his 1922 book *Language: Its Nature, Development and Origin* to sound symbolism. He noted the interest in sound symbolism shown by Classical writers, considered various types of iconic sounds and words, and finished with a swipe at Saussure: 'I hope this chapter contains throughout what is psychologically a more true and linguistically a more fruitful view.'[16]

That view gained support in 1929 from two experimental studies. Edward Sapir, an American linguist specialising in Native American languages, described an experiment in which partici-pants were told that the nonsense words *mil* and *mal* referred to tables of a different size.[17] Which was the larger table? Whether the participants were English or Mandarin Chinese speakers, adults or children, the majority (about 80 per cent) intuitively chose the word *mal*. This was explained by saying that *mal* requires a larger oral cavity to be made than when saying the word *mil*. In the same year a similar experiment was reported by Wolfgang Köhler, a German psychologist, who was exploring cognition in general rather than language per se.[18] He gave participants a rounded shape and a spiky shape and asked which should be called *maluma* and which *takete*. The overwhelming opinion was that *maluma* was appropriate for the rounded shape. This was explained by the rounded movements of the tongue and mouth when required to say *maluma* and the sharp movements of the tongue needed for the word *takete*. Sapir and Köhler's experiments demonstrated that humans are prone to cross-modal perception – an interaction between what they perceive in the visual and auditory sphere such that one can influence the other.

Following the seminal contributions by Jespersen, Sapir and Köhler, research on iconic words continued in a piecemeal fashion, but remained outside the linguistic mainstream.[19] That gradually began to change from the 1970s and then quite dramatically during the last two decades.

The rediscovery of iconic words

In 1971 the American linguist Roger Wescott collated the accumulating evidence for iconic words in a short but striking journal article entitled 'Linguistic iconism'.[20] Wescott took a broad view of language, arguing that many animal signals, such as the dances of honeybees and the alarm calls of finches, can be interpreted as iconic. He posited that the arbitrary signs of alphabets had evolved from iconic forms, as had most signs of American Sign Language.

Turning to speech, Wescott proposed that iconism is pervasive throughout language. He noted that onomatopoeias are the most familiar type of iconic words. These capture the sound of their referent, as in *plop* and *bang*, although many onomatopoeias are quite arbitrary – such as *oink* for the sound made by a pig. Other iconic words use vowels, consonants or syllables to represent the physical properties of their referent, indicating a link between auditory and visual perception. In languages throughout the world, high, front vowels, as represented by the IPA symbols /aɪ/, /ɪ/ and /iː/, are associated with small, light and fast-moving things at a higher frequency than one would expect by chance, as found in the English words *tiny*, *little* and *bee* respectively. Conversely, low front and long back vowels such as /æ/, /ɑː/ and /uː/ are used in words referring to or which describe large referents, as in *mammoth*, *vast* and *huge* respectively.

Wescott noted a similar association with consonants: laterals (in which the tip of the tongue blocks the passage of air, such as /l/) connote smallness, as in *little*, *low and light*, while labials (made using the lips, such as /b/) connote largeness, as in *big* and *bulbous*. Wescott argued that words representing muffled sounds and blurred images frequently use muffled or blurred syllables, as in *muffle*, *mumble*, *hum*, *blur*, *dull* and *dust*. He noted that iconism is found not only in European languages, being especially prominent in Spanish, but also in languages throughout the world, and stressed the role of iconism in tonal languages. Within Bini, spoken in southern Nigeria, words meaning high, tall or loud have uniform high tones; those meaning low, short, faint or dull have uniform low tones, while things that are irregular in shape, movement or behaviour have words with variable tones.

Wescott continued by citing further iconic roles for consonants as evident from a variety of languages: stops (e.g. /p/, /b/ and /d/) signal brevity or discontinuity; voiceless consonants (e.g. /f/, /s/ and /p/), especially when combined with high vowels, signal inaudibility or at least the lack of sonority (as in *sizzle* and *whisper*); dental obstruents (e.g. /d/ and /t/) are prominent in words that denote projections from the earth or body, such as Proto-Indo-European *dent* = *teeth*, Efik (Nigerian language) *ot* = *head*, *eto* = *tree*, and Mixtec (Mexico) *tu* = *tail*, *duti* = *mountain*. Note that the asterisk against *dent* indicates that it is a reconstructed word.

According to Wescott, words for male and female often have an iconic element. They either use vowels that denote size, with the contrast between *vixen* and *fox* in English matching that between *little* and *lot*, or are denoted by consonants. Languages throughout the world use hard (or stop) consonants for father, as in *dad* and *pa*, and soft, vowel-like (or continuant) consonants (e.g. /l/, /m/ and /n/) for mother, as in *mommy*, *mum*, *ma* and *nan*.[21]

Wescott also noted an alternative source for mother-type words, one that had been proposed by Roman Jakobson in 1960. He suggested the pervasive use of the labial, nasal consonant /m/ in these words may derive from breastfeeding. Jakobson had written:

> Often the sucking activities of a child are accompanied by a slight nasal murmur, the only phonation which can be produced when the lips are pressed to mother's breast or to the feeding bottle and the mouth full. Later, this phonatory reaction to nursing is reproduced as an anticipatory signal at the mere sight of food and finally as a manifestation of a desire to eat, or more generally, as an expression of discontent and impatient longing for missing food or absent nurser, and any ungranted wish. When the mouth is free from nutrition, the nasal murmur may be supplied with an oral, particularly labial release; it may also obtain an optional vocalic support.[22]

This implies that the /m/ iconic word for mother has effectively been reinvented by every infant.

Wescott went beyond the sounds of words to consider their morphology and the syntax of language. When we wish to emphasise a word or enlarge its meaning by adding tense or possession, we tend to make it longer, as in *book* vs *books*, *jump* vs *jumped*, *long*, *longer* and *longest* (and its Latin cognate *longus*, *longior* and *longissimus*). He gave an example from Russian to illustrate how words referring to an earlier time exhibit stress on their first syllable of the word, and those for a later time place emphasis on the second or third syllables. Word order in most languages is structured by subject-verb-object, which often provides an iconic representation for the order of events.

Wescott's 'linguistic iconism' can be faulted as being a hotch-potch of cherry-picked examples of iconic sounds and words, each of which might be contradicted by counter examples (e.g. in the Bahnar language spoken in Vietnam, /i/ is used in words for big, and /a/ in words for small).[23] I have dwelt on his contribution, however, because Wescott concluded with some bold propositions that recall the ideas of Classical and Enlightenment thinkers while pre-empting twenty-first-century research on iconic words. He proposed (among other things) that: (1) all languages have a degree of iconism; (2) although less precise than arbitrary words, iconic words are more easily understood by those who do not know the language; (3) many arbitrary words within a language had once been iconic, such as the English word *laugh* that derives from the Proto-Indo-European word **klak*.

Wescott concluded that language might have once consisted exclusively of iconic words, but this pan-iconicity was 'shattered' (which seems a strong word to use) by the introduction of arbitrary words. These subsequently replaced iconic words at a constant rate until language became what some of his contemporary linguists had prematurely assumed to be wholly arbitrary.[24]

Hybrid signs and words

As Wescott noted, iconic words are more prevalent in some languages than others. Sign languages have a far greater degree of iconicity than spoken languages. As with spoken words, the signs have a sliding scale of iconicity, ranging from the entirely iconic to the entirely arbitrary, with hybrid signs in the middle. Transparent signs are those for which the meaning is obvious to anyone with a shared cultural background; translucent signs have meanings that cannot be guessed by a non-signer but are easily understood once the context and motivation of the signer becomes known; obscure signs have lost the iconicity they once possessed by the signs becoming conventionalised; opaque signs are and were always arbitrary.[25] All the signs, even the transparent ones, are partly conventionalised, setting them apart from the iconic gestures that accompany our speech.

Spoken words can also be hybrids, a combination of sound-symbolic and arbitrary sound segments. As such, there is a sliding scale from absolute to relative iconicity in words, from those are that are most explicitly iconic and easily described as imagistic, to those in which iconicity can be subtly buried within the word such that it can be mistaken for being entirely arbitrary.[26]

Words known as phonoaesthemes are interesting examples. These have an initial morpheme followed by an arbitrary morpheme, although one that lacks any meaning (which is unusual for a morpheme).[27] The initial sound is repeated in several different words that have similar meanings. For instance, English words that relate to unhurried movement consistently start with *sl-*, such as in *slow, slide, slur, slouch* and *slime*. In this case the movement of the tongue over the palette to make *sl-* captures the essence of those words – we can only describe the tongue as moving slowly and sliding. Another English example are those words that start with *fl-* that refer to a sudden or moving light, such as *flame, flash, flare* and *flicker*, with that general meaning captured by the initial fricative, *fl-*, providing a sudden puff of air which is similar to the suddenness of a light coming on. English words referring to unmoving bright lights consistently use *gl-*, as in *gleam, glisten, glitter* and *glint*. In this case, it is less easy to identify *gl-* as having an iconic role.[28]

Sound symbolism enters the linguistic mainstream

Following Wescott's 1971 article, evidence for iconism in language continued to accumulate. In 1994 a heavyweight book entitled *Sound Symbolism* indicated the study of iconic words was entering the linguistic mainstream.[29] Based on an international conference, this volume had contributions from twenty-four distinguished linguists on the role of sound symbolism and drew on languages from all continents. The book provided a new typology for sound-symbolic (iconic) words, and a comprehensive coverage of the many ways in which they capture meanings in non-arbitrary ways.

Two of its chapters were devoted to the ideophones found within the languages of indigenous Australia and sub-Saharan Africa.[30] These words provide a particularly vivid, expressive and direct form of imagistic iconism, often appearing to capture a multimodal sensation – sight, movement, smell and so forth in one prosodic expression.[31] In both chapters, they were described as being prominent while sitting outside the normal rules of language, having unusual sound properties, little if any morphology, and using the repetition of sound segments to indicate movement and/or the passage of time. Those of the Yir-Yoront Indigenous Australians of Cape York Peninsula include *karrkvr-rkvrr* (which could be continued ad lib) to represent the spearing of an object in water multiple times; *kitkitkit*, for the darting of a snake's tongue in and out; *chichichi*, for a dog running; and *porrl*, for the dumping of a water lily on the ground. They were summarised as constituting a part of speech that belongs in an entirely different realm from the familiar nouns, verbs, and particles, and from bound morphemes. The members of this class are phonologically aberrant in certain patterned ways, are apparently exempt from regular sound change, tend to be onomatopoeic and sound-symbolic in certain limited ways, and constitute the only vocal communicative noises that are permitted in social contexts where 'speech' is forbidden.[32]

Sound Symbolism concluded that iconic sounds and words play a considerably larger role in language than scholarship had hitherto recognised. While it was the definitive work for the time, the ideas and data within *Sound Symbolism* were soon outdated. Most notable was a 2001 publication by the cognitive scientists V.S. Ramachandran and Ed Hubbard. This was ostensibly about synaesthesia but contained a proposal for how iconic words had kicked off the evolution of language.

Synaesthesia, cross-modal perception and iconic words

Synaesthesia is the condition by which a sensation experienced in one modality, such as vision, causes another to be stimulated, such as hearing, typified by those who always see red when they hear the number two. It is caused by heightened levels of connectivity within the brain – sometimes referred to as 'leakage' between cortical areas. Synaesthesia is common in young children but is usually lost after the age of ten as the neuronal connections with the cortex become pruned.[33] This is likely related to increased functional specialisation of cortical areas of the brain, as domain-general learning processes are supplanted by domain-specific processes, specialised mechanisms relying on specific types of inputs for undertaking specific types of cognitive tasks.[34] If the pruning does not occur, some degree of synaesthesia will remain into adult life – just as Johann Gottfried

Herder had speculated in his 1771 essay. Synaesthesia is now recognised as an exaggerated form of the cross-modal perception that all people experience to varying degrees, rather than as a qualitatively different phenomenon.[35] Cross-modal perception had been the key interest of Wolfgang Köhler in 1929 and cited by Herder in 1771 when he wrote that 'we are full of such interconnections of the most different senses'.

Ramachandran and Hubbard repeated Köhler's *maluma* and *takete* experiment, using similar shapes called *bouba* and *kiki*. They found the same result: 95 per cent of participants named the curvy shape as *bouba* and the spiky shape as *kiki*. They argued that both the movements of the tongue against the palate when saying *kiki*, and the sharp changes of the emanating sound contours, mimic the jagged changes in the lines in the spiky figure. Sensory-to-motor synaesthesia was invoked as an explanation for this chain of mimicry: an external object's visual appearance stimulates a visual region of the brain that stimulates an auditory region, which in turn stimulates a motor region that controls the movement of the tongue and lips. It is not by chance, they argue, that to make the high, front vowels that denote small objects the mouth itself becomes small; the converse happens to make vowels that signify large objects – the oral cavity is opened wide.[36] They could have mentioned Plato / Socrates from 383 BC who noted how the tongue becomes agitated when talking about movement.

Ramachandran and Hubbard took another step by suggesting that language can involve the cross-activation of two motor areas of the brain. They recalled how Charles Darwin wrote that we often unconsciously clench and unclench our jaws when using a pair of scissors for cutting. By so doing, our mouth movements are mimicking the actions of our fingers with the scissors. This is paralleled, they suggest, by how the small size of the oral cavity and position of the tongue against the palette when saying *tiny* and *teeny* mimics the pincer-like opposition of the thumb and forefinger to denote small size. Similarly, the outward pout made with my lips when I say *you* (or *tu* in French and *thoo* in Tamil) mimics my outstretched finger when I point to you; conversely, my lips and tongue move inwards when I point to myself by saying *me* (or *moi* in French and *nann* in Tamil). In 1971 Wescott had provided a further example: the asking of a question is often denoted by both a rising intonation and a rising of the eyebrows.

How ancient is cross-modal perception?

According to Ramachandran and Hubbard, a web of synaesthetic links within the early human brain connecting the appearance of objects, motor movements and vocalisations created iconic sounds and provided the initial kick-start to language – a rediscovery of Herder's 'throng of converging sensations as the origin of language'.

If so, when did that web of synaesthetic links arise, and when did the kick-start begin? Were those links in the brain of the 6-million-year-old last common ancestor of chimpanzees and humans? Or did they evolve within the *Homo* lineage after 3 million years ago, giving *Homo habilis* the kick-start to language?

As before we can look to experimental studies with living primates. Rhesus monkeys come from southern, central and southeastern Asia where they live in large groups. The males are larger than the females, and the largest male is the dominant individual with the group. They communicate with a variety or calls and gestures; one of their calls is a 'coo' sound. Its formant – a measure of its pitch and resonance – changes with the monkey's size, reflecting the length of its vocal tract. The coos of larger monkeys have a greater resonance and lower pitches.

An experiment tested whether monkeys spontaneously associate the vocal resonance, or formant, of a coo with body size. Do they associate large-sounding coos with large monkeys, which might be comparable to humans associating large back vowels of /o/ and /u/ with large objects?[37] Monkeys were seated in front of a video display showing the face of a large, sexually mature monkey and a small, juvenile monkey. They were played a simulated coo from either a long or a short vocal tract and then tested whether they preferentially looked towards the large or the small face (the size of the face correlating with overall body size).

The monkeys were able to discriminate between the coos and spontaneously associated the large coos with the large face and vice versa. This is open to two competing interpretations. The monkeys might have previously learned from experience that large coos are made by large conspecifics. Alternatively, their brains might be susceptible to the cortical leakage that creates the cross-modal perception in humans which underlies the use of iconic words. To differentiate between these interpretations, further experiments are needed to test whether monkeys look at larger objects of an unfamiliar type when hearing 'larger' coos.

Chimpanzees also exhibit cross-modal perception relating sound to visual stimulus.[38] Six chimpanzee and thirty-three human participants undertook a near-identical experimental test that required them to classify black and white squares according to their colour while hearing background sounds that were either high pitched or low pitched. Both groups spontaneously associated the high-pitched sounds with luminance (white) and low-pitched sounds with darkness (black). Further confirmation of cross-modal mapping came from an experiment that discovered chimpanzees can spontaneously detect the similarity between shapes and sounds which are structured in the same manner, referred to as being isomorphic.[39] The sounds consisted of three discrete pitches, the first and last being identical, while the shapes consisted of a row of three images, also with the first and last being identical.

These experiments indicate that cross-modal perception

might have been present in the last common ancestor with humans of 6 million years ago. Two other studies, however, suggest that the extent of such perception is limited. Both involved variants of the *maluma / takete* experiment, first undertaken by Köhler in 1929.

In the first of these, chimpanzees, gorillas and humans heard either a *maluma*-type ('round') or a *takete*-type ('spiky') word and were then asked to touch either a rounded or a spiky shape that appeared on a touch screen. Only the humans identified the association.[40] The 'failure' of the apes to do likewise might have reflected their unfamiliarity with words – the idea that a sound can represent a visual image. Consequently, a second experiment was undertaken with Kanzi, the language-trained bonobo (*Pan paniscus*) who had already demonstrated a high rate of success at matching words and images.[41] The outcome was the same; even with training in the use of words, this bonobo neither selected round shapes to go with *maluma*-type words, nor spiky shapes with *takete*-type words.

Overall, this evidence suggests that the type of synaesthetic links within the early human brain that Ramachandran and Hubbard suggest kick-started language, evolved after the time of the 6-million-year-old last common ancestor, likely with the *Homo* genus itself.

A new wave of studies

Sound Symbolism of 1994 and Ramachandran and Hubbard's 2001 proposal laid the basis for a wave of studies about iconic words during the past two decades.[42] This has placed iconicity into the mainstream of linguistics and provided a key part of the puzzle for the invention of words and evolution of language.

This new phase of research has been distinguished by a greater interest in real words as used in real life rather than with *kiki, bouba, takete, maluma, mil, mal* and other nonsense words used in laboratory settings.[43] Of most interest are the studies undertaken by the anthropologist Brent Berlin that found iconism plays a key role in the naming of birds and animals by hunter-gatherers and subsistence farmers.[44] Within Amazonian languages, for instance, nasal consonants are preferred for naming fish, with their size marked by vowels and their movement indicated by the acoustic character of consonants. Rather than anecdotal reporting Wescott-style, huge corpuses of words from languages from throughout the world have now been subjected to rigorous statistical analysis to identify the extent of iconicity.[45] Experimental research has accompanied this concern with real words, now undertaken on a far larger scale and with more rigorous controls than had been used by Edward Sapir and Wolfgang Köhler.[46]

Four major outcomes have emerged from this wealth of research: (1) iconic words are universal; (2) iconic words are easier to learn than arbitrary words; (3) iconic words constrain vocabulary and concept development; and (4) iconic words likely played a key role in the evolution of language. Each of these require some consideration because they are all important pieces of the language puzzle.

Iconic words are universal

The first outcome is confirmation of the cross-linguistic and universal nature of iconicity. Köhler's *maluma/takete* experiment has been replicated with speakers from a wide variety of languages, consistently with the same outcome.[47] The speakers include toddlers and younger infants, who return the same result that Köhler observed in 1929. It only fails to appear in infants of six months and younger, suggesting their degree of brain maturation may not have attained the level required for the cortical leakages that lead to synaesthetic experience.[48] A greater sensitivity to sound symbolism for *maluma*-type words than *takete*-type words has been detected, with the latter appearing at a later age and possibly requiring some exposure to language.[49]

Further experimental studies have tested whether people can identify the meaning of iconic words within languages for which they have no knowledge – such as English speakers listening to words in Japanese. These experiments asked the participants to match antonym pairs of iconic words (i.e. those with opposing meanings, such as big/small, fast/slow) from their own and the languages unknown to them. The outcomes of the experiments were summarised as 'demonstrating consistent mapping between sound symbolic words and their meanings or meaning dimensions across multiple semantic domains [i.e. size, shape, motion] and across ten unrelated languages with distinct phonologies and language histories'.[50]

Another striking demonstration of the universal propensity to associate specific sounds with specific meanings has come from a statistical study of words from more than 6,000 of the world's languages.[51] This demonstrated that a considerable proportion of 100 basic vocabulary items show persistent sound-meaning associations irrespective of language families, environment or culture. Some of these have already been noted in the more anecdotal studies, such as that of /i/ with words for small, and /r/ with words for round. Surprisingly, however, words for big and large lacked a consistent association with the long back vowel of /o/ that has been often reported. Strong associations of consonants with body part words were confirmed: that for tongue with /l/ and that for nose with /n/, these associations reflecting the distinctive sounds made by these organs. Words for breasts were consistently associated with the consonant /m/ and the high back vowel /u/, supporting the proposition that these words derive from the sound of suckling.

Numerous other associations were found for which an iconic explanation is not readily evident, although cannot be entirely rejected: *fish* with /a/, *dog* with /s/ and *ash* with /u/. Similarly unexplained are negative associations. While the word for *bone* was strongly associated with /k/, it was negatively associated with /y/, as was the word for *dog* with /t/, *nose* with /a/ and *skin* with /m/ and /n/.

Iconic words are easier to learn than arbitrary words

The second outcome from the post-2000 research is confirmation that iconic words are easier to learn than those with arbitrary meanings and play a key role in the initial stages of language learning by infants and children. Non-arbitrary words, which may have variable degrees of iconicity, are learned earlier and in greater numbers than arbitrary words.[52] They dominate the lexicons of two- to six-year-olds, after which a gradual shift occurs so that by the age of thirteen, lexicons are dominated by arbitrary words. During the early stages of language learning there may be feedback to cross-modal perception – by learning iconic words, children become more susceptible to synaesthetic experience. For instance, while sensitivity to iconic words that map onto round objects (e.g. *bouba*) appears to be innate, that to words with iconic association to spiky objects (e.g. *kiki*) develops with age. This might reflect a perceptual bias or greater experience of infants with round rather than spiky objects, or indicate an element of learning regarding the associations between word sounds and qualities of their referents.[53]

Iconic words are easier to learn because their meaning is grounded in the sensations experienced by the child – the sound, size, shape, texture, movement and other properties of the object or action being named. As originally argued by Ramachandran and Hubbard, there is a biologically endowed ability to integrate input from several different senses so that iconic words are automatically associated with the correct referent in the world, this having been confirmed by new brain imaging studies.[54] In this regard, iconic words alleviate what is known as the *gavagai* problem of word learning. This has troubled linguists ever since it was identified in 1960 by the philosopher Willard Quine.[55] He noted the challenge faced by a child when an adult points to a scene and says: 'Look at the *gavagai*.' How does the child know what the adult is pointing to? If a rabbit is present, and we assume that is the *gavagai*, then how does the child know that the word refers to the whole rabbit, rather than one of its properties (e.g. hopping, big ears, fluffy tail) or the complete scene, which might be a field containing buttercups and holes in the ground? By using an iconic word that captures the size, shape, movement and other qualities of the referent, the child can more easily identify its meaning than if an arbitrary word has been used.[56] When an adult says, 'Look at the balloon,' a child will instinctively be drawn to a large round object.

By acquiring iconic words, infants come to understand that speech sounds refer to entities in the world, preparing them for the harder task of acquiring the meaning of arbitrary words. As such, iconic words scaffold the entire process of language acquisition. Parents are intuitively aware of this by littering their infant-directed speech with iconic words, often replacing those with an arbitrary meaning or repeatedly using the two in direct association: *woof-woof, this is a dog*.[57] Similarly, word-like sounds such as *oops, shh!, oh-oh* and *tsk-tsk* are often used when speaking to pre-linguistic children. While these may be entirely or partly iconic, their meaning is primarily indicated by context – an accident preceding *oops* or shouting that of *shh!* These word-like sounds provide the idea of words, supporting the more challenging task of learning the meaning of entirely arbitrary words. The relative degree of iconicity is also relevant to learning sign language: the signs that are most iconic are the first to be acquired by deaf children and they facilitate sign learning by hearing adults.[58]

Iconic words constrain vocabulary and concepts

The third outcome of the post-2000 research is that while iconic words might be easier to learn, they impose constraints on language that can only be overcome with arbitrary words, whether the word is spoken or signed.[59] Iconic words are most effective for general categories rather than specific referents, and are especially poor for proper names. How can iconism alone distinguish between two, equally small and fast-moving types of birds? That is not a problem for a young child who simply needs a word for the category of bird (or even small, quick bird), but it would be for an adult birdwatcher who needs to communicate about different types of small, quick birds they had seen. This problem can be alleviated by making slight variations in the quality of the vowel sounds to enable graded contrasts in size and other properties to be indicated. Rather than just using vowel sounds to refer to small and large items, they might be nuanced to refer to the very small, medium-sized and the very large.[60] But even if there are subtle differences between the sounds of two iconic words, there is always a risk of confusion and an inherent ambiguity. Consequently, there is a threshold on the size of lexicon that can be met by iconic words alone. To get beyond that threshold, arbitrary words are required to remove ambiguity and avoid confusion.[61] This would begin by adding arbitrary sound segments to otherwise iconic words, such as extending *fl-* by adding either *-ame*, *-ash* or *-are* to discriminate between a *flame*, a *flash* or a *flare*, before creating entirely arbitrary words.

The limited ability of iconic words to refer to specific referents within a general category may influence the prevalence of iconicity in different types of words. In both English and Japanese, for instance, iconism has a greater presence in words for actions and the properties of objects, than in the names of objects themselves.[62] This is probably because objects need to be more finely discriminated between than do actions. Within the general category of objects, iconic words are more frequent within those categories for which there are fewer members and hence less chance of confusion. There are, for instance, more concepts similar to *apple* than there are for *balloon*, and hence *balloon* can enjoy the benefit of being an iconic word without the risk of it being confused, whereas arbitrary words are required to avoid confusing an apple with a pear.[63]

A reliance on iconic words not only constrains the size of the lexicon and risks confusion as to their referents, but also inhibits the types of words that the lexicon contains. The grounding of iconic words within one or more bodily sensations ties them to the real world and to concrete rather than abstract meanings. Ideas about freedom, justice or any of the multitude of abstract concepts that pervade our thinking are expressed by arbitrary words. These are defined by reference to other words rather than to the entities in the world that impact on our bodily sensations. A reliance on iconic words not only makes it difficult to express abstract concepts but to acquire them in the first place.[64]

Iconic words may have played a role in the evolution of language

The fourth outcome from post-2000 research is a catch-up with Plato and Johann Gottfried Herder by appreciating that iconic words are likely to have played a key role in the evolution of language. Ramachandran and Hubbard did so in 2001, by proposing that synaesthesia had kick-started language evolution. That was elaborated in a later review of synaesthesia and cross-modal thought by Christine Cuskley and Simon Kirby which proposed that 'the use of sensory sound symbolism would have allowed us to express and understand a variety of elementary concepts, from sharing the visual details of our surroundings to valuable information about food sources'.[65] These might have been simple onomatopoeias for types of animals, words that mimicked the calls of birds or the roars of lions, or expressions that blended multi-sensory impressions, such as a hissing and slithering type of word for a snake, accompanied by a similar type of gesture. Another study suggested that those individuals who had a heightened level of cortical leakage – Stone Age synaesthetes – 'might have been effective conduits for generating or propagating proto-words that might be mutually intelligible within early speech communities'.[66] 'Mutually intelligible' is the key phrase, because such proto-words would not need to have been learned and shared by consent – they would have been intuitively understood by virtue of shared sensory systems for sound, sight, taste and touch.

Other researchers have implicitly drawn an analogy between language learning and language evolution – if iconic words are the easiest to learn by infants, they might also have been the easiest to learn by human ancestors with relatively small brains. Brent Berlin suggested that iconic words 'had enormous adaptive significance for our hominid ancestors as they began to play the naming game in earnest'.[67]

The psychologists Pamela Perniss and Gabriella Vigliocco proposed that iconic words provided the means to communicate about entities that were not visibly present. This is known as displacement and has been proposed as a significant step in our evolutionary past.[68] How, for instance, could hominins have used language to plan an antelope hunt, some to chase the herd and others to lie in ambush, if they lacked the ability to use the word for antelope when the animal was absent from sight?

Primate vocalisations, such as those of chimpanzees and vervet monkeys, are only known to refer to the here-and-now, being responses to events that are in the field of vision (or hearing, smell, touch). The case of displacement within the orangutan predator alarm calls that we noted in Chapter 4 is likely to be a derived rather than ancestral trait, reflecting the ecological and social circumstances of orangutan life. At some time during human evolution after the last common ancestor with the chimpanzee, our ancestors also began to refer to events of the past, to referents that could be neither seen nor heard. Not only that, but also to events of the future. Such reference is required for planning cooperative behaviour, whether big game hunting, plant gathering or carcass scavenging. Perniss and Vigliocco suggest that

> The use of iconicity, i.e. of imagistic, imitative representations of real objects and actions with objects, would be a key component in achieving displaced reference. For example, in attempting to communicate to someone else the intention to go hunting, one could rely on conceptual traces of previous sensori-motor experiences in hunting, using the face, hands, body and vocal cords to imitate what can be retrieved of these previous sensori-motor experiences to convey the intention to go hunt.[69]

If these linguists and psychologists are correct that iconic words provided a vital first step in the evolution of language, we then need a mechanism for the transition to lexicons dominated by arbitrary words, as all languages are today. Earlier in this chapter we came across a possibility: Gottfried Leibniz had argued that arbitrary meanings for words arose through a long-term social process that gradually transformed what he termed 'initial sound impressions' and we would now call iconic words. Hybrid words, those with both iconic and arbitrary elements, would likely have provided a critical bridge in this transition. As such, the socially driven processes of sound change now emerge as a crucial fragment of the language puzzle, one that we will assemble in Chapter 13.

Once arbitrary words had emerged, they would have proliferated, removing the constraints on language imposed by iconic words alone. Although arbitrary words would come to dominate the lexicon, the value of absolute and hybrid iconic words would have remained essential to language, as they do today: they provide a means to facilitate language acquisition and often enhance the effectiveness and ease of communication. Because we all enjoy a good onomatopoeia – *crash, bang, wallop* – some iconic words are strongly resistant to the processes of change.[70]

Catching up with the past

Plato, via Socrates, had suspected that the first words were likely iconic words and Johann Gottfried Herder recognised these were likely to have emerged as expressions of synaesthetic experiences. Having been waylaid by Ferdinand de Saussure and Charles Hockett's conviction that arbitrary words are a defining property of language, modern-day linguistics has caught up with Plato and Herder. We now know that iconic words are pervasive in languages throughout the world. Young children find them easier to acquire than arbitrary words, but their lexicons eventually become overwhelmed by arbitrary words because iconic words are inherently limited in their communicative power. A similar trajectory appears viable for the evolution of language, with iconic words providing a bridge between the barks and grunts of the chimpanzee-like calls of the last common ancestor and the dominance of arbitrary words within fully modern language.

What then of *Homo heidelbergensis* with whom we started this chapter? As we discovered in Chapter 3, this most critical species in our evolutionary past, the ancestor for both *H. sapiens* and *H. neanderthalensis* had a vocal tract that enabled an equivalent range of vocal sounds to those of modern humans, those we use in our languages today. It seems unlikely that *H. heidelbergensis* would have been using that vocal capacity for no more than barking and grunting in a chimpanzee-like manner, especially when its brain was up to three times larger than that of the last common ancestor. But was *H. heidelbergensis* making iconic, hybrid or arbitrary words? Iconic words would have obviated the need for learning because their meanings can be intuitively grasped. Reliance on iconic words alone, however, would have severely constrained the range and precision of its speech. Had the social processes of sound change already delivered hybrid and fully arbitrary words into the human lexicon of 500,000 years ago? Or was that yet to come?

None of the linguists and psychologists whose work we have followed in this chapter can provide an answer – they deliver their scenarios for language evolution with no consideration for chronology. Neither can the fossil evidence for vocal tracts and brain size be of help. We must now turn to another source of evidence for language evolution, the stone artefacts and other debris discarded by our human ancestors. What can that tell us about the evolution of language?

7

MAKING TOOLS

From 2.8 million years ago to the present day there is a continuous record of material culture made by our human ancestors and relatives. That provides a welcome contrast to the sporadic finds of human fossils that leave intermittent evidence for how the vocal tract and other aspects of human anatomy evolved. As described in Chapter 2, the earliest traces of material culture are chips and flakes detached from nodules of stone.

Today we make different types of chips, those on which our electronic devices depend. That is because ever since the start of farming at 10,000 years ago, and possibly long before, fully modern language has been a driver of technological change. It enables each generation to build on the technology used by the previous one, enhancing the complexity, efficiency and/or capability of the tools we make and use.[1] Language delivers this so-called 'ratchet effect' by allowing us to talk about the tools we use and how they might be improved and enabling the social structures for the passage of technical knowledge from generation to generation. When did the ratchet effect begin? Had earlier types of language also been able to drive technological change? More generally, what can the stone tools and other artefacts made by our human ancestors tell us about their linguistic capabilities?

Our focus is with the stone tools made between 3 million and 40,000 years ago, coming from Africa, western Asia and Europe. Those from central, south and eastern Asia are insufficiently documented and understood to be of value for exploring what language capacities might have been possessed by *H. erectus* and the later Denisovans who inhabited those regions before the arrival of modern humans. We can be confident that all the material culture from the Americas and Australia was made by people after modern language capabilities had evolved.

Chimpanzees, the tool makers

In 1949 Kenneth Oakley (1911–81), an anthropologist at the British Museum, coined the epithet 'Man the tool-maker', using this as the title for a slim volume outlining the evolution of technology as then understood.[2] It was a fabulous read, having seven print editions over thirty years and formative in my own decision to study prehistoric archaeology. Oakley wrote that 'the most satisfactory definition of man from the scientific point of view is probably man the tool-maker'. This was a widespread view, influencing Louis Leakey who coined the name *Homo habilis*, or 'handy-man', for the fossils he found in Olduvai Gorge in 1964. In the same year, however, man the tool maker fell off his evolutionary perch because Jane Goodall, a young anthropologist supported by Leakey, published her observations of tool making by wild chimpanzees in Gombe, Tanzania.[3] She described how the chimpanzees had selected blades of grass, trimmed them to an appropriate size and shape, poked them into termite mounds, and then pulled them out to lick off the termites clinging to the grass blade. Goodall termed this 'termite fishing'. Oakley had to recant but did so reluctantly, in 1969 changing the epithet to 'man the skilled tool-maker'.[4]

Our knowledge of chimpanzee tool use has greatly expanded since Jane Goodall first saw termite fishing at Gombe. In addition to making sticks to fish for termites, chimpanzees use twigs to probe for ants, collect honey, extract marrow from bones, eyes from skulls and kernels from nuts; they use leaves to sponge up water for drinking, and branches for throwing.[5] Different groups of chimpanzees make use of different tools, responding to what resources are available. Each group also has its own cultural traditions about what tools to use and how, with such habits transmitted from generation to generation by social learning.[6] We should attribute the same range of ecological and cultural variation in tool use to the last common ancestor of humans and chimpanzees dated to 6 million years ago.

In West Africa, chimpanzees use stone hammers and anvils to crack nuts. They select hammerstones of specific sizes and mass to match the hardness of the nuts they need to crack open. The nuts are placed onto anvils, often depressions in the exposed tree roots of the nut trees themselves but sometimes boulders or stones that can also be carried around. By repeatedly using the same location, their nut-cracking creates its own archaeological record of discarded hammerstones. These can be directly compared to those of the Oldowan, some of which have similar pitting and wear traces indicating their use for pounding plants and breaking open animal bones to extract the marrow.[7] Oakley would be interested to learn that the levels of skill exhibited by chimpanzee and *H. habilis* hammerstone-use appear similar. But no wild-living chimpanzee has ever been observed to deliberately detach flakes from nodules of stone or to use the sharp edge of a found flake as a cutting tool.

There is no evidence for active teaching of tool use by chimpanzees, although the close association between mothers and infants certainly facilitates learning by the infants about tool use. Mothers have been observed to assist their infants in nut-cracking, sometimes letting them use their hammerstones. As such they scaffold the infant-learning experience, one that is otherwise achieved by passive observation.[8] Kanzi, the captive bonobo who demonstrated a capacity to learn the use of symbols, was taught by his human carers to make stone flakes and use these to cut rope and leather. He grasped how to break apart cobbles using a hammerstone but even after many years of experience could not match the Oldowan tool makers in the use of sharp-edged flakes for cutting.[9]

Chimpanzee tool use has been observed for more than sixty years, but instances of innovation are sparse while technological progress is absent. In the equivalent period, humans have lived through the digital revolution, a dizzying pace of technological change that has transformed how we live. While we became computer- and smartphone-savvy, chimpanzees continued fishing for termites.

Stone tools, from Oldowan to Acheulean

Oldowan tool-making involved the production of sharp-edged flakes by striking one stone, the core, with another, the hammerstone, referred to as knapping.[10] That simple description hides some complexities: the stones used for cores must be of a type that will flake in a predictable manner when struck, rather than just shatter into amorphous fragments; the hammerstone must strike an edge of a core with an angle of less than 90 degrees if a flake is to be removed; the blow must be delivered with appropriate precision and force; once a flake has been removed, the core might be turned to find another suitable place to strike, perhaps provided by the scar of the previous flake removal. All these tasks, from selecting the stones to removing the flakes, required visual-motor coordination, muscular control to grasp and rotate the core, and ongoing evaluation of core morphology. Each nodule of stone would have been a different shape and size, providing different opportunities, challenges and knapping problems to solve. Despite the Oldowan knappers' skill at removing flakes, they made no attempt to produce a tool of a particular design; their intent was simply to make sharp-edged flakes and cores rather than to impose a predetermined form (Figure 8).

Oldowan core tool from Olduvai Gorge, Tanzania, made from basalt, as used by *Homo habilis* c.2.0 million years ago

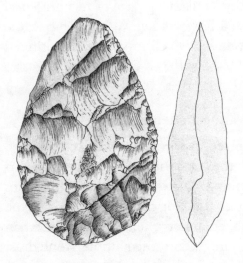

Acheulean handaxe, from Boxgrove, England, as made by *Homo erectus* and *Homo heidelbergensis*, c.500,000 years ago

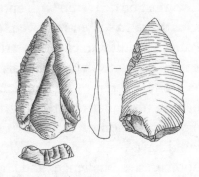

Levallois point from Tabun Cave, Israel, as made by *Homo neanderthalensis* and *Homo sapiens*, c.300,000 years ago

cm

Figure 8 Early stone tools

Collections of Oldowan tools vary from site to site in terms of the numbers of cores, hammerstones and flakes, the extent to which single cores have been flaked from just one, two or more faces, and the range of stone types that were used, including basalt, quartz, quartzite and chert.[11] Such variability can be explained by a combination of ecological factors, such as stone availability and the resources exploited, along with cultural traditions.[12] There is, however, a remarkable degree of unity and stability: Oldowan technology remained unchanged across large regions of Africa for more than a million years: a chimpanzee-like rather than human-like pace of change.

Technological innovations are evident from *c.*1.6 million years ago, broadly coinciding with the appearance of *H. erectus / ergaster* and the enlargement of the brain to about 1,000 cm^3. Flake removal became more elaborate, indicating planning and preparation before the removal of flakes.[13] Large flakes were now detached from cobbles and then shaped by the removal of smaller flakes and chips around their circumference; in other cases, cores were prepared by the removal of small flakes so that a larger flake could be struck off. Such innovations coalesced into the production of the deliberately shaped tools we call handaxes, cleavers and picks. These are collectively known as bifaces because they were made by removing flakes alternately from one side of the core and then the other to impose a specific shape and make a sharp edge. Handaxes were shaped like a flattened pear; cleavers had a straight rather than pointed tip, while picks were triangular in cross-section. There is no evidence that these tools were hafted. As the name 'handaxe' implies, they were held in the hand, sometimes with the thick end of the handaxe in the palm so pressure could be applied to the tip, and sometimes pinched between the fingers so the sharp edges could be used for cutting and slicing. Collectively, these bifacially made tools are known as Acheulean technology.[14]

The Acheulean lasted for more than a million years, again

quite unlike the constant and cumulative change of modern human technology. The earliest handaxes were predominantly made from large flakes and were typically the size of an adult's hand, although they can be both much smaller and larger. To produce the starting flake, known as the blank, the knappers required much larger cores than used in the Oldowan, along with heavier hammerstones and new body postures, arm and hand movements. Once a suitable blank had been detached, smaller hammerstones were used to remove flakes from alternate faces. Care had to be taken to maintain edge angles to enable further flakes to be removed while also imposing an overall shape onto the artefact. Handaxes were also made directly from cores, some of which were turned into picks.

Many handaxes exhibit a high degree of symmetry, not only when seen face on but also when looked at from the side. The imposed form often went beyond the functional requirements of a portable butchery tool.[15] This suggests that such handaxes were used to display technical ability because making a highly symmetrical handaxe required a combination of both cognitive and manual skills. Such display may have been to attract mates – those looking for 'good genes' for their offspring; if someone can undertake the complex tasks of making a symmetrical handaxe, they are also likely to be good at the equally complex tasks of finding food, protecting from predators, and building social alliances in the group. As such, highly symmetrical handaxes may have sometimes been the cultural equivalent of a sexually selected trait in the animal world such as a peacock's tail or a stag's antlers.[16]

By 1 million years ago *Homo erectus* had dispersed from Africa into Europe, where it evolved into one or more poorly defined and problematic species, known as *H. antecessor*, followed by *H. heidelbergensis*, the latter name used for fossils dating to after 600,000 years ago. That date also marks a further expansion of range in Europe, likely involving more effective hunting and technological elaboration.[17] Bifaces become especially refined and often abundant at sites. Their material debris is well known, providing outstanding evidence for the highly skilful knapping process. At Boxgrove, England, for instance, the waste flakes from single manufacturing episodes can be refitted to reconstruct the precise sequence of minute-by-minute knapping actions that were undertaken half a million years ago.[18]

Acheulean handaxes after 700,000 years ago were often made by a two-stage process. The first was to produce a rough-out, an approximation to the final shape. The second involved thinning the rough-out to make it lighter to carry, with sharper edges and a more regular form. They come in a variety of shapes: some pointed with straight sides, some pointed with concave sides (ficrons) and some with two rounded ends (ovates). Thinning was achieved by detaching long, thin flakes that travelled across the surface to reduce the overall thickness of the handaxe. To remove such flakes, robust platforms on the edge of the artefact were required, made by removing small flakes or grinding. Experimental replication has found that soft hammers, those made from antler or bone, are most effective at such thinning. Examples of soft hammers shaped from deer antler and with distinctive wear patterns have been found at Boxgrove. Their acquisition and preparation introduced another stage in the manufacturing process (Figure 8).

Throughout the handaxe roughing out and finishing stages, errors may have occurred: a miss-hit, a fracture arising in the hammerstone, unexpected faults in the stone, such as crystals, fossils and fractures, that caused flakes to snap, and so forth. Consequently, those making handaxes had to be constantly modifying their plans and never looking too many steps ahead. While a form was deliberately imposed, it may have emerged only during a late stage of the knapping process rather than having been planned from the start. A meticulous study of Boxgrove handaxes concluded that:

despite the variables of blank size and shape, together with the problems of frequent breakage, the Boxgrove hominins were still able to arrive at a common goal: elongated ovate handaxes with convex distal ends and wider proximal butts. The hominins clearly had a deep knowledge of their raw material and near-perfect control over the shaping technique. They were therefore able to solve technical problems and physical differences between nodules to create forms that were standardized in shape.[19]

There are important similarities and differences between Oldowan and Acheulean technology. Both involved a multi-stage process of acquiring the stone nodules and hammerstones, a high level of visual-motor coordination and dextrous motor skills. The Oldowan primarily involved a chain of repetitive actions: inspect the core, strike with a hammerstone, inspect the core, strike, and repeat until enough flakes have been detached or a new core is required. In contrast, the actions required to make bifaces were more structured, with one flake removal always being taken with a view towards the next. They also involved more stages to impose a form onto the tool, often requiring a roughing-out stage and then a thinning stage using different types of hammers and blows. Some presence of hierarchical organisation may be evident from biface manufacture, in which repetitive sequences of actions such as preparing an edge for striking, the removal of a flake, assessing the biface, and further preparation, are embedded within each other.[20]

Despite the greater technical demands of making bifaces, the Acheulean shares an important feature with the Oldowan – immense stability though time. The archaeologist Robin Dennell recently provided a succinct summary of more than a million years of human technology, stating: 'the Acheulean is unimaginably conservative. For over a million years – or 50,000 generations (assuming a generation length of 20 years) – lithic technology continued to be centred on bifaces and unstandard-ized flake tools. On the surface, the Acheulean appears to be a record of unbelievable monotony.'[21] That is, of course, a 'big picture' view. It must be balanced against considerable variation in the Acheulean when the specific details of knapping tech-niques are examined, these reflecting responses to raw materials, ecological need and cultural traditions. Nevertheless, these are minor variations within a tediously monotonous theme.

Implications for language and the brain?

It is easy to draw analogies between making stone tools and language, whether that is speaking or signing. Grasping a ham-merstone and rotating a core to make a strike involves motor control of the hand comparable to that of the vocal tract when forming phonemes and syllables. Sequences of blows can seem like a flow of words, each contributing to a final product, whether that is a spoken utterance or a completed handaxe.

Another view is that knapping is like having a conversation with a stone: a hammer blow makes a statement, and the core responds by losing a flake. Sometimes the answer is unexpected, with the flake refusing to leave or an inclusion being exposed. The knapper then pauses to reflect, just as we might do when hearing a surprising answer to a question. Like us, the knapper might then 'change the conversation' such as by tackling another face of the core or changing the type of hammer. Recursive sequences of platform preparation are like embedded clauses in speech – both are required to achieve the intended meaning of the final statement, whether that is an Acheulean biface or a spoken utterance.

A third view is to note similarities between bifaces and words that distinguish them from both Oldowan tools and the vocali-sations of apes. One is the deliberately imposed form. Handaxes have this, and often show a striking consistency in their shapes and sizes when coming from a single site or region. Words also have that quality: while varying in their pronunciation they need a minimum level of consistency to be understood. In contrast, both Oldowan tools and chimpanzee calls can entertain much higher degrees of variability while serving their purpose to the best effect. Another similarity is that bifaces were carried around the landscape, ready for use for whatever task was required, including digging for tubers, butchering carcasses, chopping wood, and providing a source of razor-sharp flakes. As such, bifaces were used in meaningful ways in a variety of contexts. Words are the same: we carry them around with us and can use them in different ways in different contexts. I can say 'Hello' in a cheerful way to greet a friend, inquisitively to answer the phone, formally to my boss, and tiresomely when arriving at a meeting I do not want to attend. Oldowan tools and chimpanzee calls are primarily situation specific: the former were usually made, used and discarded at the same spot; the latter are made in response to specific situations. Neither are 'carried around' and used in a variety of ways.

Are these analogies between bifaces and words, between the Acheulean and language, just fanciful and potentially misleading? Or is there a relationship between the neural networks for making stone tools and those for language? A collaboration between Dietrich Stout, an archaeologist and expert knapper, and Thierry Chaminade, a neuroscientist, found that when modern humans have their brains scanned while making Oldowan and Acheulean tools, there is significant activity in regions known to be used for speech and language.[22] Oldowan knapping activates the pre-motor cortex as if it were being used to manipulate breathing and the vocal folds for speech. When making handaxes, areas of the prefrontal cortex that are known also to process language are activated, possibly reflecting common problems of coordinating elements into hierarchically structured sequences. An experiment found comparable patterns of blood flow in the brain when a knapper is at the planning stage of making a handaxe and when asked to undertake a standard word-generation task.[23]

Although these overlaps are undeniably exciting, we must be cautious. Only a few studies have been undertaken and a small number of individuals tested. Generalising from a small sample is always problematic, and we know that language exhibits considerable interpersonal variability in patterns of brain activity. Moreover, these studies are by necessity testing modern humans with fully evolved brains: they illustrate the brain areas activated in a language-using modern human to make Oldowan and Acheulean tools rather than those in a *H. habilis* or *H. erectus/ergaster* brain. We also know that language processing is pervasive throughout the brain with each of its regions having multiple functions (as described in Chapter 11). With the relatively low-resolution brain imaging of Chaminade's experiments (undertaken more than a decade ago) compared with what can be achieved today, it would be more surprising if areas associated with language were *not* activated.

If we are generous to these brain scanning experiments and sympathetic to Stout and Chaminade's conclusion that tool making exhibits 'multiple levels of overlap with cortical language circuits', we might consider an evolutionary relationship.[24] These circuits might have evolved to enable tool making and at a later date became used for language because speaking has similar requirements: fine muscular control, hierarchical processing of information, forward planning and so forth. The converse – neural circuits evolved for speech becoming adopted for tool making – seems unlikely because that would require language to be present before 3 million years ago when brains were no larger than those of chimpanzees today. Alternatively, language and tool making may have co-evolved, bootstrapping each other as enhancements in one created the conditions for improvements in the other. A further possibility is that these circuits are general purpose, used for a range of complex, goal-oriented tasks of which tool making and language are simply two examples.[25]

A co-evolution of tool making and language might have arisen if spoken language had been used to facilitate the transmission of knapping skills from one generation to the next. There have been several experimental programmes to explore the effectiveness of spoken language for this role, testing how it enhances skill acquisition over that gained from the passive observation of an expert at work or when instruction is provided by gesture.[26] As with the brain scanning experiments, I am cautious, and in some cases sceptical, about the experimental designs. Is it possible to ask language-using modern humans to 'switch off' inner speech and pretend they are learning by observation or from gesture alone? I wouldn't be able to help myself from silently translating gestures into words and talking myself through the task. Also, those who have learned to replicate Oldowan and more particularly Acheulean tools will say the only way to learn is through many hours of practice, ideally spread over several years. The experimental programmes are very short term: in one case novice participants were given just five minutes of instruction, whether by observation alone, gesture or speech, and then a mere twenty minutes to practise, before being asked to teach others.[27] This seems so divorced from reality that drawing conclusions about evolutionary relationships between language and tool making is problematic.

Nevertheless, if one is generous to such experiments the outcomes are of interest. The most elaborate was conducted by a team from the University of St Andrews and involved novices learning to replicate Oldowan flake production and then teaching others in a chain of cultural transmission.[28] By measuring skill acquisition by the quality and quantity of flakes detached from a core, it found that performance improved with teaching, particularly when spoken language was added to gestural instruction. Active teaching appears to facilitate the acquisition of subtle but crucial information about knapping that is difficult to grasp from observation alone, such as attending to edge angles. The use of labels – words for 'edge', 'platform' and so forth – may have been particularly important in breaking the learning task into constituent parts and enabling someone who was newly taught to go on to teach others.

Curiously, having supposedly demonstrated the role of speech, the St Andrews team concluded that language had not been used to transmit Oldowan technology from one generation to the next. They forced themselves into that position by suggesting that the absence of teaching, whether by gesture or speech, explains why chimpanzee tool use has not changed over time – it has reached the threshold of complexity that can be transmitted by observational learning alone. Because Oldowan technology also remains unchanging for over a million years, that must have been under the same constraint.

The overall outcome from experiments involving tool making, brain scanning and various forms of instruction is inconclusive. The idea that stone tool technology and language co-evolved, with spoken instruction facilitating technical tasks which then selected neural circuits that also enhanced the capacity for speech, is appealing. But the relative roles of gesture and speech remain unclear. The St Andrews experiment conflated these by adding speech to gestural instruction. When a later study tested the value of speech without gesture, it was shown to be of less value than gesture alone.[29] A final concern is about the nature of speech and language itself. All these experiments have been undertaken by modern humans with language as we know it today, having a variety of word classes, extensive lexicons and grammatical structures. If language evolved gradually, perhaps beginning with few words and no grammar, could it have provided the subtle instructions for how to find and strike the edge of a stone cobble?

The need for such verbal instructions might have been greater for the next stage in stone tool technology – the Levallois technique. This emerged between 400,000 and 350,000 years ago, being found in Africa, Europe and Asia although whether it had a single origin or was independently invented in different regions is unknown.[30] While often identified with the Neanderthals, it was extensively used by the early *H. sapiens* in Africa and western Asia before 40,000 years ago.

Levallois technology involved preparing cores with convex surfaces, sometimes called tortoise cores. By carefully creating a platform at one end of the core, a single hammer blow removed a flake of a predetermined size and shape. The principle was the same as removing a thinning flake from a handaxe but the detached flake rather than the core was now the desired product. The Levallois cores were sometimes carefully shaped so that the detached flakes were pointed or had long parallel sides referred to as blades (Figure 8). The flakes, points and blades were either used as they were with razor-sharp edges or chipped into specific shapes to make them easier to hold or attach to a handle or spear shaft. A second removal, and potentially more, could often be detached from the Levallois core, either by striking the same platform or one at the opposite end of the nodule.[31]

The Levallois technique is more complex than making an Acheulean handaxe because the hierarchical organisation of flake removal is significantly more pronounced: many think it represents the most complex stone technology from throughout human evolution. I am inclined to think that a few instructive words on how to prepare the core might have been required in the transmission of this technology from one generation to the next.

The innovations made by modern humans after 40,000 years ago enhanced efficiency by securing more cutting edges from the same quantity of stone but did not require a higher level of technical and cognitive skill.[32] Those innovations included producing long blades from pyramid-shaped cores, held point down to expose a flat platform for striking, using an antler punch rather than directly striking the platform with a hammerstone, removing flakes by pressure rather than percussion, and making lots of tiny blades to use in multi-component tools.

The 'missing majority' and composite tools

Archaeologists focus on stone tools because they are often all that is available for study. There can be no question, however, that even from the earliest times wood would have been extensively used, possibly more so than stone – chimpanzees use an assortment of sticks for a variety of tasks while wood is pervasive in the toolkits of modern hunter-gatherers. Not just wood, but also sinews, skins, bone, horn and other perishable materials. Artefacts made from such materials are appropriately referred to as the 'missing majority' of Stone Age tools.[33]

In 1911, the broken point of a wooden spear was found at Clacton-on-Sea, UK, later dated to c.400,000 years ago. Further insights into the missing majority had to wait until 1997 when complete wooden spears were discovered at Schöningen, Germany, dated to between 337,000 and 300,000 years ago. A further discovery came in 2023 with evidence for the earliest structural use of wood dating to at least 476,000 years ago from Zambia.[34]

Unique lakeshore conditions at Schöningen had preserved at least ten spears and two double-pointed sticks made from the trunks and branches of spruce trees by early Neanderthals. The spears ranged from 1.8 to 2.5 metres in length and had similar designs to modern javelins, suggesting that they had once been thrown although they could equally have been used for thrusting. The likely target had been wild horses, with collections of their bones found nearby.[35] The pointed sticks were about one metre in length and might have been used for throwing, stabbing or digging.[36]

As with Levallois points, making these tools had required a chain of actions: selecting the type of wood to be used; acquiring suitable small trunks and branches; making stone tools and using these for initial shaping and then finishing the tool.[37] The spear-making chain would have taken a longer time, most likely spread over several days rather than the few minutes to make a Levallois point (after the nodule of stone had been acquired). But although the spears and pointed sticks provide new insights into Neanderthal lifestyles, they offer no further information about the likely character of Neanderthal language.[38]

Further insights into Neanderthal technology are available from 171,000-year-old wooden artefacts excavated at Poggetti Vecchi, Italy.[39] These are fragments of shaped sticks with traces of handles that had been over one metre long, made from boxwood, pointed at one end and rounded at the other. Their surfaces had been burnt. Described as 'digging sticks', they were likely used for a variety of tasks with boxwood having been selected because it is particularly hard. As with making a Schöningen spear, preparing the digging sticks had required a chain of actions: selecting an appropriate type and sample of wood; removal of branches and the shaping of handles by stone tools; the use of fire to help remove the outer bark after scraping with stone tools; final shaping and the rounding of an end by abrading with coarse stone. Like the spears, I find it difficult to imagine that words were not used to label such tools, but have no direct evidence that was the case.

We should not doubt that the Schöningen spears and Poggetti Vecchi sticks are the tip of an iceberg of tools made from organic materials that have perished. A glimpse of what must have been comes from a tiny piece of twisted plant fibre, found adhering to a Levallois flake from the site of Abri du Maras in France.[40] The flake had been made by a Neanderthal at c.41,000 years ago; the fibre fragment is no more than 6.2 mm long and 0.5 mm wide, made from the inner bark of a conifer. The fibres were twisted together to make what may have been a piece of string. Whether that was once attached to the stone flake or is the remnant of a piece of rope, net or a bag that had been discarded below the stone flake is unclear. The Abri du Maras find suggests that woven items had been made by Neanderthals. Even without it, however, there is no reason to doubt that was the case nor about the likely role of tools and tool components made from twisted fibres in many aspects of daily life.

A key difference between the Abri du Maras twisted fibres and the Schöningen hunting spears is that the fibres represent a composite technology, one made by combining items together. Neanderthals in Europe, *Homo sapiens* in Africa and both species in western Asia were also combining stone and wood to make spears with hafted points and scrapers with handles. The evidence for this is threefold.[41] First, stone points, flakes and blades often have scratched and/or polished surfaces indicating where they had been slotted into a haft; second, pointed flakes, notably those from Levallois cores, often have signs of impact damage – small chips and striations indicating the point had forcibly struck the bone of an animal; third, several examples of birch bark tar have been found adhering to stone flakes having been used as an adhesive.[42] These imply that hafts and points were separately made, birch bark tar was prepared, probably mixed with bees' wax to make a strong resin, and these were joined together, possibly using sinew or cord from plant fibres.

Does such composite technology have implications for language beyond those evident from a reductive technology such as flaking stone and carving wood? One might argue that combining different materials together to make a single tool is analogous to how we connect words to make a meaningful utterance, both outcomes being more than the sum of their parts. The wooden shaft, adhesive and stone point of a spear play different functional roles just as a noun, verb and adjective do in an utterance. Embedded clauses might be a better comparison than words for the component parts of a composite tool, because each needed to be separately prepared before being combined with the other parts. The case for words as labels for both component pieces and the finished items is compelling – but that might come from my own language-using bias.

We must be cautious about over-interpreting the significance of wooden artefacts and composite tools. Because of their rarity, archaeologists can get overexcited and make extravagant claims. The tiny piece of twisted fibre from Abri du Maras was claimed to demonstrate that Neanderthals had 'mathematical understanding of pairs, sets, and numbers' (whatever that means).[43] Birch bark tar was initially claimed to indicate that Neanderthals had mastered its manufacture using the same complex dry distillation process we know from Bronze Age and Roman times, whereas later experiments demonstrated it can simply be produced by burning bark close to cobbles in a hearth.[44] Moreover, despite such composite artefacts and those made from single pieces of wood, there is further evidence from stone artefacts that suggest that both Neanderthals in Europe and early *H. sapiens* in Africa, at least before 150,000 years ago, were still behaving, thinking and talking in a quite different way from language-using modern humans.

The linguistic implications of technological change – and its absence

The shift from hand-held bifacial core tools to a predominance of hafted smaller flake tools between 400,000 and 300,000 years ago is denoted as the start of the Middle Stone Age (MSA) in Africa and the Middle Palaeolithic (MP) in Europe and western Asia.[45] The shift was initiated by *H. heidelbergensis* (or whatever multiple species that fossil category represents) and maintained by its descendants, Neanderthals in Europe, *H. sapiens* in Africa and both species in western Asia. Having considered the evolved vocal tracts and large brains of all three species we can be confident they had some form of linguistic capability. That also seems a reasonable conclusion from how bifaces and Levallois flakes were made. Can further insights into the linguistic capabilities of Neanderthals and early *Homo sapiens* be acquired from how their technology varied across space and time?

Before considering this question, we should recall how fully modern language is a driver of technological change via the rachet effect – the cumulative increase in technological complexity, efficiency and / or capability from generation to generation. This has been happening ever since fully modern language evolved, at least 40,000 years ago and notably during the final stages of the Ice Age and when farming was invented: whenever fully modern language is present, technology cannot stand still. We cannot, however, exclude the possibility that earlier forms of language, perhaps entirely reliant on iconic words, had also driven technological change with their own form of the ratchet effect.

Let's start with the Neanderthals. How had their technology changed through time? Is there evidence for the ratchet effect?

After 350,000 years ago, the Levallois technique became prevalent throughout Europe, with many variations in how cores were prepared and flakes removed. Several other core preparation and flake removal methods were used. A discoid technique detached thick flakes from around the circumference of a core, each removal producing a platform for the next strike, and a bipolar technique removed flakes from both ends of a core by striking it when positioned on an anvil. The Neanderthals also used the so-called Quina technique which sliced up cores, producing flakes that were thick at one end and thin at the other, and could be repeatedly resharpened. Cores were sometimes prepared to remove long blades – once thought to be the preserve of modern humans. Bifaces continued to be made, becoming prominent within a culture known as the 'Mousterian of Acheulean Tradition' that dates to around 50,000 years ago in southwest France.[46]

The Neanderthals often carried unworked nodules, cores and flakes around the landscape, using, resharpening and discarding them as required. Other than bifaces, tools were rarely chipped into deliberate shapes: the differences in those we excavate tend to be an incidental consequence of the extent to which they have been resharpened.[47] However produced, flakes and blades of diverse shapes and sizes were used for the same range of tasks: cutting and shaping wood and other materials, chopping and pounding plants, as points for hunting weapons and to provide sharp edges for butchery.[48]

Changing temperatures, rainfall and ecosystems altered the availability of stone, the patterns of Neanderthal mobility and the resources that could be found. These in turn influenced the tool-making techniques that were employed. The Quina method, for instance, is associated with colder conditions, when reindeer dominated the fauna. But there is little impression that the Neanderthals deliberately adjusted their technology to adapt to their new environments. They are best described as being reluctantly responsive, forced to change their technology because of changing availability of raw materials and their own mobility patterns that required either more or less resharpening of the tools they carried around. Bifaces were made in temperate, forested, cold and open environments, their variations in shape and methods of manufacture reflecting available types of stone and cultural traditions of Neanderthal groups rather than adaptive responses to the natural world.[49]

Techniques came and went, often leaving alternating layers of Levallois, discoid and other types of cores in the stratified cultural deposits of single sites. Leading experts on Neanderthal technology describe it as showing 'stochastic [i.e. random] variation' through time and space, and 'not going anywhere in particular'.[50] Even when Europe enjoyed warm and wet conditions between 130,000 and 80,000 years ago, Neander-thals continued to draw from the same limited repertoire of tool-making methods they had used during the colder periods; when these were insufficient for coping with the new wood-land, they moved to live elsewhere.[51] The only bone tools that Neanderthals made were minimally worked pieces for use as hammers for flaking stone, 'retouchers' for further chipping those flakes, and (possibly) bone scrapers for cleaning hides.[52]

The absence of innovation in Neanderthal technology is striking because their skeletal remains display many injuries from hunting large game.[53] If a population was ever in need of new technology to enhance hunting efficiency and safety, it was the Neanderthals. Despite being able to haft points and twist fibres, there is no evidence from the huge number of animal bones, antlers and worked stone excavated from their sites that they invented spear throwers and bows and arrows; such tech-nology would have significantly enhanced their personal safety and hunting success.[54]

In summary, despite the Neanderthals' evolved vocal tracts and large brains, and despite their abilities to make Levallois stone points, wooden spears and whatever the twisted fibres from Abri du Maras represent, there was an absence of tech-nological innovation for more than 300,000 years. This and the seemingly random, directionless variation in tool-making techniques, indicate that whatever type of language the Nean-derthals possessed, it was quite different to that of modern humans – it did not drive cumulative technological change.

Homo sapiens in Africa appears to have been little different, especially before 150,000 years ago. Their stone technology also exhibits a diverse range of prepared core techniques without evident geographical patterning and directional change. Bifa-cially made handaxes had also remained part of the technological repertoire, being found up until 160,000 years ago.[55] As with the Neanderthals, functional specialisation appears absent: Middle

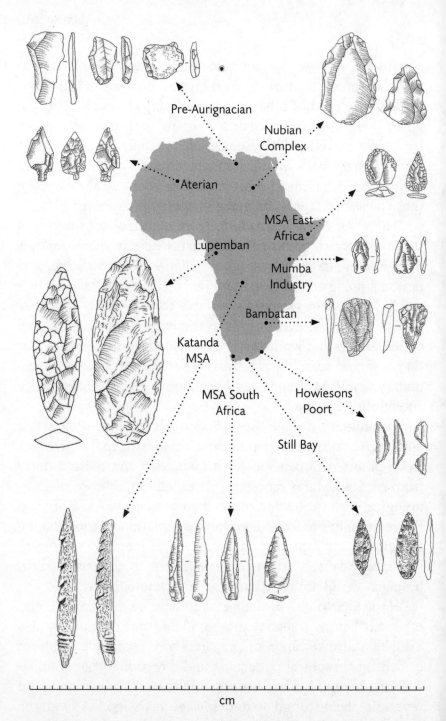

Figure 9 Technological diversity in the African Middle Stone Age

Stone Age points have evidence for scraping activities while those which archaeologists call scrapers have impact damage from projectile use.[56]

After 150,000 years ago, however, *H. sapiens* appears to become more innovative, creating a mosaic of cultural traditions across the continent.[57] (Figure 9). There is a greater extent of imposed form onto stone tools throughout Africa, representing geographically based stylistic variation and/or adaptation to specific environmental conditions. Long blades were made in North Africa where a distinctive type of point looking like an arrowhead had appeared by 150,000 years ago, known as Aterian points.[58] In West Africa, elongated bifaces, small axes, blades and points date to at least 150,000 years ago, collectively known as the Lupemban culture. The earliest known barbed bone points come from Katanda, central Africa, at 90,000 years ago, and were likely used for fishing. Bone points are also found in South Africa, where the first use of pressure flaking occurs at *c.*75,000 years ago. This technique removes small flakes from a stone tool by pressing on its surface with a bone or wooden point, enabling a precise form to be imposed. It was used to make symmetrical, elongated stone points known as Still Bay points, which had a point at both ends. Impressive examples are found at Blombos Cave, a coastal site in southern Africa dating to 70,000 years ago, where they are associated with equally impressive bone points, incised ochre and shell beads.[59]

Contemporary with or a little later than the Still Bay points, small blades began to be produced in the same region, representing the Howiesons Poort culture. These are similar to those that would proliferate throughout the continent after 40,000 years ago. They were steeply chipped along one side, and had likely been inset into composite projectile weapons.[60] Both these and the Still Bay points disappear at or soon after 60,000 years ago, as technology returns to the production of large flakes which then remain until *c.*40,000 years ago.

Quite what these new types of stone tools, varying from Aterian points in the north to Still Bay points in the south, represent is unclear. As with Neanderthal technology, there is neither evidence for directional change nor trace of one innovation sweeping through the continent. As in Europe, we find sequences in which different techniques come and go, perhaps reflecting the movement of populations or communities, or maybe technological drift.[61]

Nevertheless, technological innovations of a type unknown in Europe occurred throughout Africa after 150,000 years ago. Those from southern Africa are the best dated and come in pulses that correlate with abrupt periods of climate and environmental change, either those of drought or of especially wet conditions with abundant resources.[62] They appear quite different to the reluctant responses of Neanderthals to environmental change: *Homo sapiens* in Africa had been proactive and innovative, indicative of new capabilities to design tools to exploit novel opportunities and cope with new challenges. How had those capabilities arisen? Might they reflect a shift in the nature of language that enhanced their abilities to talk about tool making and how their current tools might be improved?

Stone tools and language?

We began this chapter hoping that the evolution of stone tool technology would provide insights into the evolution of language, providing us with a well-defined fragment of the language puzzle. It hasn't worked out like that.

We certainly found several analogies between words and stone tools, notably handaxes: they both have an imposed form, are carried around, and are used in different ways in different situations. Similarly, the making of a handaxe seems comparable to the making of an utterance, the first needing specific types of hammer blows made in the right order, the second requiring the same but using words. Composite tools are likewise: combining a wooden haft, vegetable fibre, pine resin mastic and a flint blade to make a spear, seems comparable to combining a noun, verb and adjective to make a statement. While these analogies might be coincidental, the overlaps in brain activity when using language and making tools suggest otherwise: there appears to be a co-evolution of stone tool technology and language. That suggests some form of linguistic transitions might coincide with that between the Oldowan and Acheulean at 1.6 mya, between the Acheulean and the flake-based, prepared core technologies soon after 400,000 years ago, and may account for the innovative tool making by *Homo sapiens* in Africa after *c.*200,000 years ago.

While this leads us to think of *H. erectus* handaxe makers as having a form of language, it must have been radically different from the fully modern languages that I described in Chapter 2. Those languages exhibit enormous diversity, are constantly changing, and are associated with a similar dynamic in technology, quite unlike the million-year monotony of the Acheulean. Following on from the previous chapter, I am tempted to suggest that *H. erectus* had remained reliant on iconic words, these providing sufficient verbal communication to transmit tool-making skills, but fell short of being able to drive technological change. With their larger brains and evolved vocal tracts, the Neanderthals must have had additional language capabilities. But their own lack of technological innovation suggests their languages had also remained quite different to those of modern humans.

Drawing such conclusions is, however, premature before further fragments of the language puzzle have been assembled. The next to consider has already been prompted by our review of handaxe manufacture: how to ensure all the hammerstone blows are of the right type and made in the correct order to produce the desired end-product. How do we manage to achieve the same with our words to convey the meaning that we intend?

8

LESSONS FROM AN ARTIFICIAL LANGUAGE

It is puzzling that we know so many words and so many rules for how to combine them into meaningful utterances, without any awareness of our own knowledge. I've often asked modern-day flint knappers who have learned to replicate handaxes how they know which type of hammer to use, where to strike their flint nodules, with what angle and force. They usually look at me quite blankly because they also do not know, or at least it is not part of what psychologists call their discursive knowledge – that which they are aware of and can talk about. It is the same with syntax; we use it whenever we make an utterance but, unless we have studied linguistics, we don't know that we know it, and we certainly can't talk about it. Very puzzling.

Just as we learn the words of our speech community during childhood, so too do we learn the rules by which those words are ordered to make meaningful utterances. The principle of compositionality enables us to use those rules to make a potentially infinite number of meaningful utterances from a finite number of words, constantly generating expressions that have never been heard before but which can be effortlessly understood by those who speak the same language. Although rules gradually change over time, just as the meaning and pronunciation of words change, each generation learns the rules used by the previous generation from whom they are learning language. As an English speaker, I learned to put adjectives before nouns from my parents, and they did the same from their parents and so forth back in time. That raises a question central to the language puzzle: how did the rules originate? Were they invented by a clever hominin in the early Stone Age, who has left a very long legacy because their rules have been copied and accidentally modified by every generation of language learners that followed? No, of course not. But what is the alternative?

The answer was discovered during the 1990s: syntax spontaneously emerges from the generation-to-generation language-learning process itself. This surprising and linguistically revolutionary finding was discovered by a new sub-discipline of linguistics that is known as computational evolutionary linguistics. This constructs computer simulation models for how language evolves by using artificial languages and virtual people. Known as agents, the virtual people have their knowledge of language, learning processes and production of utterances defined by computer algorithms. They learn and transmit their language from generation to generation while the model monitors how the language evolves. The outcomes of such models have provided a new fragment of the language puzzle.

Simulating language change

Computer simulations have several advantages over other methods of study.[1] By translating a theory into a set of algorithms, one is forced to define precisely what one means by terms that are frequently used too casually, such as 'learning' and 'meaning'. One can also manipulate the population size of a speech community at will, along with how the agents learn and transmit their language to the next generation, these controlled by setting parameters in the computer code to different values. Moreover, there is no limit on the number of generations across which language can be transmitted, whereas historical studies of language change are limited to 200 at a maximum because, with twenty-five years per generation, that takes us back to the invention of writing at 5,000 years ago in Mesopotamia. Perhaps the greatest advantage of a computer simulation is that by removing all the messiness of the real world and changing each of the parameters in turn, one can disentangle what is a causal factor, what is an influence and what is irrelevant in language evolution, a task that is virtually impossible when dealing with real people speaking real words and phrases over a few generations.

The weakness of simulation models is the mirror image of their value. By removing the messiness of the real world, one risks simplifying the process of learning and language transmission to such trivial conditions that the outcome has no bearing on what really goes on.

Simon Kirby, Henry Brighton and Kenny Smith of the University of Edinburgh pioneered simulation models for language evolution, blending expertise in linguistics, artificial intelligence and computer modelling.[2] From their innovative research, the field has expanded to encompass several types of approach ranging from formal mathematical models to robots playing language games.[3] To gain a flavour of the methods used and types of outcomes, we will consider one of the earliest and least complicated simulations developed by the Edinburgh team. This is known as the iterated learning model (ILM).

The iterated learning model

The basic form of the ILM consists of a single agent in each generation.[4] The simulation starts with an agent referred to as A_1 that has been primed with a start-up language from which it produces utterances. The second-generation agent, A_2, 'hears' these utterances and forms a hypothesis for the nature of language that underlies them and produces its own utterances. You can think of this by imagining learning a new foreign language: you go to classes and hear your teacher say a number

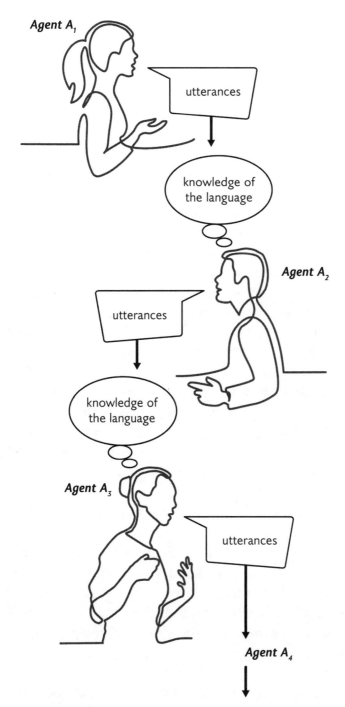

Figure 10 The language-learning bottleneck

of phrases in the new language and are then required to make your own phrases based on what you have heard. In the ILM, the next agent, A_3, hears A_2's utterances and makes their own for the next agent – it is as if you have become the teacher of the new language. This continues for however many generations one wishes to explore (Figure 10). The key question is: how does the language of, say, A_{100}, or however many iterated generations one desires, differ from that of A_1? It is like a very clever game of telephone (once known as Chinese whispers) concerned with the structure rather than content of language.

The artificial language consists of two components: meaning and signal. These are equivalent to concepts and words in the real world. Both are represented by strings of symbols in the computer model – there is no attempt to relate them to actual concepts or words. The character of the artificial language is represented by the relationship between meanings and signals, just as real language is dependent on the relationship between concepts and words. When there is no matching up between the symbols in the strings used to represent meanings and those for signals, the language is referred to as being a holistic language. The closest to a holistic phrase in spoken English is *abracadabra*. This means *let something magical happen* but neither the *abra*, *cad* nor any other sound segment in the phrase has its own meaning relating to a part of *let something magical happen*. Monkeys and apes rely on holistic phrases. Their alarm calls, for instance, have multiple sound segments but meaning only resides in the whole call. This type of communication is assumed to have been used by the common ancestor, pre-dating the invention of words.

An artificial language in which there is a consistent matching-up between symbols in the strings representing meanings and those in the strings for signals is referred to as being a compositional language. All known languages in the world today have this quality: it is the foundation of syntax. For instance, in the phrase *Steven walked home*, each of the sound segments,

Steven, walked and *home*, has their own separate meanings which are used to construct the overall meaning of the phrase. The value of compositional language is that previously unencountered utterances can be interpreted as to their likely meaning by drawing on knowledge of what its component elements mean, these having been learned from previous utterances. By knowing the meaning of *Steven walked home*, a good guess can be made to the meaning of *Zinkey walked home* – despite never having heard the word *Zinkey* before, you can guess it is the name of a person. Such interpretation is impossible with a previously unencountered holistic utterance: knowing that *abracadabra* means *let something magical happen* has no value for knowing the meaning of *elcocadabra*.

Having described how the ILM represents language by representing meanings and signals, the next component of the model is what the Edinburgh group call the 'association matrix'. This is used for three tasks: it defines how agents represent their knowledge of language; how they use the utterances they hear to enhance that knowledge; and how they use their knowledge of language to make their own utterances. The matrix defines the known strength of association between symbols in the meaning and signal strings. Each agent has their own matrix with its rows corresponding to every possible symbol within the meaning-strings and its columns to every possible symbol in the signal-strings. Each cell of the matrix has a number that defines the strength of the association between the two symbols. Prior to learning, every cell has a value of zero; the agents have no knowledge of what any of the signals mean. Again, think about learning a new language. Before you start you have no idea what any of the words mean. If the language is Swahili (which I am assuming you do not know) you would be unable to associate the words *nguruwe*, *nyumbani* and *akaenda* with any meanings, and hence they would have a score of zero in your mental association matrix for all the concepts (meanings) in your brain – *book, pig, banana, house* and so on for everything you know.

When an agent in the ILM hears some utterances, its association matrix will be updated: the value of cells for which meaning and signal symbols have coincided will be made more positive, indicating learning has taken place. This is equivalent to you finding that the Swahili word *nguruwe* is used whenever you see a picture of a pig. At first you may only suspect *nguruwe* means pig, because you can also see a cow and goat when it is spoken, and hence *nguruwe-pig* has a low positive value in your own mental association matrix. As more examples accumulate which lack evidence for cow and goat, that number becomes more positive – you learn that *nguruwe* definitely means pig. At the same time, you learn that *nguruwe* definitely does not mean *cow, goat, book, banana* or *house*, and hence those concepts will be given negative values in your association matrix with the word *nguruwe*.

The ILM simulation prompts agents with randomly selected meanings and requires them to produce the paired signal. This is similar to your language teacher giving you an English phrase to say in Swahili. The agent in the ILM searches every cell in its association matrix to find the highest scores for each of the symbols in the meaning string with those in signal strings, just as you will make your best guesses for which Swahili words to use. The signals produced by the ILM agent as their 'best guess' for the meanings are passed to the next agent in the simulation, which it uses to update its own association matrix – by so doing, the new agent improves its knowledge of the language. This is like you telling a classmate that *nguruwe akaenda nyumbani* means *the pig walked home*, and the classmate learning about the meaning of these words. You might, of course, have made an error and said that *nguruwe akaruka nyumbani*, which would lead your classmate to mistakenly think that *akaruka* means *walked* when it really means *flew*. The next stage of the ILM prompts the new agent with a set of meanings requiring that agent to produce a paired signal, that agent being able to draw on its updated association matrix. Similarly, your classmate might be asked by the language teacher to translate *the cow walked home*. He would be able to have a good guess at the words for *walked* and *home* from what you had just told him but might be less confident about the word for *cow*, having to search memory (his association matrix) for any possibility and possibly having to choose a word at random.

Each utterance made by an agent in the ILM is analysed for its degree of compositionality – the extent to which its meaning is derived from the meaning of its parts, the words, and the way they are combined. The mathematical algorithm used for this is applied to all of the utterances made by each agent to generate a single statistic that measures their overall compositionality, this varying between zero for a holistic language and 1 for a perfectly compositional language.[5] This is like your language teacher giving you a test as to how much you have learned about Swahili. To what extent are you simply combining random words together to make phrases rather than composing phrases by drawing on the learned meanings of individual words and the rules for how they can be combined?

To summarise so far, the ILM has four key components: (1) the representation of language by a combination of meanings (concept-like) and signals (word-like); (2) an association matrix for each agent which is used for representing their linguistic knowledge, learning and production of utterances; (3) the iterative process for one agent to transmit their meaning-signal pairs to the next; and (4) a measure of the compositionality of language at each iteration. Once these have been encoded, the simulation provides a tool for exploring how language evolves under varying conditions.

The initial interest of the Edinburgh group was the impact of the quantity of linguistic knowledge – as measured by the number of paired meaning-signals that are passed from generation to generation. In the real world this is limited: children only hear a tiny fraction of the language of their speech community from which they derive the meaning of words and the rules by which they can be combined to make meaningful utterances. This is what Noam Chomsky called the poverty of the stimulus and the Edinburgh group refer to as a learning bottleneck. It can be mimicked in their ILM model by only allowing each agent to transmit a fraction of its linguistic knowledge to the next agent in the simulated generational chain by providing a small number of paired meaning-signals. Conversely, the ILM model can also allow each agent to transmit the entirety of the artificial language – every possible meaning paired with an associated signal – to the next generation. This is a 'no-bottleneck' scenario, one that never occurs in the real world; it is impossible for you to hear and memorise every possible phrase that could ever be spoken in Swahili.

When these two conditions were explored by multiple runs of the simulation a dramatic contrast emerged in how language evolves. In both the bottleneck and no-bottleneck scenarios, the first agent, A_1, began with no linguistic knowledge by having a value of zero for all cells in its association matrix. As such, its utterances for each of the meanings it was required to produce were random strings of symbols, which gave a compositionality score of zero – a holistic language. Your equivalent would be having to select Swahili words at random when not knowing what any of them mean.

Under the no-bottleneck condition, this persisted in every generation of the simulation. In contrast, under the bottleneck scenario when only a fraction of the possible paired meaning-signals are passed on in each iteration, a compositional language always evolved, reaching the maximum score of 1 within twenty generations.

The reason for the difference is that in the no-bottleneck

condition, the agents can simply memorise the complete language of the previous generation and therefore pair each meaning with its associated signal. As such, the language remains stable through time. If you could hear and memorise every possible phrase in Swahili, you would simply repeat these when prompted. When there is a bottleneck, however, the ILM agents need to produce signals for meanings that had not been previously heard, causing a change in the language. This is equivalent to your classmate needing to translate *the cow walked home*, despite never having heard this phrase in Swahili before. He may have heard the words *cow, walked* and *home* used in other phrases and hence be able to undertake the translation.

The Edinburgh group describe the bottleneck as introducing a pressure for language to become generalisable – meanings of whole utterances are constructed by combining meaning components drawn from other utterances that have already been learned. Over time, the bottleneck causes language to become highly generalisable, compositional, and stable. In other words, it acquires syntax.

The emergence of linguistic structure

The discovery that syntax-like structures can emerge within language by the process of cultural transmission alone was transformative. The existence and rules of syntax had long been thought to be biologically driven, incrementally constructed during human evolution by natural selection.[6] Each additional step was assumed to enhance the communicative efficiency of language, as a reliance on word order was supplemented with that of hierarchical phrase structure, and then recursion. The Edinburgh group's ILM demonstrated that some and perhaps all forms of syntax can emerge from cultural transmission with neither biological intervention nor intentionality by the speakers of the language. Before the ILM, the potential consequence of the learning bottleneck for linguistic structure had not been appreciated.

It might be argued that this discovery is merely an inevitable outcome of the design of the model itself – it was effectively written into the algorithms and computer code. That is correct, but the outcome was non-intuitive and unexpected: once it occurs, it can be readily understood with hindsight. It only becomes a significant outcome, however, if the design of the model is thought to be sufficiently close to how people learn language and generate their utterances. That a bottleneck exists in language acquisition is unquestionably the case – children only ever hear a fraction of the possible utterances that their language can generate, just as you could only ever hear a fraction of Swahili phrases when learning the language. Moreover, the outcome of the ILM was validated by laboratory-based experiments using real human subjects that mirrored the structure of the computer model. By providing the participants with an artificial language-learning task that had to be learned and passed on through a chain of participants, linguistic structure emerged without any intentional design on the part of the participants.[7]

Once the ILM had been constructed, Kirby and his colleagues undertook experiments to explore how the width of the bottleneck – how much of language is passed on from generation to generation – influences the emergence of compositionality and other features of language.[8] The impact of the learning process was also explored, represented in the ILM by how the agents update their association matrices once they have heard a set of paired meanings and signals from the previous agent in the chain. Agents were made to learn in different ways by altering the extent to which a matching between a meaning and signal symbol influences the score in the appropriate cell of their association matrix. Such experiments mimic how people learn at different rates and how some are better at generalising from the use of a word in one context to another. The ILM agents were also made to make mistakes, as happens in the real world when we mishear a word or misunderstand what it means – such as when telling a classmate that the Swahili word *akaruka* means *walked*. The agents were also made to minimise effort by always choosing the shortest string of symbols to convey a meaning.

One of the Edinburgh team's experiments explored how the process of learning and bottlenecks influence the communicative accuracy of utterances. To what extent does one agent decode the signal it receives to extract the same meaning that the previous agent had sought to encode into the signal? These experiments demonstrated how communicative accuracy can only be achieved if the agents are biased against acquiring what the Edinburgh team call 'one-to-many meanings' – the same signal (i.e. word) pairing with multiple meanings; similarly, agents must also be biased against the converse, 'many-to-one meanings' – multiple words all meaning the same thing. Having identified these biases, the Edinburgh team were able to look at the evidence in the real world for how children learn language and found evidence for such biases, the significance of which had not been previously appreciated. Other key experiments explored how not only compositionality but also recursive syntax can emerge via the bottleneck effect.[9]

The significance of ILM experiments was summarised by Simon Kirby. 'Counter to intuition, the [language] system appears to adapt. In even the simplest instantiation of the model, structure emerges in the meaning-signal mapping. Words / morphemes spontaneously emerge that correspond to subparts of the meaning, and regular rules evolve for combining these into complete sentences.'[10]

A key phrase of this statement is that 'the system adapts'. Before the ILM, the emphasis had been on how people must adapt, either culturally or biologically, to meet the demands of the language they are acquiring. Kirby and his colleagues turned that on its head: it is language that must adapt to pass through the learning bottleneck. Unless it has words and structures that allow the complete language system to be generated from the samples heard and processed by language learners, the language will not survive, becoming replaced by an alternative version that is better adapted to the human brain. As implied by the ILM, the brain needs only a generalised learning mechanism, rather than a dedicated language acquisition mechanism, for such linguistic structure to emerge.

This view has been further developed by the psychologists Morten Christiansen and Nick Chater.[11] They suggest that language should be seen as 'a complex and interdependent "organism", which evolves under selectional pressures from human learning and processing mechanisms. That is, languages themselves are shaped by severe selectional pressure from each generation of language users and learners.'

Central to this phenomenon is the effect of the learning bottleneck. Consequently, our next task in this book is to explore the bottleneck as it exists in the real world. What further fragment of the language puzzle can we find by exploring how children find words and learn their meaning in the constant stream of sound they hear?

9

FINDING AND LEARNING THE MEANING OF WORDS

How children learn language has long been of interest to those concerned with its evolution. The idea that 'ontogeny recapitulates phylogeny' has been promoted, which means the stages of children's development on their way to adulthood replicate those of our human ancestors on their way to becoming modern humans. This idea has been applied to language acquisition and its evolution, but I've never been persuaded.[1] It is intellectually problematic because our human ancestors were never 'on their way' to anywhere other than being themselves. My interest in language acquisition is different and twofold.

First, is language acquired by specialised mental processes that are dedicated to this task or learned by general-purpose processes used for a variety of learning tasks? Second, can we project the processes of language acquisition/learning that we observe in the present into the prehistoric past to gain insights about the evolution of language?

The language-learning challenge

Let's remind ourselves of the problem that language learners face. When we write, we leave spaces between words. Reading wordswithoutsuchspacesisdifficult. When we speak, however, our words flow from one to another without any pauses, generating a continuous stream of sound. Yet we hear and understand each word without difficulty – we automatically 'parse' the utterances we hear into their component parts. We can do this because we know not only the words and grammar of our language, but also its prosody – how syllables at the start or end of words are stressed and how pauses are used between phrases. Without any awareness of such knowledge, we use it effortlessly both when speaking and when listening; we use it to identify words as they appear within the continuous stream of sound we hear, just as if the words had been written down interspersed with spaces. One can appreciate this with a little reflection. Just think how you pronounce the syllable *ham* when referring to a piece of meat and when talking about a furry animal – a long *ham* and a short *ham*-ster.[2] Think about listening to a language unknown to you, one with different words, grammar and prosody. You will be at an utter loss to identify its words, let alone their meaning.

How then, can babies learn language? Although they are born into the world without any knowledge of the language their caregivers will speak, some learning takes places within the womb.[3] The reception of external speech sounds prepares the foetal brain for the processing of language once born. Languages have different rhythms and infants a mere three days old prefer listening to the rhythm of their caregivers' language, indicating that they became familiarised to it before birth via the sounds heard or the vibrations felt within the womb.[4]

Nevertheless, babies are born with no specific knowledge of the language they will hear. Yet whether that is English, Mandarin, Swahili or any other of the 7,000 or so languages that remain today, by around one year of age infants are saying their first words, and within another year they are stringing words together to make simple sentences by using the correct order for their native language – those acquiring/learning English or French will place verbs before objects, and those learning Turkish or Japanese will do the converse. Without any formal teaching, by a mere four years of age, most children have acquired more than 1,000 words, the rules for how words can be combined to generate meaningful utterances, and the ability to change the tone of an utterance to further flex its meaning.

Such language acquisition is truly remarkable. How it is achieved has been one of the major questions facing linguists ever since the study of language began, generating a constant flow of new ideas and evidence. We have already dismissed arguments about Universal Grammar in Chapter 3 and can now move on to more interesting and viable ideas. Within the voluminous literature there are two questions of particular interest for the language puzzle. The first is how infants find discrete words in the continuous stream of speech they hear. The second is how they learn the meaning of those words.

Finding words

How can babies possibly discover where a word begins and ends within a continuous sound stream? A breakthrough in our understanding came in 1996 in a three-page article in the journal *Science* entitled 'Statistical learning by 8-month-old infants', authored by the psychologists Jenny Saffran, Richard Aslin and Elissa Newport.[5] This explained that infants use 'transitional probabilities' (TPs) between syllables to identify which syllable strings recurrently go together, and hence are likely to constitute words, and which syllables have low probabilities of following each other and hence are likely to mark the break between words. The phrase *pretty baby*, for instance, has four syllables (*pre-ty-ba-by*) and three transitional probabilities between syllable pairs. In English the probability that *ty* will be followed by *ba* is lower than that of *pre* being followed by *ty*, and of *ba* being followed by *by*. That eight-month-old infants can calculate and use such transitional probabilities came as a surprise.

Saffran and her colleagues had exposed infants to no more

than two minutes of continuous speech that contained four three-syllable nonsense words, such as *tupiro* and *padoti*. These 'words' were repeated in random order by a monotone speech synthesiser that created a continuous sound sequence, such as *bidakupadotigolabubidakupadotigolabubidakutupiro*. . . The sequence contained no pauses, variations in stress or any other acoustic cues between word boundaries. The only cue available to the infants were the transitional probabilities (TPs) between syllables. Those within words were 1.0, because the first syllable was always followed by the second, and the second by the third, while the TPs of syllables between words was always 0.33. After a mere two minutes of listening, the infants were tested as to whether they differentiated between words (syllable strings that had TPs of 1.0) and non-words (syllable strings that contained TPs of 0.33). For this the infants were presented with words and non-words and found to have longer listening time for the non-words. This indicated they had already become familiar with the words by listening to the continuous sequence of syllables within which they had been embedded. The only way that could have happened was by monitoring the TPs between syllables – the infants were capable of statistical learning.

Although this demonstrated that infants could identify recurrent strings of syllables, a question arose about whether they understood these were 'words' – sounds that could have referents in the world. To explore this, a follow-up study repeated the segmentation task and then tested whether infants were better at learning the association of high TP syllable strings ('words') with visual images, than they were at learning associations between low TP syllable strings (non-words) and images.[6] That proved to be the case, indicating that the infants were able not only to extract the syllable strings but also were ready to attribute meaning to those strings.

Saffran and her colleagues had excluded prosody from their artificial language, but this is pervasive in the sound streams heard by infants in the real world. When we speak, we pause at the end of a sentence while we tend to lengthen the syllable and drop our pitch at the end of a clause. As adults, we rely on prosody to comprehend speech. It is essential for segmenting identical strings of syllables, such as inferring the difference between *crisis turnip* and *cry sister nip* and inferring the intended meaning of an otherwise ambiguous utterance, such as *Raoul murdered the man with a gun.*[7] (Did Raoul use the gun or did the man he murdered have the gun? The intonation of the spoken utterance provides the answer.)

For infant language learners, prosody provides further cues as to where words begin and end. For instance, the last syllable spoken before a pause must mark the end of a word. Experiments have shown that such 'edge' words are easier for infants to acquire than the words that come in the middle of a phrase.[8] They have also shown that pauses are the most helpful prosodic cues for segmenting speech, followed by the lengthening of syllables, while changes in pitch have little or no effect.[9]

It soon became evident that infants integrate cues from both TPs and prosody when segmenting sound streams into words. Jenny Saffran and her colleague Erik Thiessen discovered that infants alternate between such cues. They initially use statistical learning to segment words from a continuous stream of speech sound. By six and a half months old, infants have acquired sufficient words to learn that English words tend to be stressed on the first syllable. They then switch from using TPs to the location of stress as their main cue for word learning. This leads to a few mistakes because some words, such as *guitar*, are stressed on the second syllable, causing an infant of seven and a half months of age, on hearing the syllable string '*guitar is*', to decide that *taris* is a likely word. By eleven months old, infants have learned that non-stressed words exist within the language and integrate the use of both TPs and stress for identifying words.[10]

Such integration of multiple cues must lie at the heart of

what is an even more impressive feat achieved by some infants than finding words within a continuous sound stream. This is finding the words of two or even more different languages within multiple sound streams, coming from either the same or different people. Infants immersed in multilingual environments keep track of and learn different sets of transitional probabilities simultaneously and hence different sets of words for different languages.[11] Remarkable!

In the real world with real language

Laboratory-based experiments using synthesised speech are critical because they control what elements of language an infant is exposed to and which aspect of language is being explored. The real world is a far more complex learning environment. Infants hear multiple voices, all with a slightly different pronunciation and prosody for the same words and phrases.[12] Moreover, the sound streams they hear contain the same words but in different morphologies, such as verbs with the addition of *-ed* to make a past tense and nouns with an *-s* to make a plural – not to mention the pervasive irregularities found within any language. Then there are the grammatical words that must also be acquired, such as *the*, *of* and *to*, along with syntax, the rules for how words must be ordered to make meaningful utterances.

Ongoing experimental studies are finding that statistical learning and prosodic cues enable such complexities of language to be acquired.[13] One set of experiments has shown that infants not only track TPs going forward, but also in a backward direction.[14] This is useful for learning relationships between both syllables and words. For instance, in the pair of words *the dog*, the backward probability (*the* preceding *dog*) is much higher in speech that the forward probability (*dog* following *the*). Other experiments have shown that twelve-month-old infants use statistical learning to first segment words from continuous speech, and then use those words as input to discover their permissible ordering. The infants can move from mere syllables to knowing a rudimentary syntax after no more than a few minutes of exposure to a novel sound stream demonstrating the power of their statistical learning processes.[15]

Further experiments have shown that consonants and vowels play significantly different roles in these two levels of language acquisition, at least by adults. By manipulating the frequencies of consonants and vowels in artificial speech streams, experiments found that participants could identify words by computing TPs between consonants but were unable to do so by the TPs of vowels. Conversely, they could identify grammatical rules for the ordering of words by attending to the patterns between vowels but not between consonants.[16] As with all studies, we must be cautious about generalising from one set of experiments. The participants in these studies were Italian university students and we should ask whether their knowledge of Italian, with the particular role that consonants and vowels might play in that language, influenced how they learned this second artificial language.

One of the challenges involved in learning syntax is that it involves relationships between words that may be some distance from each other, referred to as non-adjacent dependencies. For instance, the tense difference between *is talking* and *has talked* depends on elements (*is . . .ing, has . . .ed*) that are separated by another element (*talk*). Similarly for subject-verb agreement, as in *the dog [down the street] barks* versus *the dogs [down the street] bark*. Such syntactical structures will not be discovered by computing transitional probabilities between adjacent syllables or words.

Experiments have shown that infants can swiftly learn non-adjacent dependencies when nonsense words are separated by pauses, such as *pel wadi rud* and *vot licey jic*, when the first word predicts the third word regardless of what word is between them.[17] In these experiments, however, the hard work of parsing the sentence (finding the words) has already been done. When learners are provided with the type of continuous syllable strings that mimic actual speech, as used by Saffran and her colleagues, their performance is quite different: they are unable to find the non-adjacent dependencies.[18]

On reflection, this is not surprising because the types of non-adjacent dependencies between syllables that arise in such sound streams, such as *pokigopoguga* in which *po. . .ga* are the non-adjacent syllables, are rarely found in spoken languages. Far more frequent are non-adjacent sound segments defined by consonants separated by vowels, such as *pagitpogut* in which *p..g..t* are non-adjacent recurring consonants. Similarly, the converse, when recurring vowels are separated by consonants. When the experiments were re-run using an artificial language structured in this manner, the participants were quite able to learn non-adjacent dependencies.

This, and various follow-up experiments, support the discovery from the computer models we examined in the previous chapter: the form that languages can take is partly defined by the constraints on what human learners are able to acquire. In other words, only some types of language can fit through the language-acquisition bottleneck. It is not surprising that non-adjacent dependencies between syllables cannot be acquired. Doing so while also tracking adjacent syllables, and doing both forwards and backwards through utterances, creates an overwhelmingly massive computational task.

While the real-word sound streams are vastly more complex than those used in laboratory experiments and computer models, caregivers provide infants with help by modulating the way they speak. So-called motherese, or infant-directed speech (IDS), deliberately exaggerates prosody by heightening pitch contours and lengthening vowels, adding pauses, repetition and emphasising consonants, as in '*You are a beauutifulll baabyeee, yooo are, yooo are, yes yooo are.*' IDS also uses shorter words and phrases, often providing words in isolation which babies are known to prefer. Such baby-talk aids the use of both TPs and prosodic cues for segmenting the sound stream into words, and most likely acquiring the grammatical structure of language.

The contribution of IDS should not be overestimated: it is an aid and may accelerate learning, but ultimately infants do it for themselves. Indeed, they learn how to help themselves. As soon as some words have been acquired, they use the end of these words to identify that a new word is beginning and hence they only need to search for its likely end – which might be the start of another already known word.[19] As they approach two years of age with a sizeable vocabulary, they become resistant to acquiring words that are phonetically untypical for their language.[20]

Speech or signing usually happens in a rich visual environment involving the speaker and listener, their facial expressions, body language, gestures, and surrounding objects and settings.[21] These provide further cues that facilitate the identification of words and learning their meanings, such as pointing to a teddy bear while also saying the two-syllable string *ted-dy*. Moreover, the contexts in which words are spoken is always changing. This reduces the ambiguity of word meaning because the co-occurrence of the word with one of several possible referents can be tracked. If the child learns the word *teddy* when playing with a teddy bear, a doll and a toy dog, he/she may not know which object this word refers to; when hearing it again when playing with a toy car and a teddy bear that ambiguity can be resolved. This is another form of statistical learning, one using so-called cross-situational statistics.[22]

We have now drifted into learning the meaning of words rather than simply their existence. Before further addressing that process, three points must be stressed. First, the statistical learning that underpins the segmentation of sound streams is a general learning process used in a variety of domains other than language. It also underpins learning about music and the visual world.[23] Second, the nature of those learning processes imposes structure onto language, causing recurrent features within linguistic diversity. Third, statistical learning is not specifically human but found in other animals for learning about the world, because extracting regularities in sensory input is a core mechanism for effective adaptation. Of most interest for our purposes is its use by non-human primates that potentially provide analogues for our early ancestors who were the first to invent – or perhaps discover – words.[24] Laboratory experiments have shown that chimpanzees use statistical learning by exploring their abilities to find recurrent patterns in sequences of shape and tones. Even more evolutionarily distinct primates such as marmosets and cotton-top tamarins have been shown to use statistical learning, evaluated by testing these primates on equivalent tasks to those Saffran and her colleagues used for newborn babies.

Learning the meaning of words

Discovering words and learning their meanings often happen together, as when an infant watches a caregiver point to a soft toy and say *ted-dy*. But we frequently learn words before we know what they mean. While researching this book, I came across words such as *apocope* and *lenition* and had to look up their meanings.[25] We can all recall the experience of hearing a word and wondering what it means. I recently overheard my wife say *rubato*. When asked, she told me it is a musical term that means playing with a degree of rhythmic freedom by slightly speeding up and then slowing down the tempo of the music as one pleases. I recall being ten years old and wondering what the word *condom* meant, along with various euphemisms of this word that I was beginning to hear spoken by my rather more developed classmates.

Wondering what words mean must be a regular experience for children as they learn language, this requiring an average rate of ten new words a day to reach the 60,000 words known by most seventeen-year-olds. How do they manage that in the 'blooming, buzzing, confusion' of a language learner's world? An answer was provided within an exceptionally good 2000 book by the psychologist Paul Bloom called *How Children Learn the Meanings of Words*.[26] Although the ideas and evidence he provides have been extended during the last twenty years, Bloom's arguments remain sound and have been validated by new research.

Do children simply learn to associate the words they hear with the things they see? That would require calculating the statistical co-occurrence between heard words and seen entities to derive the meanings of words on a probabilistic basis. We know that children can use statistical learning in a variety of learning tasks including finding recurrent syllable strings (i.e. words) within streams of speech. The cross-situational learning experiments referred to above show that statistically based learning for the meanings of words does indeed occur. Although we should not neglect its importance, Bloom argues that associative learning is supplementary to the principal ways in which word meanings are acquired.

If associative learning was the main process, children would be unable to learn the meaning of a word they hear when they are looking at something else. Nor would they grasp the meanings of words that have no visual representation, words such as *think* and *know*. By needing to calculate the statistical co-occurrence between words heard and entities seen, children's word learning would be much slower than it is; they would be hesitant when using words and prone to error. Children are neither of these: they display 'fast-mapping', grasping the meanings of new words after a few incidental exposures and rarely make mistakes. How do they do it?

There are three key methods. First, children are born into the world with perceptual biases that facilitate their learning. Second, they exploit the mental states of others by sharing their attention and inferring what speakers intend and are thinking about when uttering words. Third, they use their initial learning of word meanings to acquire those of more challenging words from language itself. Let's look at each of these in turn.

Perceptual biases

Children have an intuitive understanding that the world contains entities, properties, events and processes. They have a bias towards thinking about objects, evident from both language learning and other areas of cognition. This encourages children to interpret novel words as referring to objects and to learn nouns more quickly than other types of words – their acquired vocabulary has more names for objects than they hear in the speech directed to them. This bias is not surprising because object names are the easiest type of word meanings to acquire. Other types of words, such as *the*, *many* and *of*, are difficult to grasp without knowledge of their semantic context, which is only provided by the meaning of nouns and proper names. Whichever language is being acquired, children's first words are similar: proper names for people and animals, and common nouns such as *ball* and *milk*.

Object bias encourages children to interpret the words they hear as referring to complete objects. When a caregiver points to a dog and says *dog*, the child will intuitively understand that word means the whole animal, rather than just part of it such as the dog's tail or ears. Nor will they mistake the word to mean the wider scene consisting of multiple objects, such as the 'dog on the lead with a lady'. Caregivers are intuitively aware of this bias. When presenting words to children that are parts of objects, they often introduce the whole object first. When looking at a rabbit (whether a toy or living) rather than saying, 'Look at the big ears,' they will say, 'This is a rabbit. Look at its big ears.'

This object bias is further channelled by the child's sensitivity to iconic words, otherwise known as cross-modal perception, as considered in Chapter 6. These are words that mimic one or more aspects of their referents, such as how they sound, their size, shape, movement or texture, the most familiar of which are onomatopoeias. Because such words are grounded in the sensations of the child, they are easier to identify and learn by association. When hearing a single-syllable word with a high, front vowel, the child will be automatically drawn to objects in its visual field that are small and quick moving, such as a fly or a bee. Similarly, words with /r/ sounds will naturally lead them to round objects such as an orange or a marble, and to verbs such as roll and rotate (note how all these words contain an /r/). As we previously noted, iconic words dominate the lexicons of children up to the age of six years old.

Children are also biased to shape. When they learn the name of an object, they are prone to transfer that name to another object of the same shape. That works, because objects which are of the same shape are likely to belong to the same category: chairs can be different sizes, colours and textures, but are of a similar shape.

That items can be placed into categories is another intuitive bias that facilitates word learning. Children are either born with or rapidly develop an impulse to place entities into categories, with each category forming a concept within their minds (as further considered in Chapters 11 and 14). Infants have an intuitive, conceptual understanding of differences between living things and inanimate objects.[27] They know that the former can move by themselves whereas the latter cannot, although children can ignore such differences when engaged in pretend play. They understand that categories are hierarchical, enabling them to swiftly learn that terriers are types of dogs, which are types of animals, which are types of living things. Hierarchical categorisation into basic (e.g. dogs), subordinate (e.g. terriers) and superordinate (e.g. animals) categories is essential for learning, problem solving and decision making.

Such categorisation is crucial because if every object, action and idea was unique, minds would become overloaded with information, having to remember the specific features that define every single entity. Some entities are unique, however, and are referred to by proper rather than by common names, such as Winston Churchill, which does require a feat of memory. For common names, entities are not only placed into categories, but each category is also believed to have essential qualities that are shared by its members: dogs have four legs, a tail and ears and bark; birds fly and lay eggs, and so forth. In fact, neither of these are absolutely true: there are breeds of dogs that do not bark; male and juvenile birds do not lay eggs. Nevertheless, such qualities are sufficiently common to allow inductive inferences from one member of a category to another.[28] As the child acquires knowledge they can loosen their concept boundaries, such as to accommodate penguins and ostriches within the category of birds, even though they do not fly.

By 'learning the meaning of a word', the child is attaching the sound of that word to a mental concept that refers to a category of entities in the world. Once the word *chair* is learned, it will be used for all chairs (i.e. the concept of chair) rather than the specific chair that may have been pointed to and labelled by its caregiver. This is particularly evident from the fact that when a child learns a word such as *table* from a drawing of that object, it has no hesitation in using that word when it sees an actual table or a tiny table within a doll's house.[29]

While children are born with some ready-made concepts, others are swiftly acquired through experience before learning their associated words, providing children with many unlabelled mental concepts, perhaps about food, toys, pets and family members, such as grandma or sister. Because their caregivers are likely to share the same concepts, children are primed to identify labels for their concepts as soon as they are used by caregivers. For instance, even before hearing the word *cat*, infants are likely to have a concept of cat and be aware that cats come in different colours, sizes and degrees of fluffiness; also as living as living creatures, toys and pictures. When they learn the word, perhaps from a picture in a book while enjoying a story, they can apply the term to all the different types of cats that fall into their concept of cat. Moreover, they are biased to treat new words as relating to a basic-level concept. The word *dog*, for instance, will be treated neither as representing a superordinate concept (such as animal) nor as a subordinate concept (such as terrier). In this way they can make maximum use of the word with minimal risk of making errors.

Children vary as to the number of words they know for different categories of objects. Some know lots of names for animals, but few names for types of vehicles; others are the converse or might have a relatively large number of words for types of food. This reflects what they are interested in, which in turn depends on factors such as their caregivers' interests and their early experiences, although inborn variation should not be dismissed. Such interests facilitate word learning, as shown in a recent experimental study with children of thirty months. Each child's interest in four types of categories was measured: animals, vehicles, clothes and drinks. This was done by asking their parents and measuring the dilation of the child's pupils when shown familiar members of each category, items for which they already knew the words. Pupil dilation is a reliable measure for cognitive interest in visual stimulus. The children were then shown novel members of the four categories and told their names, such as *pangolin* for an animal and *rickshaw* for a vehicle. Later they were tested on their abilities to correctly name the novel items. Children were found to be best at learning the words that belonged to the categories they were most interested in.

With categorisation playing such an important role in the process of word learning by children, we might ask whether this is a unique mental process to humans or whether, like statistical learning, it is shared with other primates, notably chimpanzees, and hence be attributed to our early human ancestors.

The answer is an emphatic yes, and not just for chimpanzees. Perceiving the similarities between entities and recognising that they can be grouped into categories is a must for all animals, with each category being represented by a mental concept. Concepts and categories are required to recognise types of food items, predators and members of their own group, such as who to be friends with and who to avoid.[30] The key difference with humans is that whereas we can express any concept we knowingly hold using words, non-human primates are restricted to expressing a tiny fraction of the many concepts their minds contain.[31] Experiments have shown how chimpanzees can sort objects into categories. They do so with ease when the categories are natural entities, such as trees and flowers, but struggle when the categories are dependent on functional qualities, such as tools and non-tools.[32] This suggests that chimpanzees and other animals have a strong bias to categorising objects by appearance alone.[33]

Mind reading

Having a perceptual bias to objects, shapes, iconic words and natural categories is the first of three key methods that children use to learn the meaning of words. The second is by exploiting the mental states of others. Understanding the beliefs, desires and knowledge of other people is referred to as having a theory of mind (ToM) – informally, we can call this mind reading. As adults we constantly use this in our day-to-day social interactions to explain and predict others' behaviour. We may have no self-awareness of doing so, but we are always thinking about what other people think and know, and what they think we know, and so forth. We base what we say and how we act on what we believe others believe, and sometimes what we believe they believe that we believe – and so on.

As far back as 1781, Jean-Jacques Rousseau had proposed a role for theory of mind in the acquisition and evolution of language, although that was long before this phrase had been coined. In his *Essai sur l'origine des langues*, he wrote: 'As soon as one man was recognized by another as a sentient, thinking Being, similar to himself, the desire or the need to communicate to him his sentiments and thoughts made him seek the means to do so.'[34]

Although children are not born with a theory of mind, they have its precursors which lead to its swift development provided they grow up with sufficient social interaction. By the age of four years old most children can appreciate that others may have different thoughts and beliefs to their own. Their capability is tested by giving them a false-belief task. In one such test, they first watch someone putting sweets into a drawer who then leaves the room. While away, another person enters and moves the sweets into a cupboard. The first person returns and the children are asked, where will she look for the sweets? If they say the cupboard, it shows they cannot differentiate her knowledge from their own; if they say the drawer, it shows they understand that someone can have a belief that is not only different from their own but which is false. Children over four years old tend to pass this false-belief task, while those younger fail it – but there is plenty of variation. There are similar experimental protocols for testing other elements of mind-reading development, such as their understanding how people can have different desires and different, but not necessarily incorrect, beliefs.[35]

The key precursors for mind reading are engaging in joint

attention and appreciating intentionality in others. Both tend to be evident from the age of six months old and continue developing until the age of two years. Joint attention involves the ability to follow another's direction of gaze or pointing gesture. This will direct the infant to an entity that is being spoken about and thereby help them grasp the meaning of the word, especially if they also understand that is the intention of the other person. Children develop joint attention skills with different rates and to different levels. Long-term studies tracking infant development have shown that differences between children in their joint attention skills at ages between six and eighteen months old predict the size of their vocabularies at thirty months.[36]

Children's understanding of their own intentionality and that of others is apparent from the names they give to pictures. Rather than relying on what the picture looks like, they refer to the intention of the artist. A two-year-old can point to their own scribble and confidently call it 'mummy'.

Once children are aware of their own and other people's minds – what they respectively don't and do know – the next step is to ask questions. At around sixteen months old, children begin to point. They are soon asking, 'What's that?', although may initially use words like 'wha', 'tha' and 'eh'. Such words are often within the first fifty words that a child learns, and sometimes within the first ten.[37] When caregivers provide the names for the things that children are pointing at, those words are preferentially retained over the words that caregivers might choose to tell the children themselves – and which might be of little interest to the children.[38] As we saw with categorisation, infants drive their own language learning.

As mind reading further develops, children appreciate that different people know different things, and will differ in their reliability as a source of word meanings. This was demonstrated in an experiment in which pre-school children were given two puppets to watch and listen to.[39] The children were first tested on a range of mind-reading tasks to find which of them were more advanced and which less developed in this ability. Both puppets were asked to name three objects that were already familiar to the children: a toy car, a spoon and a toy cat. One of the puppets always gave the correct names and the other always got them wrong. Then the puppets were given three objects that the children had never seen before and were asked to name them. Each puppet labelled them with different nonsense words – one would call an object a *mirp* and the other call it a *preek*. The children were then asked to choose the correct name of the novel objects. Those who had been measured as already having more advanced levels of mind reading were more likely to select the words that had been given by the puppet that had accurately named the objects already known to the children (the toy car, spoon and toy cat). To check that the children's use of these puppet role models for word meanings was specifically about their apparent knowledge of vocabulary, the experiment was repeated with new puppets and making one appear much stronger than the other by showing them lifting heavy objects. The nonsense word test was repeated and the children with more advanced levels of mind reading showed no preference for the names provided by either of the puppets.

The value of understanding the beliefs, desires and knowledge of other persons for acquiring language can be appreciated from children whose development of mind reading is inhibited. This is the case for children with autism spectrum disorder (ASD), who have problems with social communication and interaction. ASD can be manifest in a variety of ways, at different levels of severity and with different consequences for language acquisition. In general, however, the extent to which a child with ASD has problems with learning the meaning of words appears to be a direct consequence of their level of impairment regarding theory of mind.[40]

Might an absence of mind-reading abilities be one of

the reasons why chimpanzees are constrained in their vocal capabilities, having word-like calls but no words with learned and shared meanings? When reviewing the research of Klaus Zuberbühler and his colleagues in Chapter 4, we found evidence that chimpanzees modify their calls depending on what their companions did or did not know about predator danger and food availability. The dilemma is that alternative interpretations are always possible. Juvenile chimpanzees refrain from giving food calls when they might attract dominant individuals who would steal their food. Is that because they have thoughts about others' cognitive states, or because they have simply learned through trial and error that being quiet enables them to keep their food?

Despite extensive laboratory experiments over the last fifty years to discover whether or not chimpanzees have similar mind-reading abilities to humans, the results have been inconclusive.[41] The task is far harder than when testing human infants because one cannot simply ask a chimpanzee what he/she thinks someone else intends, knows or believes.

By 2008, a sufficient number of experiments had been undertaken to allow the psychologists Josep Call and Michael Tomasello to undertake a review and decide whether chimpanzees have a theory of mind.[42] They concluded that chimpanzees are quite able to understand that others have knowledge, goals and intentions. Chimpanzees engage in gaze-following and joint attention, and evidently appreciate that others see, hear and know things. However, Call and Tomasello were unable to find any experimental outcomes to indicate that chimpanzees can understand that others may have false beliefs.[43] This is a skill that four-year-old children master with ease and one of the reasons why they know the meanings of words and chimpanzees do not.

Learning language from language

Perceptual bias and mind reading provide two of three key methods by which children learn the meanings of words. The third is by drawing on cues from within language itself. This only becomes possible after some words have been acquired. They provide clues to the meanings of other words, some of which may be impossible to learn in any other way.

Paul Bloom gives the example of *the*. It is difficult to explain the meaning of this word to someone without using the support of other words. Once a child has acquired object labels, such as *dog, chair* and *toy*, the meaning of *the*, as in *the dog* (*chair* or *toy*) will become evident.

Children tend to think that words can have only one meaning, known as the mutual exclusivity principle.[44] Although that is not entirely correct, as evident from homonyms such as *bat* (both a flying mouse and an instrument for playing cricket), it is a helpful guide to learning words by the use of words that are already understood. When a child knows the words *apple* and *red*, and is shown what is described as a *round, red apple*, he/she will be able to guess the meaning of *round*. If a child hears *Do you want me to cook some porridge for breakfast?*, they can use their knowledge of words for *cook* and *breakfast* to infer that the word porridge is a type of food. That would be confirmed when the porridge arrives in the bowl with a spoon – children are always integrating multiple cues for the meanings of words.

A host of experimental studies has shown that children use a wide range of syntactic cues for the meanings of words, especially grammatical words such as *of, on* and *for*, and verbs. When shown an unfamiliar object and given a name, children use syntax to decide whether the name belongs to that specific object or is a category name. What does the word *flopsy* mean when heard in association with the first sight of a small furry animal with long ears? Syntax provides the clue: *this is Flopsy*, rather than *this is a flopsy* (NB it is a rabbit). Once the word *rabbit* is known, the placement of a word immediately before it will suggest a property word, as in *this is a big rabbit*, whereas a word positioned between *the rabbit* and another object word will suggest a spatial relationship, as in *the rabbit is in the hutch*.

Paul Bloom explains that a child hearing *I would like you to clean your room* could infer from the syntax that *clean* refers to an action that one entity does to another. Another example from Bloom involves combining cues. When hearing *I'm really annoyed that you kicked the dog*, the child might use the speaker's quivering body language to guess that the verb *annoyed* has a negative connotation. The syntax of this utterance is known as a sentential complement because it has an embedded clause (*you kicked the dog*) under the main verb (*annoyed*). The main verb of such phrases can only refer to mental states, such as *think, guess, annoy* and *believe*, or a means of communication, such as *say* and *tell*. As such, sentential complements provide particularly important cues for acquiring the meanings of these types of verbs, which are difficult to grasp outside the context of language itself.

The use of sentential complements provides us with another important lesson: there is a two-way street between the acquisition of language and the development of mind reading. The embedded clause in such phrases can be a false statement within an otherwise true statement as denoted by the main verb, such as *Lucy thinks the moon is made of green cheese*. *Lucy thinks* and *I'm annoyed* in the previous example are true while their embedded clauses might be either true or false. As such, when hearing sentential complement phrases, one is invited to entertain the idea that other minds might contain false beliefs.[45]

An experimental study took a group of children aged between thirty-six and fifty-eight months old and evaluated their mind-reading capabilities using false-belief tasks.[46] The group was then split into two sets. One set was trained in the use of sentential complements, using phrases such as *John said that Fred went shopping*. The other set of children were trained in the use of a phrase that also had an embedded clause, but which lacked reference to beliefs, such as *the boy that had red hair* . . . Following this training in language use, the children were re-tested on the false-belief tasks. The first set, those trained in the use of sentential complements, were found to have enhanced their ToM skills, whereas there was no change in the second.

This study suggests a bootstrapping effect between the acquisition of language and the development of mind-reading skills. Joint attention initially enables the meanings of some words to be acquired; those words will then enhance the attribution of mental states to others; the developing mind-reading skills enable further word meanings to be acquired, which in turn provide access to sentential complements. These further improve the child's understanding that other minds can hold false beliefs while helping them to learn the meanings of more words about mental states. Bilingual children appear to have an advantage, often passing the false-belief task at the age of three. The linguist Viorica Marian suggests this is because bilingual children must learn to pay extra attention to the language of the person they are interacting with, improving appreciation of their perspective of a situation.[47]

A similar bootstrapping effect exists between language acquisition and concept formation. While many concepts are acquired through observation of the world and relate to what can be termed natural categories, such as *cats*, *trees* and *toys*, other concepts are reliant on words themselves. It is unlikely that a child, or even an adult, can acquire the concept of *freedom* without having the word or a suite of related words to first install this concept into his or her mind. Such concept formation via exposure to language has been termed 'category-assembly' to contrast it with 'category-recognition' when a pre-existing category is labelled by exposure to a word, such as the *cat* example I gave above.[48] But once that category-assembled concept exists, it facilitates the learning of other words that fall into the same category, such as *liberty* and *autonomy*, just as a child finds it easier to learn the meaning of *pangolin* once it already has some names within the category of animal.

An accident rather than a miracle

Parents are often awestruck when their children say their first few words, which so swiftly become a stream and then a flood of words. How do they ever learn so many words so quickly and know how to use them? It was once easy to think that children needed highly specialised and dedicated language-learning mechanisms in their brains to achieve that feat – the Universal Grammar idea that dominated linguistics for so many years. That was wrong, because general-purpose mechanisms such as statistical learning, supported by a few perceptual biases and a little mind reading, is all children require. Traces of these mechanisms are found in our primate relatives, indicating they had evolved long ago for reasons unrelated to language. Paul Bloom suggests their suitability for learning the meaning of words may have been no more than a lucky accident.[49]

Now that we have covered language acquisition, the interior of the language puzzle is filling up. A picture of how language evolved has yet to emerge, but some hints are appearing: even our earliest human ancestors would have communicated using word-like entities, with some degree of control over when and how to make their vocalisations (Chapter 4); a form of language requiring a modern-like vocal and auditory tract was present in *H. heidelbergensis* by 500,000 years ago (Chapter 5); the earliest words are likely to have been iconic, with arbitrary words emerging from a gradual process of sound change (Chapter 6); whatever type of languages were used by *H. habilis*, *H. erectus* and *H. neanderthalensis*, they did not drive technological change and cultural diversity (Chapter 7); the transmission of language from one generation to the next through a learning bottleneck plays an active role in shaping languages (Chapter 8); general-purpose learning mechanisms, traces of which are found in the minds of our living primate relatives, are used for language learning by the child.

With several gaps still to fill within the puzzle, it remains unclear how these fragments can be joined together. Maybe the next fragment will throw some light on this matter: firelight.

10

FIRE

Modern-day hunter-gatherers make fire on every possible occasion, using it for a variety of tasks, including cooking food, providing warmth, deterring predators, preparing tool-making materials, and for light – a means to extend the day. Most of all, fire provides a social hub, a focus for talking, telling stories and singing together. Sitting by the fire with their parents, siblings and elders provides a stimulating environment for children when finding and learning the meaning of words.

One might compare the skills of gathering fuel, building, lighting and tending a fire to those of making composite tools and putting words together to make a meaningful utterance – all three outcomes are more than the sum of their parts. Like language, controlling fire is a unique attribute of humans and is found in all modern human communities. Does the trajectory of discovering and mastering the use of fire parallel that of words and language in human evolution?

My interest in fire goes beyond its use as a locus for where children can learn the meaning of words and how making fire might be a proxy for making spoken utterances. I am also concerned with the types of words and linguistic phrases that fire entices out of human minds.

This interest originates from a study of 'night-talk' among hunter-gatherers undertaken by the anthropologist Polly Wiessner.[1] Having made meticulous recordings and analyses of talk by the Ju/'hoansi Bushmen of the Kalahari, Wiessner identified their daytime talk to be about economic matters and social relations. When seated around the fire at night, however, the Ju/'hoansi told enthralling stories about people in the recent and distant past, engaged in singing and dancing, and evoked the supernatural. Wiessner explained why there is such difference:

> Sufficiently bright firelight represses the production of melatonin and energizes at a time when little economically productive work can be done; time is ample. In hot seasons, the cool of the evening releases pent up energy; in cold seasons, people huddle together. Fireside gatherings are often, although not always, composed of people of mixed sexes and ages. The moon and starlit skies awaken imagination of the supernatural, as well as a sense of vulnerability to malevolent spirits, predators, and antagonists countered by security in numbers. Body language is dimmed by firelight and awareness of self and others is reduced. Facial expressions – flickering with the flames – are either softened, or in the case of fear or anguish, accentuated. Agendas of the day are dropped while small children fall asleep in the laps of kin. Whereas time structures interactions by day because of economic exigencies, by night social interactions structure time and often continue until relationships are right. Foragers make use of daytime efficiently and night-time effectively.[2]

Not restricting herself to the Ju/'hoansi Bushmen, Wiessner reviewed evidence from hunter-gatherers around the world. She found striking commonalities regarding night-time activities, to conclude that:

> When the night appears to have really mattered was for the extension of cultural institutions over time and space to link individuals from different bands into larger 'imagined

communities' beyond village limits, an enterprise that involved complex cognition and time-consuming information transfer. In most hunter-gatherer societies, firelit hours drew aggregations of individuals who were out foraging by day and provided time for ventures into such virtual communities, whether human or supernatural, via stories and ritual. Stories conveyed unifying cosmologies and charters for rules and rites governing behavior. These stories also conveyed information about the nature of individuals in the present and recent past, their experiences and feelings, as well as factual knowledge about long-distance networks, kinship, and land tenure. Stories told by firelight put listeners on the same emotional wavelength, elicited understanding, trust, and sympathy, and built positive reputations for qualities like humor, congeniality, and innovation.[3]

Experimental studies provide a physiological basis to these observations by demonstrating that sitting by fires induces relaxation as part of a multi-sensory, absorptive and social experience.[4] As such, the warmth, light and aromas of the fire create emotional and physiological states that are especially receptive to new ideas and information. Although it is possible to talk about the supernatural and tell stories during daylight hours and conversely discuss mundane matters while sitting around flickering flames in the dark, the history of fire use might provide an insight into the prevalence of language that both drew on and excited the imagination.

From the taking to the making of fire

The use of fire to cook food has played a prominent role in a theory about the enlargement of the brain in early *Homo*.[5] Because of its energetic demands, for the brain to increase in size the gut had to get smaller. The only way that could happen was by cooking food to reduce the time required for its digestion while also increasing the calorific return. It's a great theory, predicting that by 1.6 mya *Homo erectus* must have had control of fire to cook meat acquired from hunting and aggressive scavenging. It has, however, run into a problem because there is a complete absence of deliberate fire use before 1 million years ago. Scatters of burnt bones, artefacts and sediments dating to *c*.1.5 mya in Kenya, notably at FxJj20 and Chesowanja, are best explained by natural savannah fires caused by lightning strikes that swept over pre-existing debris from hominin activity.[6] We must be cautious, however. Remember making a fire on a beach or on a scouts' camping trip? Even a substantial blaze can leave no more than a small patch of ash, one that is swiftly blown away. That may have happened to the earliest evidence for the human use of fire that had been critical to the evolution of the brain.

The earliest evidence for fire use that has survived dates to 1 million years ago and comes from Wonderwerk Cave in South Africa, followed by that from 0.8 mya at the site of Gesher Benot Ya'aqov in Israel.[7] Both have burnt bones, stone artefacts and sediments indicating that fires had been deliberately started and controlled at these locations – not just once but on several occasions. Constructed fireplaces are absent, however, and it seems most likely that the flames had originated in natural fires and were then taken to these locations. If fire could have been made at will, we would expect greater evidence for its use.

The earliest known use of fire in Europe is at *c*.800,000 years ago at Cueva Negra in southwest Spain, represented by burnt bone and stone.[8] This is an exception because no other evidence is known until after 400,000 years ago, suggesting that Cueva Negra is another example of transporting fire from a natural combustion.[9] The absence of further evidence is surprising because it indicates that *H. erectus* and its descendants had colonised northern latitudes of Europe and Asia without the use of fire, despite the often glacial conditions. There is no convincing evidence for fire use at the largest and best-preserved early Stone Age sites in Europe, such as Terra Amata in France, Boxgrove in England and Gran Dolina in Spain. The first known trace of fire use in Asia is not until *c.*300,000 years ago, found at Zhoukoudian Cave in China, despite a record of occupation reaching back beyond 1 million years ago.[10]

Recurrent traces of fire use only appear after 400,000 years ago, broadly associated with larger-brained species of *Homo heidelbergensis*, *H. neanderthalensis*, and a little later, early *H. sapiens*, although *H. erectus* remained in eastern Asia. This date appears to mark when humans had learned how to create fire for themselves. There are two ways this could have been done. One is by friction to create smoke which is then coaxed into a flame. A 1979 account of making fire by Ju/'hoansi Bushmen describes this method:

> Two different kinds of wood are used: a hard wood for the drill and a soft wood for the base. The operator cuts a notch near the tip of the base stick held flat on the ground, with a knife blade to receive the coal, and places the tip of the drill stick in the notch. He twirls the drill stick rapidly between his hands with a firm downward pressure, taking care to keep the drill tip from slipping out of the notch . . . Drilling fire looks deceptively easy in the hands of a skilled operator yet real muscular strength and control are required to get the fire started. Even experts in the task are sweating with exertion after a minute's drilling.[11]

The second way to make fire is by striking two stones together to create a spark, which must then be caught to create a flame. Flint, or other types of stone containing silica, are ideal for making sparks and hominins must have noticed accidental sparks when making their tools. Dry grass can be used as tinder, but the best natural tinder is a tree fungus, *Fomes fomentarius*, sometimes called the hoof fungus because of the way it looks. Although claims have been made, there are no verified discoveries of this fungus at archaeological sites – not surprising, because it is unlikely to be preserved.[12]

The earliest evidence for making fire comes from the 25-metre-deep sequence of deposits in Tabun Cave, Israel.[13] This has almost 100 archaeological layers dating from *c.*500,000 to *c.*150,000 years ago, beginning with Acheulean handaxes and ending with Levallois tools. Burnt artefacts only appear at around 350,000 years ago, after which they are found in all layers interpreted as indicating a habitual use of fire. There is a marked increase in the frequency of burnt artefacts at around 200,000 years ago, coinciding with the first trace of fireplaces – thin layers of deep black-and-white sediment containing fragments of burnt wood. This evidence is matched by that from Qesem Cave in Israel which shows repeated use of fire between roughly 400,000 and 200,000 years ago.[14] Much of that evidence comes from microscopic analysis of sediments which can identify recrystallised wood ash and fragments of burnt bones. A particularly large hearth was located, centrally positioned in the cave with associated stone artefacts, butchered animal bones and evidence for repeated use at around 300,000 years ago.

The first evidence in Europe is dated to *c.*400,000 years ago, coming from Beeches Pit, England, and Gruta da Aroeira, Portugal. In both cases it consists of burnt sediments, bones and artefacts rather than constructed hearths, and could reflect a continuing use of fire taken from natural events.[15] That cannot explain the impressive sequence of hearths from within Bolomor Cave, Spain.[16] These were first made at *c.*350,000 years ago and continued to be made as deposits accumulated in the cave throughout the next 250,000 years. The Bolomor hearths are quite simple: circular patches of ash directly on the floor usually less than one metre in diameter, with occasional use of stones to delimit the areas. Fourteen hearths have been excavated, sometimes found in small groups, but many more are suspected in the unexcavated areas of the cave.

Similar fireplaces are found at Neanderthal sites throughout Europe, with notable examples from the Pech-de-l'Azé and Roc de Marsal in France, and Grotta di Fumane in Italy.[17] They are often found in association with pieces of mineral that contain manganese dioxide, which had sometimes been scraped to look like crayons. Experiments have shown that these minerals and the powder they produce promotes the ignition and combustion of wood, and hence they are likely to have been used with sparks from struck flints to start the fires.[18]

Whether the ability to make fire was known by all Neanderthal communities is unclear. Fireplaces are absent at many sites in Europe including several dated to the coldest conditions of the last ice age. Does that indicate a lack of knowledge or simply that fire played a less significant role in daily life than we find among modern-day hunter-gatherers? Maybe there had there been a dearth of woody fuel to burn in the vicinity of these sites.[19] The most impressive Neanderthal fireplaces come from Kebara Cave in Israel, dated to *c.*60,000 years ago.[20] This cave has a long sequence of stacked hearths, represented by fine laminations of black charcoal and white ashy layers throughout the stratified deposits. They had been repeatedly made in the centre of the cave, while ashes cleared from the fireplaces had been dumped by the side walls.

The most intriguing and unexplained use of fire by Neanderthals comes from Bruniquel Cave in southwest France, dating to *c.*176,000 years ago.[21] Rather than being close to the entrance, this is found deep underground in a chamber where Neanderthals are claimed to have arranged broken stalagmites into circular arrangements on the floor, although the possibility of these arising from bears or water action cannot be excluded. There were no artefacts, footprints or Neanderthal remains. Signs of burning on several stalagmite fragments, a small patch of carbonised organic material, and pieces of burnt animal bone indicate that several fires had been lit. For what reason is unclear. Some archaeologists have jumped to conclude this reflects a site of Neanderthal ritual; my strong preference is towards a practical explanation concerning generating light and warmth, combined with an appreciation that Neanderthals were as curious about the underground world as we are today.

The two most notable sites for the use of fire by *H. sapiens* in the Middle Stone Age of Africa are Border Cave and Blombos Cave, both in South Africa and with stratified sequences of burnt and ash-rich deposits, similar to those found in Kebara Cave. The hearths in Border Cave were made at around 200,000 years ago, placed in the centre of the cave and probably used to deter predators and to cook food.[22] Microscopic analysis of sediments and plant remains has shown that ash from the fires had been raked out and spread over the floor at the rear of the cave. When cool, sheaves of grass were placed over the ash, probably to use as bedding. The ash was likely to deter crawling and biting insects because they cannot easily move through fine powder which also blocks their breathing and biting apparatus. The fireplaces in Blombos Cave were made a little later, between 100,000 and 70,000 years ago.[23] They vary in their thickness, the extent to which they are stacked above each other, and their preservation with some being heavily trampled. This variation has been used to infer changing patterns of occupancy in the cave, but their overall presence indicates fire had always been essential to activities at Blombos.

Our knowledge of fire use by the modern humans who dispersed from Africa after 70,000 years ago is best represented by that from Europe, this simply reflecting the quantity of research that has been undertaken. A recent review of Upper Palaeolithic fire use considered it to be a ubiquitous and fundamental part of daily life.[24] Many of the earliest Upper Palaeolithic hearths were similar to those made by the Neanderthals, no more than circular patches of ashes and burnt settlement on the floor. Hearths became larger and more complex and diverse over time, using intentionally dug depressions, circular rings of stone and prepared surfaces of stone or clay. Grotta di Fumane has evidence for this change within a stratified sequence of hearths with those of the earliest Upper Palaeolithic being multilayered and better defined than those made by the preceding Neanderthals.[25] Between 32,000 and 28,000 years ago hearths throughout Europe became notably larger and often surrounded by features perhaps for storing fuel or cooked items. This period coincided with developments in symbolic art, notably the manufacture of female figurines. Kiln-like structures were made at the eastern European site of Dolní Věstonice where clay models of animals and women were baked hard.

Sparking the imagination?

The evidence of fire use suggests that long before humans had learned how to make it for themselves, fire was valued as a source of heat and light, causing our ancestors to capture and transport smouldering embers from natural fires to their living sites where they ignited gathered fuel. Lightning strikes and bush fires are also likely candidates for early iconic sound segments and words, these expressing the multi-sensory experiences of seeing, hearing and smelling fire. The sparse presence of fire residues and the complete absence of hearths before 400,000 years ago suggests that humans before then were not engaging in the type of night-time talk described by Polly Wiessner for modern-day hunter-gatherers. Would it have been possible to tell stories about the distant past and invoke the supernatural without the physiological states induced by fire? Without the flames, hominins likely wanted to keep quiet at night to avoid attracting predators, sleeping in trees for their protection.

The evidence for making fire at will appears to arise soon after 400,000 years ago. This required the execution of hierarchically organised skills as complicated as those required to make composite tools and potentially using the same neural networks that support language. The date is significant, broadly corresponding with the demise of bifaces as the dominant technology and the rise of prepared cores to produce flakes and blades, notably the Levallois technique. An impression is emerging that a linguistic threshold may have been crossed soon after 400,000 years ago.

What type of night-time talk occurred around the hearths in caves from Bolomor in Spain to Blombos in South Africa? If the imaginative capacities and linguistic tools had been available, the hominins might have swapped stories about ancestors and the spirit world. If not, daytime talk may have extended into the night, but unlikely for long because further words about the mundane would have induced boredom and sleep. Might that be reflected by the Neanderthal hearths that were small, appearing not to have burnt for long?

A third scenario is that sitting around night-time fires began for practical ends, using the fire's light and warmth to continue daytime talk, making tools and preparing food. But slowly, through time and across generations, perhaps in some parts of the world but not others, the multi-sensory impact of the fire provided a new stimulus for the brain. Did the flickering flames entice new ideas to form – those involving abstract concepts about time, space and imaginary beings which needed to be anchored in the mind by the use of new abstract words? Was it within the heady aromas of wood smoke where metaphors were first spoken to convey such ideas to others? Did those individuals who, by chance, had most proficiency at such talk gain status, not only having their ideas and words copied but also their genes eagerly acquired? Did their capability for night-time talk, that which enabled them to tell the best stories, invent new words and create the most effective metaphors, begin to pervade all of language, whether spoken during the day or night? If the human imagination and a new phase of language were being sparked into life by fire, then we had better look at what happens within the human brain.

11

LANGUAGE AND THE BRAIN

The capacity for language resides within the brain. Although language requires the body for a conduit, whether that is via the hands for signing or the vocal tract for speaking, to understand how language evolved we must delve inside the brain. Not just the brain of *Homo sapiens*, but also the brains of our living relatives and extinct ancestors – as far as we are able to do so.

The australopiths had brain sizes of up to 500 cubic centimetres (cm³). After the first *Homo* appears by 2.8 mya, brain size began a gradual increase reaching the lower range of its modern size of c.1,100–1,700 cm³ by 500,000 years ago. The brain also changed shape, with the most marked development occurring after 300,000 years ago within the *Homo sapiens* lineage while evolving in Africa. This led to the globular shaped brain we have today, one quite different to the relatively flat and elongated brain of the Neanderthals.

What happened within the brain during those 2.8 million years of change to deliver the capacity for language? When did that occur? We can only answer these questions by knowing where and how language is instantiated in the brain today. We can then begin to interrogate the evidence from the past.

The human brain

The typical modern human brain weighs between 1.2 and 1.4 kg, with an average size of about 1,260 cm³ in men and 1,130 cm³ in women, this reflecting differences in average body size. There is substantial variation around these figures, some men having brains smaller than 1,000 cm³ and others larger than 1,600 cm³, with no apparent variation in intelligence, language or any other cognitive ability. Although representing only 2 per cent of total body weight, brains require 20 per cent of the body's energy to keep them running. They are expensive to have.

I've read that a brain feels like a blob of jelly. It contains blood vessels and many cells: 86 billion neurons, connected by thin extensions called axons which send information via electrical impulses around the brain. Neurons are supported by a similar number of glial cells. Their name derives from the Greek word for glue, which is fitting because they support the neurons by holding them in place. They also supply oxygen and nutrients, destroy pathogens and insulate neurons from each other.[1]

The cerebrum is the largest part of the brain, partitioned by a deep groove into left and right cerebral hemispheres (Figure 11). The two hemispheres have a degree of asymmetry – the left extends further back and the right further forward. Their outer region is known as the cortex, made up of six layers of neurons known as grey matter. Below this is a mass of axons known as white matter. The axons are insulated by membranous sheaths made from fats and proteins, called myelin, which helps the transmission of electrical impulses.

Each cerebral hemisphere is divided into four lobes: the frontal lobe, parietal lobe, temporal lobe (at the base) and occipital lobe (at the rear), all named after the skull bones that overlie them. Specific areas of the lobes are referred to by their relative position: anterior or posterior, inferior or superior (as in the anterior temporal lobe). The surface of each hemisphere is folded into ridges and grooves, each ridge known as a gyrus and each groove as a sulcus. These are denoted according to their position on the cerebrum, such as the superior occipital gyrus and inferior frontal sulcus.

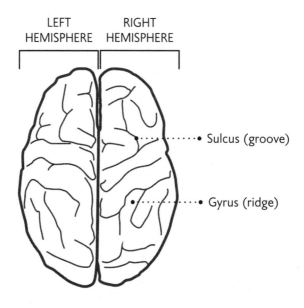

LEFT HEMISPHERE   RIGHT HEMISPHERE

Sulcus (groove)

Gyrus (ridge)

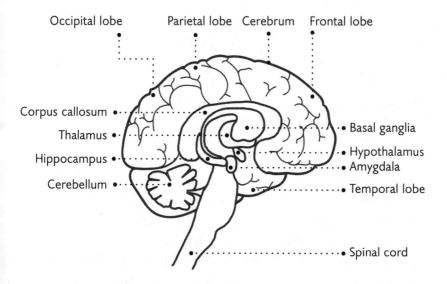

Occipital lobe   Parietal lobe   Cerebrum   Frontal lobe

Corpus callosum
Thalamus
Hippocampus
Cerebellum

Basal ganglia
Hypothalamus
Amygdala
Temporal lobe

Spinal cord

Figure 11 The human brain

In addition to the lobes, the cortex is divided into several areas relating to one of their known functions which are evident from their names: the motor cortex (within the frontal lobe), and three sensory cortices known as the visual cortex (within the occipital lobe), auditory cortex (within the temporal lobe) and somatosensory cortex (within the anterior parietal lobe, processing tactile sensations such as touch and temperature). The remaining parts of the cortex are referred to as the association areas, combining information from the cortices and other parts of the brain to undertake processes such as perception, thought and decision making. An area of the frontal lobe known as the prefrontal cortex is recognised as playing a key role in planning and decision making.

Below the cortex and not visible when looking at the surface of the brain are several subcortical structures. These include the corpus callosum, a large tract of white matter that connects the two hemispheres, the cerebellum, hypothalamus, basal ganglia, thalamus, amygdala and the hippocampus. All these have multiple functions. The amygdala, for instance, is sometimes described as the emotional centre of the brain but is also significant for memory; the basal ganglia are important for motor control and habit formation.

The cerebellum, positioned behind the occipital lobes, is like a mini brain, being divided into two hemispheres connected by a central region known as the vermis. The hemispheres are formed by grey matter over white matter, with tightly folded surfaces. Although approaching only a tenth of the size of the cerebrum, the cerebellum contains 69 billion neurons, 80 per cent of those within the entire brain. It contributes to many aspects of the body's muscular and cognitive functioning.

Neurons and networks

During foetal development brain cells proliferate at a remarkable rate, at times reaching 15 million per hour. Under the influence of chemical signals arising from inherited genes, they migrate to predetermined locations within the developing brain, becoming specific types of neurons and glial cells. Once at their location, each neuron grows long, thin extensions known as neurites. One of these develops into an axon which provides the channel along which the neuron will transmit information. The other neurites become dendrites which will receive signals from other neurons (Figure 12).

Axons grow towards specific targets in the brain, either nearby or at some distance from their neuronal bodies. They connect with dendrites from other neurons by coming very close to form small gaps, each referred to as a synapse. A neuronal body will send a signal by an electrical impulse along its axon to the synapse. That causes a chemical change within the gap that transmits the signal to the adjoining dendrite, from where

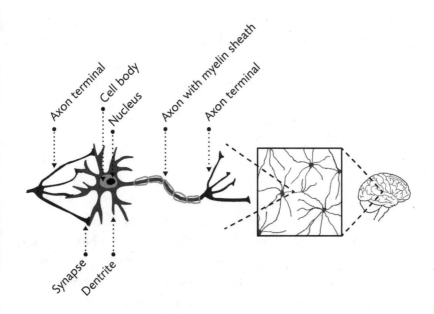

Figure 12 Neurons in the brain

it travels to that dendrite's own neuronal body, and then via its axon to another synapse. In this manner, signals flow through a web of interconnected neurons, referred to as a neural network.

Such networks can be relatively localised or extend throughout the brain. A single neuron can have more than 1,000 dendrites, making connections with tens of thousands of other neurons. With 86 billion neurons in the brain, there are trillions of connections.

While some of these connections may be fixed, others develop and change over time. Our routines and new experiences may either strengthen or weaken synaptic connections, influencing which neurons are networked together. By doing so memories are built and new skills are acquired – both of which can equally be lost by inactivity.[2] As such, neural networks within the brain are products of both inherited genes and life experience – it is difficult and unwise to separate the two.

Such synaptic plasticity plays a key role in the developing brain, shaping it by the physical, social and cultural influences of the infant's environment. At birth each neuron has about 2,500 synapses; by the age of three, these will have increased to around 15,000 as the infant is exploring its new world. By adulthood, that number will have halved by a process known as synaptic pruning, removing those that have been little used.

Our patterns of neural connections are continually modified as we learn new activities, skills and languages through life. Repetition is important – the 10,000 hours required to fully master a new skill is about building the required networks within the brain and throughout the nervous system of the body. Synaptic plasticity also allows a damaged brain to be repaired – within limits – because neural pathways can be forged to undertake functions that have been lost from elsewhere in the brain.

Such remodelling occurred in the brain of every individual from the human past: it enabled *H. erectus* children to learn to make bifaces and acquire whatever form of language they possessed. In the previous chapter, I wondered about the impact of fire on the human brain – the extent to which flickering flames and the aromas of smoke encouraged new connections in the brain, leading to the invention and adoption of new words, metaphors and imaginary worlds. Quite how the billions of neurons and trillions of connections translate into such imagination and conscious awareness, such as the sensation of how a word sounds or the thought of what it means, remains unknown.

Where does language reside?

Since the 1990s we have had a range of sophisticated techniques for identifying which parts of the brain are associated with which bodily and cognitive processes, including language. Most notable is fMRI (functional magnetic resonance imaging) scanning. This detects the locations of changes in blood flow in the brain of subjects when they perform designated tasks, with increased flows indicating brain activity. Other techniques include electroencephalography (EEG) and magnetoencephalography that rely on detecting electric signals in the brain and their magnetic fields respectively.[3] Before the development of such imaging methods, we were reliant on discovering how the brain works only after parts of it broke down.

During the mid-nineteenth century, the French anatomist Paul Broca undertook post-mortem studies of the brains of people who had suffered aphasia – the loss of all or some aspects of speech. He concluded their disorders had arisen from damage to an area of the frontal lobe that became known as Broca's area. This is also known as Brodmann areas 44 and 45, Korbinian Brodmann being a German neuropsychiatrist who gave each area of the brain a specific number. Inspired by Broca's work, the German anatomist Carl Wernicke identified another area of the brain that impacted on language, one located within the temporal lobe. When damaged, comprehension rather than the production of speech becomes disrupted.

Broca's and Wernicke's areas are both in the left hemisphere and were later found to be connected by a bundle of nerve fibres called the arcuate fasciculus. These dominated our understanding and further research into the neuroanatomy of language until the 1990s, encouraging a view that language is entirely located in the left hemisphere of the brain.[4] Even before the advent of fMRI, however, the three-component system for language provided by Broca's area, Wernicke's area and the arcuate fasciculus had been questioned. Why was it that some patients with aphasia were found to have undisturbed Broca's areas? Conversely, why did those with damaged Broca's areas not necessarily suffer a loss of speech function?

The answer was provided by fMRI and other brain-imaging techniques that showed language is distributed throughout the brain rather than being located in a small number of specific areas. Moreover, by developing a closer association with linguistic science, neuroscientists have begun to identify where specific aspects of language are processed. Broca's and Wernicke's areas remain important, with the former identified as critical to processing hierarchical phrase structures, such as embedded clauses.[5] The cerebellum is recognised as playing key roles regarding language perception, processing and production, with a notable contribution to overall verbal fluency.[6] Although the basal ganglia do not appear to be involved in basic linguistic functions, when damaged or diseased they can have a negative impact on language capability.[7] The connection between Broca's area in the frontal lobe and Wernicke's area in the temporal lobe remains critical. But rather than the single-fibre tract of the arcuate fasciculus, two pathways are now recognised, a ventral pathway supporting how we connect the sounds of words to what they mean and a dorsal pathway responsible for speech production.[8]

The global network view of language is exemplified by

semantic mapping. This is a method to identify where the brain stores and processes words or, more accurately, the concepts attached to the sounds designated as words.[9] This research began by using fMRI to identify which areas of the cortex are activated when subjects hear different types of words, such as those relating to objects, actions and social narratives. It found these were processed at different locations in the brain, but predominantly in the left hemisphere. Further work derived a 'semantic atlas' for the brain. This stretches across both hemispheres, with some preference for narrative-related words on the right.[10] Rather than being stored at single cortical locations, abstract concepts are represented by spatially distributed networks, thought to reflect the open-ended nature of such concepts. To quote Yizhen Zhang, the lead scientist, the take-home message from semantic mapping is that 'the human brain encodes a continuous semantic space'.[11] This exemplifies the new understanding of how language in the brain relies on extensively distributed neural networks, extending into all four lobes, both hemispheres, the cerebellum, and most likely every anatomical part of the brain.

Two other key lessons have been learned. The first is that no area of the brain appears to be entirely dedicated to language – they are all multifunctional. Broca's area, for instance, consists of at least ten sub-regions, supporting both language and non-language roles, including complex hand movements.[12] While the cerebellum has a significant role in language processing, it also has a controlling role for balance, movement, motor skills and memory.[13]

Second, there is considerable variability in the neural networks for language found in different people. While this can have a genetic cause and lead to language dysfunction, some people develop different networks for no apparent reason and with no impact on their linguistic ability, such as having a greater prominence of their language-significant neural networks in the right hemisphere.[14] The language that one speaks, signs or reads also influences the neural networks that develop: a study comparing Italian and English found that the words of these languages influence the extent to which areas of the temporal gyrus and frontal gyrus are activated when reading, probably because English words have a more complicated mapping of letters to sounds.[15] Such studies have mainly been undertaken using spoken English or other languages of the Indo-European family, and hence share many linguistic features. We have no idea how the neuroanatomy of language would vary for a language that has a much greater reliance on word morphology than English, such as indigenous Australian and North American languages with their extremely long words. How would the brain respond to a language that lacks word order entirely?

Only a few studies have been undertaken to explore the brains of people who speak more than one language.[16] Bilingual and multilingual people have been found to have a greater quantity of grey matter in frontal regions of their brains and of white matter that provides connections between different brain regions. Rather than switching between languages, their brains keep all their languages activated at the same time, although how this occurs remains unknown.[17]

Human and chimpanzee brains compared

Learning that the capacity for language is distributed throughout the modern human brain removes any easy neuroanatomical target for seeking to document when language evolved. We can, however, make a comparison between the human brain and that of the chimpanzee to consider what differences may have given rise to language.

Although the modern human brain is more than three times the size of the chimpanzee's, the two brains remain broadly consistent in shape and anatomy. As the neuroscientist Suzana

Herculano-Houzel explained, the human brain, is 'remarkable, but not extraordinary'.[18] We can nevertheless find some areas of the human brain which are larger and some smaller than would be expected if it was no more than a threefold evenly scaled-up version of a chimpanzee brain.[19]

The visual cortex within the occipital lobe is only 60 per cent of the size expected for an evenly scaled-up chimpanzee brain. The olfactory bulb, providing the sense of smell and located in the frontal lobe, is even smaller, just 30 per cent. This may not be surprising because we know our capacity for smell is limited compared with other animals.[20] Similarly, our muscle control is not noticeably better than other primates, reflected in our motor and pre-motor cortices (located within the frontal lobe) being only 33 per cent of the size expected. The cerebellum and frontal lobe are a better fit, both approximately three times larger than those of the chimpanzee.

Because the human frontal lobe contains relatively small motor and pre-motor cortices, its other part, the prefrontal cortex, must be proportionately larger than would arise from a simple tripling in size of the chimpanzee brain. Although difficult to precisely measure because of its convolutions, the prefrontal cortex may be twice as large as would be expected. The temporal lobe is also larger than expected, which is likely significant for language considering the presence of Wernicke's area within this lobe.

Beyond these gross differences, there are more subtle variations. Chimpanzee brains are also asymmetrical, but to a lesser degree than in humans, and differ in the patterning of the ridges (gyri) and the grooves (sulci) in the frontal lobe. Like all other great apes, chimpanzees have a sulcus at the rear of the occipital lobe, referred to as the lunate sulcus. This is pushed further back in the human brain, possibly reflecting an expansion of frontal regions that may have related to language. Another subtle contrast probably related to language is that Brodmann areas 44 and

45, known as Broca's area in humans, are six times larger in the human brain than in the chimpanzee.[21]

We have considered the visual appearance of the brain. What about its inside? The human cerebral cortex contains twice as many cells as that of the chimpanzee.[22] Its frontal cortex has a proportionately greater quantity of white matter over grey matter – the white matter having expanded 4.6-fold and the grey matter threefold.[23] This indicates a higher degree of neuronal connectivity within the human brain, providing greater links from the prefrontal cortex to other regions of the brain.

The pattern of connectivity can be mapped by using a brain imaging technique called tractography.[24] This technique relies on how water molecules within the brain diffuse along the white matter fibres that connect neurons. By monitoring their diffusion, and then using a package of neuroscience statistical and modelling tools, maps of connectivity known as connectomes can be constructed.

In 2019, Dirk Jan Ardesch, from the aptly named Connectome Laboratory in Amsterdam, and his colleagues from Emory University in Atlanta, United States, published the first comparison between the modern human and chimpanzee connectomes.[25] The human brain was found to have significantly higher degrees of connectivity between the multimodal association areas than in the chimpanzee, those areas that receive and merge information from different sensory systems (sound, touch, smell and so forth). Such connectivity complemented a more pronounced modular organisation in the human brain, indicative of increased functional specialisation. Connections observed in humans but not in chimpanzees linked areas of the brain known to be important for language processing.[26] Overall, the human brain is distinctive by having long-range connections that link areas of functional specialisation into a global network. The significance of such long-range connections has also been identified in studies that compare the development of the chimpanzee and human brain. While both species begin with similar levels of integration between different regions of the brain, this is only maintained into adulthood by humans, likely reflecting enhanced functional specialisation of multiple areas within the human brain.[27]

Functional specialisation and domain-specific thought

We tend to think of functional specialisation of the brain as being for cognitive skills such as memory, attention, visual, language and auditory processing. There is, however, another dimension of functional specialisation that concerns how we store and access different concepts and categories in the brain. This is of considerable significance for our language capability today and how it may have evolved because concepts are associated with words, and we typically combine words drawn from several different categories when we engage in speech.

Functional specialisation for categorical knowledge is indicated by people who suffer from brain damage that impairs their ability to recall the names and comprehend entities of one category, while leaving those of another unaffected. Alfonso Caramazza and Jennifer Shelton, Harvard-based psychologists, provide examples of people who lose such abilities for naming living things while remaining unaffected for inanimate objects such as tools. Other people have undergone the converse, and hence this phenomenon is known as a double disassociation.[28] One person known as E.W. was disproportionately impaired at naming and talking about animals compared with other living things such as fruits and vegetables, and relative to all inanimate categories, such as tools and food items. Another person was impaired at defining musical instruments, precious stones, clothes, materials and other inanimate objects but was unaffected regarding body parts.[29]

Caramazza and Shelton argue that these deficits relate to

categories of knowledge for which evolutionary pressures led to the development of specialised neural mechanisms for their perceptual and conceptual distinction, such as those of conspecifics, animals, plant life and artefacts. Such evolutionary-driven domain-specific knowledge is precisely the conclusion I had reached in my 1996 book *The Prehistory of the Mind* based on archaeological evidence, phrased in terms of social, natural history and technical intelligences. I had argued that these remained relatively isolated from each other in the early human mind (notably the Neanderthals) but became interconnected in modern humans.

Fifteen years after that publication, Alfonso Caramazza and the neuroscientist Bradford Mahon were able to provide a neurological description of such domain-specific knowledge systems:

> A domain-specific neural system is a network of brain regions in which each region processes a different type of information about the same domain or category of objects. The types of information processed by different parts of a network can be sensory, motor, affective or conceptual. The range of potential domains or classes of items that can have dedicated neural circuits is restricted to those with an evolutionarily relevant history that could have biased the system toward a coherent organisation.[30]

Camarazza and Mahon argue that domain-specific knowledge arises from patterns of connectivity between regions of the brain that process sensory and conceptual information relevant to that domain. For instance, specialisation for faces is in the lateral occipitotemporal gyrus because that region of the brain has connectivity with the amygdala and the superior temporal sulcus which are important for the extraction of socially relevant information. In contrast, knowledge about tools and objects that can be manipulated is located in the medial occipitotemporal gyrus which is connected to the parietal cortex.[31]

Domain-specific knowledge systems develop during brain maturation. Because even babies appear to have an intuitive understanding of the difference between animate and inanimate objects, along with concepts about living things and human minds, the neural networks supporting these knowledge systems must develop rapidly and largely under genetic control.[32] This is not surprising because those domains of knowledge are significant for living within any cultural environment, whether that of a prehistoric hunter-gatherer or modern city-dweller. Other domains of knowledge are more specific. Reading and writing, for instance, are also functional specialisations of the brain underpinned by neural networks but are too recent to have their own evolved genetic basis. They appear to have 'invaded' evolutionarily older neural networks that had developed for different, but similar functions, seeking out their own 'neuronal niche'.[33]

Considering the development of multiple areas of domain-specific knowledge in the brain, the psychologist Annette Karmiloff-Smith argued (in 2015) that 'global information sharing would therefore depend on a set of interconnected high-level cortical regions forming . . . a "global workspace", i.e. a distributed network of cortical areas tightly interconnected by long-distance axons, which sends information back to specialized processors'.[34] This structure for the brain is precisely what Dirk Jan Ardesch and his colleagues found in their connectome study published in 2019, and what I characterised in my 1996 archaeological study of the modern human mind as cognitive fluidity.

Brain evolution from 3 million to 800,000 years ago

The chimpanzee brain is frequently used as a model for that of the 6-million-year-old common ancestor. We must be cautious, however, because the chimpanzee brain has also evolved over the same 6 million years and may contain derived features that were absent in the brain of the common ancestor. Nevertheless, using the chimpanzee brain as a model, we can ask: when did the size, structure and connectivity of the human brain arise, especially the neuroanatomy that underlies language?

Asking such questions is easier than finding the answers: we cannot examine our ancestors' brains because they do not survive in the fossil record. Although, as the next chapter will explain, advances in ancient genomics are providing insights into brain evolution during the last 500,000 years ago, before that we remain reliant on the fossilised skulls of our early ancestors. By measuring the volume of their cranial vaults, we can estimate that of their brains, and the interior surface of the vault may carry an impression of the exterior surface of the brain it had once contained. These impressions are called endocasts. They can either arise naturally after a skull has become full of sediment that solidifies with an impression of the cranium's interior or be made in a laboratory. That was once done by using rubber moulds but is now undertaken by digital methods that capture the tiniest bumps and fissures of ancient brains. While such undulations record the past gyri and sulci, endocasts cannot tell us about the internal structure of the brain.[35]

There has been a gradual increase in brain size from between 300 cm³ and 500 cm³ before 3 million years ago, as found in our australopith ancestors and relatives, to an average brain size of about 1,250 cm³ at around 500,000 years ago, as found in *Homo heidelbergensis*, the ancestor of *H. sapiens* and *H. neanderthalensis* (Figure 13).[36] Since that date, brain size has remained stable, with a slight decrease in *Homo sapiens* following the end of the Ice Age and possibly occurring within the last 3,000 years.[37] Just as we find in any species, brain size varies with body size and gender. Consequently, we should give a range for previous human species rather than a single figure, which incidentally shows how much these overlap:[38] *H. habilis* 550–800 cm³; *H. ergaster* 700–900 cm³; *H. erectus*, 600–1,250 cm³; *H. heidelbergensis*, 1,100–1,400 cm³; *H. neanderthalensis* and *H. sapiens* 1,100–1,700 cm³.

The rate of increase in brain size appears to have been steady over the past 2.5 million years, amounting to no more than 0.008 cm³ per generation (assuming generations of twenty years' duration) to achieve the 1,000 cm³ increase from the earliest to the most recent *Homo*. Two small but relatively recent types of humans provide an exception to the ever-increasing trend: *H. naledi* living in southern Africa between 330,000 and 240,000 years ago had a brain size of 465–560 cm³ and the brain of *H.*

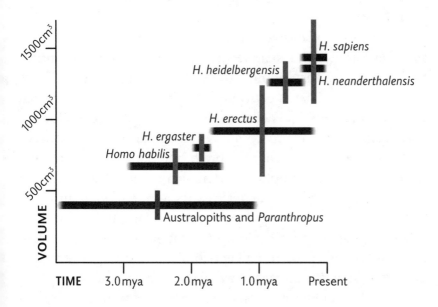

Figure 13 The evolution of brain size in *Homo*. The horizontal black bars indicate the duration of the species in the fossil record and the vertical grey bars the variation in its brain size during that period as measured by the volumes of fossil crania.

floresiensis from Flores Island, Indonesia, between 100,000 and 60,000 years ago was a little smaller at 417 cm³. These relatively recent but small human brains provide particularly challenging questions about the cognitive capabilities of their owners.[39] They illustrate how human evolution did not always involve an increase in brain size – in some ecological situations it went in the other direction.

All living primates have the same density of neurons in their brains, whether they are tiny monkeys, chimpanzees or modern humans. Assuming this holds for our extinct ancestors, we can convert their brain sizes into numbers of neurons: australopiths 35–37 million; early *Homo*, 50–60 million; *H. erectus*, 62 million; and 76–90 million neurons for most recent humans and ourselves today. While such increases provided a wide range of new abilities, most notably language, the additional neurons came with a cost.

Brains are expensive organs. To feed the extra 60 billion neurons found in the Neanderthal or *H. sapiens* brain compared with that of an australopith, an extra 360 kcal/day would have been required. That is equivalent to a meal of two scrambled eggs, two rashers of bacon, toast, butter and coffee. Cooked breakfasts were hard to come by for our ancestors, but their diet had to change to fuel the brain, most likely by increasing meat and fruit consumption, noting there is no evidence for the regular use of fire until 400,000 years ago. Larger brains also impose costs by needing longer pregnancies and periods of infant dependency, both requiring changes in social relations; large brains also need to be kept cool and that might require a change in activity patterns to avoid the midday sun.

Throughout human evolution, these costs were outweighed by the reproductive benefit gained from having a larger brain. That benefit is likely to have partly derived from an enhanced level of vocal communication that a larger brain could provide, this enabling more effective cooperation, such as for scavenging and hunting, and the ability to live in larger social groups for protection from predators. In the cases of *Homo naledi* and *H. floresiensis*, however, the costs of having a larger brain were not worth paying. *Homo floresiensis* most likely derives from a population of *H. erectus* that colonised Flores Island which had no large predators to threaten them or to compete for food. Without such pressures, over time this island-bound hominin population reduced both body and brain size from the norm of *H. erectus* to become the diminutive *H. floresiensis*.

Size is not everything when considering the brain; we also need to consider its shape, reflecting the size of its component parts. When did the brain evolve from looking like that of a chimpanzee to that of a modern human? There are two clues we can look for on endocasts. First, the position of the lunate sulcus, located further forward in the chimpanzee than human brain; second, a bulge relating to Broca's area, covering Brodmann areas 44 and 45. In chimpanzee endocasts, this is relatively slight and restricted to Brodmann area 44. The base of the human and chimpanzee bulges are delimited by different sulci – that of the chimpanzee by the fronto-orbital sulcus and the human by the lateral orbital sulcus.

Endocasts from *Australopithecus afarensis* of 3 million years ago, with adult brain sizes of 350–530 cm^3, have the characteristic features of a chimpanzee-like brain.[40] A well-preserved endocast from an infant specimen (DIK-1-1), from Dikika in Ethiopia, has an unambiguous impression of the lunate sulcus in the anterior, ape-like position, and lacks any bulge indicative of Broca's area.

There is, however, a key difference between *A. afarensis* and the chimpanzee that may relate to an evolving capacity for language, or at least provide one of its foundations. The dentition of two *A. afarensis* infants from Dikika indicate they were between two and two and a half years old, with brains two-thirds the size of the adults (infants averaging 294 cm^3, compared with the adult 445 cm^3). That is relatively small compared with the brain size of an infant chimpanzee of the same age when growing its brain to the adult size (an average of 369 cm^3). This suggests that even before substantial brain enlargement had occurred and before any sign of human-like reorganisation, a prolonged period of brain growth had appeared. That implies a longer period of infant dependency on its caregivers, which would in turn allow more social learning, potentially encompassing what types of calls to make, when and to whom. Even at this date of 3 million years ago, therefore, a typical ape-like brain might have been developing new patterns of brain connectivity, not from genetic change but from new life experiences acquired during an extended infancy.

One of the australopith species had given rise to *Homo habilis* by 2.8 mya. While the lunate sulcus cannot be detected on *H. habilis* endocasts, suggesting a more human-like brain, there is no impression of Broca's area. That also remains indistinct on five well-preserved *H. erectus* crania from Dmanisi, Georgia, that date to 1.85–1.77 mya, deriving from an early dispersal out of Africa. The chimpanzee pattern remains, with bulges on the Dmanisi specimens largely contained within Brodmann area 44 and delimited underneath by the fronto-orbital sulcus.

From what evidence exists, the same chimpanzee-like pattern is found on African *H. erectus* endocasts before 1.7 mya.[41] The levels of brain asymmetry for these early members of *Homo* are also in line with that of chimpanzees. Fossil skulls dating to between 1.7 and 1.5 mya have a greater degree of variability, while after that date the bulge takes on the modern human appearance and indicates the expansion of Brodmann areas 44 and 45 that can now be confidently denoted as Broca's area. This indicates brain reorganisation had occurred within the frontal lobe between 1.7 and 1.5 mya, a period when average brain size increased from 650 to 830 cm^3. The same distinctive human pattern is found on *H. erectus* endocasts from East Asia, notably at Zhoukoudian dating to 770,000 years ago. This indicates a second dispersal from Africa had occurred after brain reorganisation.

Although Broca's area is just one part of a widely distributed neural network for language, its expansion suggests a neuroanatomical step towards language. We must remember, however, that Broca's area has multiple functions, including motor control. It may be significant, therefore, that 1.7–1.5 mya coincides with the appearance of a new mode of stone tool technology, known as the Acheulean. As was covered in Chapter 7, making Acheulean bifaces required greater manual dexterity than making Oldowan choppers, and consequently may have been dependent on an evolved Broca's area. The selective value of enhanced motor control for accurate throwing may also have been significant. The shape of *H. erectus* shoulder bones and its mobile waist suggest that proficient throwing had evolved by 2 million years ago, this requiring at least as much hand–eye coordination as making an Acheulean biface.[42]

Neanderthals, Denisovans and modern humans

By 800,000 years ago, the fossil record contains the skeletal remains of a species designated as *Homo heidelbergensis*. The sample of skulls and cranial fragments available indicates a more pronounced level of asymmetry than seen for the chimpanzee and the earliest *Homo*.[43] *Homo heidelbergensis* is recognised as the common ancestor to *Homo sapiens* and *Homo neanderthalensis*, with the genetic divergence of these species dated to between 800,000 and 630,000 years ago. They may have continued to exchange genes before full population divergence, estimated at between 440,000 and 270,000 years ago.[44] The Neanderthal lineage further split at around 400,000 years ago, with the Denisovans evolving as a species in central Asia – a species without a

Latin name and principally known from its genome rather than its skeletal remains.

Although Neanderthal and modern human brains remained equivalent in size at between *c*.1,100 and 1,700 cm³, their shapes and surrounding crania diverged (Figure 14). At birth, Neanderthals and modern humans had the same shape of skull and hence the same shape of the brain: relatively elongated from front to back and flat on top – the ancestral shape. Soon after birth, however, and as with newborn infants today, the modern human brain began to change.[45] The parietal and cerebellar regions expanded and the angle between the cranial base and the cranial body increased.[46] Once fully mature, the Neanderthal and modern human crania had similar volumes but different shapes, as did the brains they contained. While the Neanderthals maintained the front-to-back elongated and relatively flat crania that characterises all previous types of humans, modern humans had more spherical crania, referred to as globular.

Endocasts provide some hints as to how the brains inside the crania may have differed, but new and penetrating insights

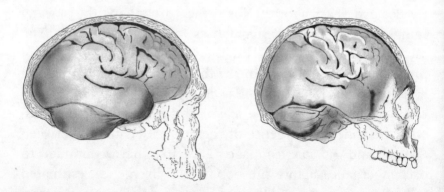

Figure 14 Neanderthal (left) and modern human (right), illustrating the different shapes of their cranium and brain

have come from a 3D digital reconstruction of the Neanderthal brain. This was achieved by deforming a digital model of the modern human brain and fitting it against Neanderthal endocasts.[47] Led by Takanori Kochiyama of Kyoto University, this study had two key findings, one previously suspected and one completely novel with potential implications for language.

The first is that the Neanderthal occipital lobe was significantly larger than that of modern humans. This lobe contains the visual cortex, which had already been identified as being relatively large in Neanderthals from endocast evidence alone. The larger occipital lobe is matched by relatively large eye sockets, indicating that the Neanderthals evolved better vision than modern humans.[48] By investing more in the visual system, the Neanderthals had less cerebral brain tissue to invest in other cognitive functions, which may have included language.

Kochiyama and his team's new finding is that the Neanderthals had a significantly smaller cerebellum. Not only smaller, but also positioned differently to that of modern humans and of a different shape.[49] Earlier in this chapter I noted the role of the cerebellum in a range of cognitive and motor processes, including language and working memory, and how it contains up to 80 per cent of the neurons of the entire brain. Variation in cerebellum size within modern humans influences performance on a range of tasks: people with a larger cerebellum achieve better results in tests concerning speech comprehension, speech production, memory and attention, all feeding into abilities for learning and reasoning.[50] Whereas the modern human cerebellum is symmetrical, the right side of the Neanderthal version is smaller than the left. Because each side of the cerebellum is connected to the opposite side of the cerebrum, this indicates limited connection between the Neanderthal cerebellum with the left hemisphere, where Broca's and Wernicke's areas are located.

Kochiyama and his colleagues concluded their study by

arguing that 'the differences in neuroanatomical organization of the cerebellum may have resulted in a critical difference in cognitive and social ability between the two species [Neanderthals and modern humans]'.

A partial fragment for the language puzzle

The evolutionary history of the human brain provides us with a fragment for the language puzzle, but one that has many missing pieces: how can it ever be complete when all we have are empty brain cases to study? The pieces we do have are, however, of immense value.

Size may not be the only significant aspect of a brain but it is undoubtedly important because of the energetic cost that comes from adding extra neurons. By 500,000 years ago *Homo heidelbergensis*, soon followed by *H. neanderthalensis* and the earliest *H. sapiens*, had brains equivalent in size to modern humans. That suggests that much of the evolutionary work for language might have been completed by that date, which would fit with the modern-like vocal tract of *H. heidelbergensis* that we identified in Chapter 5.

Brain shape, indicated by endocasts, is also important. The fossil record suggests a shift from an ape-like to a human-like shape occurred between 1.7 and 1.5 mya with the enlargement of Brodmann areas 44 and 45, known in modern humans as Broca's area. While we know that Broca's area has multiple functions, it is certainly key to the language network suggesting that a significant evolutionary step in language evolution occurred at this date. This is coincident with the emergent of Acheulean technology that, as we noted in Chapter 7, provides similarities between its bifaces and words: both have an imposed form, are carried around the landscape, and can be used in different contexts.

A more dramatic change of shape occurred within the *H.*

sapiens lineage in Africa after 300,000 years ago: the brain became more globular in shape. This involved a relative decrease in the size of the occipital lobe compared to the Neanderthals, which may have freed up space for other tasks, including language. It also resulted in a significantly larger cerebellum – the mini-brain within the brain – which is known to play multiple roles in the modern-day language capacity.

Neither size nor shape can reveal a third crucial feature of the modern brain that likely plays a key role in the fully modern language capacity: its pattern of connectivity.[51] When compared with the chimpanzee brain, that of modern humans is characterised by neural networks that provide long-distance connections between different areas of increased functional specialisation. When did that global network arise in human evolution? Fossil skulls have not been able to provide an answer. Might it be found within the pieces of the next fragment we need to assemble: the genetics of language?

12

THE GENETICS OF LANGUAGE

We are living through a revolution in our understanding, use and manipulation of genes, those inherited strands of DNA that encode instructions about how to build a living organism. Medicine and agriculture are being transformed, as is our knowledge about the past by the extraction of ancient DNA from skeletal and fossilised remains. So too is our understanding of language.

The genetics of language provides a key fragment of the language puzzle, but it is a devilish one to assemble. This is not only because of the inherent complexities of genetics but also because the field is so fast moving. Indeed, a greatly valued geneticist colleague advised me to skip over genetics entirely because whatever I write will be out of date by the time of publication. With apologies to him, and at risk of providing an oversimplified account with a short-use life, I cannot neglect the genetics of language. Within the existing literature on language evolution there are claims about a 'language gene' and that a single genetic mutation at c.100,000 years ago instantiated language in the brain.[1] Such ideas need replacing by a more informed understanding of genetics, language and evolution.

Moreover, the last decade has witnessed a flood of new information about the differences between the Neanderthal and modern human genomes, along with evidence for interbreeding. The implications for their comparative capabilities are profound and need to be captured in this fragment of the language puzzle.

The human genome

A human genome is the set of instructions required for an individual to develop and function; these are encoded within a molecule known as DNA (deoxyribonucleic acid). As members of *Homo sapiens*, our genomes are highly similar and collectively referred to as the human genome, this being distinguished from, say, the chimpanzee or Neanderthal genome, both of which also contain individual-level variability.

DNA is formed by two interlinked strands of chemical compounds known as nucleotides that wind around each other to create a double helix. Each nucleotide contains one of four bases: adenine (A), cytosine (C), guanine (G) and thymine (T).[2] The sequence of these 'letters' provides instructions for how to build a human being, everyone having a slightly different order of letters, except for identical twins, triplets and multiple births that come from the fertilisation of the same egg. More than three billion of these letters are spread across twenty-three paired strands of DNA, each referred to as a chromosome and found within every cell in the body. Genes are discrete stretches of the As, Cs, Gs and Ts, varying in length from a few hundred to over a million letters.

Genes provide instructions to build proteins. They can be switched on or off, or modified to become either more or less active, referred to as being up- or down-regulated. Proteins are built by a process known as transcription during which the gene is converted into a molecule of messenger ribonucleic acid, mRNA. This molecule has sequences of three bases known as exons, such as ATG or ACG. Each sequence codes for one of the twenty amino acid molecules that are the building blocks of life; the amino acids created by each gene combine to form the protein. Many different types of proteins can arise. Some give cells their shape, others act as enzymes to facilitate chemical reactions, antibodies to protect the body, or hormones that coordinate bodily functions. Other proteins, known as transcription factors (TFs), influence the expression of further genes – whether they are switched on, off, up- or down-regulated.

That regulation depends on strands of DNA in the vicinity of the gene. These act as binding sites for the TFs that either prevent or allow the transcription of DNA to mRNA. A single TF can target hundreds of genes spread throughout the genome. As such, it is the interaction between genes and proteins, rather than the genes themselves, that is most critical to both the development and functioning of humans, and any living organism.

Genes control the development of the brain. They determine which foetal cells become neurons, where those neurons migrate to within the developing brain and how their neurites grow. Genes guide axons from the nucleus of the neuron to other parts of the developing brain; they influence the formation, maintenance and the plasticity of synapses which are fundamental to the processes of learning and memory.

The most we can say is that there are genes that influence the construction of the neural networks and patterns of global connectivity that are essential for language. Because the same genes may play several other roles in the human body, we cannot call them 'language genes' or even 'genes for language' but can use the term 'language-significant genes'. With no more than 20,000 structural genes in the human genome – those which build proteins – language-significant genes can only play their role by operating in combination with other genes and by serving multiple functions in the development and operation of the whole range of bodily and cognitive processes. Moreover, genes cannot do any of this by themselves: they rely on environmental stimuli. Place a newborn baby into a silent world and the neural networks that enable language will either not develop or not function for this purpose; begin to stimulate that deprived child with language, and some capability might emerge as the brain begins to remodel itself.[3]

Cell division, mutation and selection

Cell division is a continuous process throughout the human life course – it is how we develop, grow and repair damaged tissue. When one cell divides to create two cells, an error can arise in the replication of its DNA. One of the As, Cs, Gs or Ts might be deleted or be replaced with an alternative letter. This is referred to as a mutation, and more specifically a single nucleotide polymorphism, often abbreviated to SNP (pronounced snip). In some cases, this mutation will alter the amino acid produced during transcription, which will then alter the protein end-product. This type of mutation is referred to as a missense substitution.

Mutations are happening all the time within each of us. In addition to those arising from cell division, mutations can occur when cells are damaged by viral infections, or exposure to radiation and some chemical substances. The majority are neither beneficial nor harmful, with many being repaired quickly before they can do any damage. Some, however, can have an impact by either being located within a gene or inside a regulatory area that influences the formation of transcription factors.

Cells can be divided into two types, somatic and sex cells. Somatic cells make up most of the body's tissues and organs, such as skin, muscles, the gut and the brain. When mutations occur within somatic cells or during their division, they remain restricted to the individual because such mutations cannot be inherited. Some can be harmful, such as by causing cells to keep dividing to form tumours, otherwise known as cancer.

Sex cells are those found within female eggs and male sperm. When the DNA within these cells mutate, there is a 50 per cent chance that the mutation will be inherited by the next generation when the DNA from the mother combines with that from the father during conception. That is because the offspring gains one version of the gene from its mother and one from its father, these referred to as alleles. If inherited during conception, the mutation will be passed to every cell within the offspring's body.

Although most mutations have no impact, some can have either a positive or negative influence on human physiology, behaviour and/or cognition, adding to the continuous range of human variability that provides the basis for natural selection. Mutations that have a deleterious effect on reproduction, perhaps by making someone less attractive or less able as a sexual partner, will not be sustained within the population. Conversely, those mutations that increase reproductive success (the number of offspring that survive to reproduce themselves) will proliferate, with the potential to make a total sweep through the population. That is what happened to those genetic mutations within our ancestors that facilitated the construction of the neural networks that delivered language – precisely when, why and how we have yet to find out. We can, however, begin to specify some of those language-significant genes.[4]

From language disorders to language-significant genes

Dyslexia is a language disorder that inhibits a person's abilities at reading and writing despite having typical intelligence and learning opportunities.[5] A considerable number of dyslexic children also have difficulties with processing the sounds of speech and consequently dyslexia is considered to be a language-related disorder. Four genes have been recurrently associated with dyslexia: *DYX1C1*, *KIAA0319*, *DCDC2* and *ROBO1*. All of these play a role in neuronal migration during foetal development, which has been identified as a likely cause of dyslexia – neurons and glial cells not arriving at their required location in the brain. These genes may carry their own mutations or be influenced by mutations within other genes.

The most detailed studies have been on *KIAA0319*. In a dyslexic person, a TF reduces the expression of this gene by 60 per cent which causes irregularities in neuronal migration. *DYX1C1* also illustrates the interaction between genes. A mutation on this gene was discovered to be associated with the dyslexia found within a particular Finnish family, but the same mutation appears to have no effect when found within other families. This suggests the expression of *DYX1C1* is influenced by one or more other genes that must vary between the Finnish and other families carrying the *DYX1C1* mutation.

Language disorders also arise from genetic mutations that influence axon guidance and the formation of synapses.[6] Mutations on the *NRXN1* gene, for instance, which is known to be influential in the formation and activity at synapses, have been found in people with various language impairments, notably delayed acquisition.[7] Similarly, the *CNTNAP2* gene has been implicated in language disorders associated with autism and epilepsy. This gene produces a protein, CASPR2, which influences, among other things, how neurons migrate and connect to one another.

While most genes associated with language have been found by investigating disorders, linguistic proficiency also provides a clue.[8] A study followed the progress made by newly arrived students from China at a US university when learning English as their second language. As soon as they began learning, the amount of white matter in their brains began to change, reflecting increased connectivity between neurons. That continued throughout the training, and then reversed when the language course was over. DNA taken from the students showed they each had one of three versions of a gene known as *COMT*. This gene makes an enzyme known to be important to the function of the prefrontal cortex. Students with either of two versions expanded their white matter – experiencing increases in brain connectivity – when taking the language class; those students with the third version had no such luck. When the English-language tests were taken at the end of the course, student performance was strongly related to which version of the *COMT* gene they possessed and the expansion of their white matter.

The most well-known gene influencing language, and which was once mistakenly called 'the language gene', is known as *FOXP2*.[9] This gene was identified after half the members of a family known as KE were found to have a severe speech disorder called verbal dyspraxia. This is a malfunction in the brain that constrains speech despite vocal tracts being in perfect working order. During the 1990s, the KE family's disorder was tracked to a mutation on the *FOXP2* gene. In other individuals, verbal dyspraxia has been associated with other genes, such as *ECR1* and *BCL11A*, providing further evidence of the complex relationships between genes and language.[10]

The *FOXP2* gene encodes a protein known as FOXP2 (note the non-italics when referring to the protein rather than the gene). This protein acts as a transcription factor that regulates the expression of hundreds of other genes within the brain and elsewhere in the body, including the lungs and gut. The gene and its protein have been highly conserved in evolution, meaning that it is found in near-identical forms in a wide range of animals including mice, bats, birds and the zebra finch.

When the human version of *FOXP2* was studied, it was found to differ from that of the chimpanzee, assumed to be the ancestral version, by two missense substitutions. The human *FOXP2* was further shown to differ by three substitutions from the mouse version and seven from that of the zebra finch. To discover the role of the *FOXP2* gene, mice were genetically modified to carry the human version with its two missense substitutions. The mice displayed higher degrees of synaptic plasticity, which increased their exploratory behaviour and rate of learning – they became quicker at solving mazes and faster at finding food. Conversely, when mice were modified to carry the malfunctioning *FOXP2* version as found within the KE family, their synaptic plasticity was reduced, resulting in motor impairments and the loss of the ultrasonic vocalisations that are elicited when baby mice are separated from their mothers. This suggests how the mutation may have been impacting on the KE family members and why they suffered from such speech impairments.

The epithet of 'the language gene' became attached to *FOXP2* in 2002 after researchers determined that the mutations on the human version had happened in the last 200,000 years and then swept throughout the human population. That suggested the mutations must have increased reproductive success by enhancing – or maybe even delivering – the language capability. The date was curiously coincidental with the appearance of modern human behaviour in Africa. It helped to impose a divide from the Neanderthals who, at that time, were often claimed to lack language. However, this theory soon ran into trouble: when the Neanderthal genome was sequenced in 2007, the Neanderthals were also found to have the *FOXP2* version with two mutations. Similarly for the Denisovans.[11]

In 2018 the original 2002 study was repeated using a much larger and more diverse sample of human genomes than the mere twenty available for the original research. The new study could draw on those from the '1000 Genomes Project', reflecting the tremendous advances made in genomic research.[12] It found no evidence for the 200,000-year-old sweep of the mutated gene through the human population. Some present-day African populations have the ancestral form of the gene without showing any language defects. This study concluded that the one undertaken in 2002 had been unfortunate in not being able to sample enough genomes. While the 2018 study undermined *FOXP2* as the 'language gene', that designation was always a media-driven nonsense because language can only be underpinned by a complex array of interacting genes, of which scientists have currently gained no more than a sneak preview.

Nevertheless, *FOXP2* remains a gene with a high level of language-significance. Although modern humans, Neanderthals and Denisovans had identical versions of this gene, it appears to have been regulated in a different manner within modern humans which may have had an impact on its language capabilities.[13]

Genetic differences between Neanderthals and modern humans

Homo sapiens, *Homo neanderthalensis* and the Denisovans are distinctive human types because of differences in their genomes with consequences for their physiology, behaviour and cognition.[14] They originated from isolated populations of *H. heidelbergensis* that accumulated genetic differences by the constant process of chance mutation and selection. These gradually pulled the genomes of isolated populations apart and differentiated their anatomy such that anthropologists would eventually designate them as three separate species. The fossil record of Europe shows progressively more pronounced Neanderthal traits appearing in specimens dated between 400,000 and their extinction after 40,000 years ago – heavier brow ridges, more protruding faces and wider nasal cavities, although Neanderthals maintained a similar degree of brain asymmetry as found in modern humans.[15] The Denisovans of central and eastern Asia probably looked similar, but there are too few fossil specimens to enable a reconstruction. Conversely, in Africa, evolution led to the less robust anatomy, flatter faces and globular crania that designate *Homo sapiens*, these appearing at different times with flatter faces by 300,000 years ago and with globular crania after 200,000 years ago.

Population movements, including the modern human dispersal from Africa after 70,000 years ago, led to the three species meeting each other (although not at the same time), resulting in multiple occasions of interbreeding – Neanderthals with *Homo sapiens*, and both with the Denisovans.[16] This introduced some of the mutations that had occurred in any one of the lineages into the other two, referred to as the introgression of DNA and resulting in genetic admixture.[17]

Our knowledge about the genetic differences between *Homo sapiens*, *Homo neanderthalensis* and the Denisovans, along with the extent and consequences of their interbreeding, has arisen from the remarkable advances in genomics of the last decade. The first complete human genome was derived in 2003. Today we have many thousands, allowing us to appreciate the extent of genetic variation and its consequences within our species. The first complete Neanderthal and Denisovan genomes were derived in 2010, and a second Neanderthal in 2012.[18]

In 2019, the geneticists Martin Kuhlwilm and Cedric Boeckx made a catalogue of the known changes in modern human and Neanderthal DNA since these species had diverged.[19] They found that more than 12,000 SNPs (single nucleotide polymorphisms) had become fixed in the modern human genome, meaning that they are found in the entire population, with nearly 30,000 fixed in over 95 per cent of modern humans.[20] While this sounds a large number of mutations affecting genes and regulatory regions of DNA, it amounts to only between 1.5 and 7 per cent of the modern human genome. That is what makes us unique as a species and within which any uniquely modern human aspect of language must find its genetic base. The rest of our genome is shared by the Neanderthals and Denisovans, primarily because of our shared common ancestor, but with a contribution from interbreeding.[21]

Having identified the list of modified genes, Kuhlwilm and Boeckx tracked down their impact on the modern human condition.

Several of the modified genes play key roles in the process of cell-cycle machinery – the process that enables cell growth, duplication of genetic material and cell division. The brain is known to be especially sensitive to such changes. Differences in the cell-cycle machinery between chimpanzees and modern humans are key to their threefold difference in brain size; it may also be critical to the neurological differences between modern humans and Neanderthals. One of the modified genes, the *CASC5* gene, is known to influence variation in grey matter volumes between modern human populations, and likely did the same for modern human and Neanderthal species.[22] Changes in *CASC5* and the regulation of other genes by modern human SNPs may have influenced the pattern of brain growth. Other mutations were on genes known to influence the size of the cerebellum (e.g. *ZIC1*, *AHI1* and *ABHD14A*), identified by 3D modelling as a key difference between the modern human and Neanderthal brain.

Another set of specifically modern human mutations influenced genes associated with neurons and how they are connected in the brain: *VCAN* promotes neurite growth; *PIEZO1* and *NOVA1* influence axon guidance; *DCX*, *SCAP* and *RB1CC1* are associated with myelination and synaptic transmission. Other modified genes (e.g. *SLITRK1* and *NOVA1*) influence the function of the basal ganglia, this also known to impact on language.

Rather than influencing the brain, a further set of modified genes impacted on the soft palette and tooth morphology, potentially modifying the vocal tract and hence the nature of verbal expression. Other modified genes are known to influence the rate of human development, which may have had consequences for language learning. These might explain why childhood became longer in modern humans than in the Neanderthals, providing children with greater opportunities to shape their language-significant neural networks during extended periods of vocal learning while their brains were still developing.[23]

Considering this entire suite of modified genes, Kuhlwilm and Boeckx concluded that 'modifications of a complex network in cognition or learning took place in modern human evolution'.[24] They stressed that no single gene modification was primary, with the 'modernity' of modern humans arising from the whole network of changes.

Kuhlwilm and Boeckx's catalogue of uniquely modern human mutations has already been extended and their functional consequences further explored – such is the pace of genomic research. A 2022 study placed greater emphasis on the importance of mutations beyond the cerebral neocortex, finding that regions such as the cerebellum have some of the most derived features of the modern human brain.[25] That conclusion chimes with the cerebellum's larger size and different shape in modern humans than in the Neanderthals, as discovered by Kochiyama's digital reconstruction. Another study identified the impact of the modern human variant of the *TKTL1* gene, finding that this generates a greater number of neurons during foetal development than does the Neanderthal variant.[26]

The globularisation of the modern human brain

As this research demonstrates, a new synthesis about brain evolution is emerging from the developments in genomics and neuroscience. This synthesis also encompasses anthropology, as illustrated by recent research on the globularisation of the modern human brain.[27] Led by the palaeoanthropologist Philipp Gunz and the geneticist Simon Fisher, this study integrated methods from palaeoanthropology, genomics, neuroimaging and gene expression. It drew on the fact that most people today carry some Neanderthal DNA, this having arisen by interbreeding between Neanderthals and *Homo sapiens*.

Gunz and Fisher's team first measured the extent of cranial globularity in thousands of people and imaged their brains. They found that people with more pronounced globularity had larger amounts of grey matter in their temporal lobes, the vermis and adjacent parts of the cerebellum, and in their subcortical structures. Conversely, they had less grey matter elsewhere in their brains, including other regions of their cerebellum and their gyri in the frontal, temporal and occipital lobes.

Next, the research team measured the globularity of over 4,000 individuals of European ancestry whose genomes contained some Neanderthal DNA arising from interbreeding by their ancestors. They tested which introgressed fragments of Neanderthal DNA were associated with relatively elongated brain shapes in this sample of modern humans. They found eleven fragments to be significant, located on two different chromosomes. Although these fragments were outside the protein-coding region of DNA – the genes – they were identified as influencing the expression of two nearby genes, UBR4 and PHLPP1, possessed by both Neanderthals and modern humans. Both genes are associated with brain development, including neurogenesis, neuron migration and myelination, with notable impacts on the cerebellum.[28]

This brilliant study shows how the distribution of grey matter would have varied between the Neanderthal and modern human brain, it being relatively higher in some regions and lower in others, with different parts of the cerebellum affected in different ways. It demonstrated that two of the many genes that influence the shape of the brain were regulated differently within the Neanderthals and modern humans, leading to different patterns of neural development and contributing to the elongated Neanderthal and the globularised modern human brain. At present, we cannot specify which cognitive functions, if any, were enhanced by the modern human expression of UBR4 and PHLPP1, but with the network for language distributed throughout the brain, it would be unwise to bet against some influence over the capacity for language.

I find it difficult to resist the idea that the genetics behind globularisation developed those extensive neural networks that connect areas of functional specialisation in the modern human brain. They result in what the psychologist Annette

Karmiloff-Smith called a global workspace and what I have termed cognitive fluidity. That would account for the slow emergence of the novel behaviours in Africa after 200,000 years ago that we reviewed briefly in Chapter 2, represented by the technological innovations described in Chapter 7 and the new use of body adornments, ochre and engravings to be considered in Chapter 15. Once those behaviours were present, they would have continually reinforced the long-distance connections in the brain, which would in turn further enhance those novel behaviours: the brain, language and material culture bootstrapping each other into modernity.

While changes in connectivity may have been a consequence of globularisation, it is possible that the change in shape alone was of most significance.[29] This arises from a study published in May 2023 by James Pang and colleagues that challenged the conventional view that patterns of connectivity are the prime determinant of brain function. They analysed thousands of brain images derived from people undertaking a wide array of tasks and concluded that the shape of the brain was more significant than patterns of connectivity in the observed brain activity. By shape they refer to the many grooves, contours and folds of the cerebral cortex that, they argue, influence how waves of excited neurons travel through the brain. To summarise their findings, Pang suggests that 'ripples in a pond may be a more appropriate analogy for large-scale brain function than a telecommunications network'.

Inevitably, their work has been challenged by neuroscientists building connectome models of the brain that emphasise patterns of connectivity (the telecommunications network). It is very early days for the approach taken by Pang and his colleagues and a great deal more research is required for their methods and findings to be validated. Common sense, however, suggests that both the shape and the internal wiring of the brain would play significant roles in turning the swirling chemicals and electrical currents within the brain into thoughts, feelings and words – although noting that common sense is not always the best guide in science.

If shape does play a role, there may be profound consequences for our understanding of how language and cognition evolved because modern human and Neanderthal brains differed not only in their surface geometry but also in their overall shape. Rather than thinking about ripples in ponds, one might think how vibrations move through differently shaped bells to create different sounds and compare these with how patterns of neuronal excitation might have passed differently through the elongated shape of a Neanderthal brain and globular shape of a modern human brain to create different types of thoughts and ultimately spoken words.

Exchanging genes . . .

Returning to our current state of knowledge, we are confident that Neanderthals had much lower levels of genetic diversity than those found in modern humans. This reflects a Neanderthal population bottleneck, a period when there had been a drastic reduction in population size that appears to have happened after their split from the lineage leading to the Denisovans at *c*.400,000 years ago, and for an unknown reason.[30] Inbreeding within their small, isolated communities and genetic drift led to the accumulation of mutations with deleterious effects. This genetic load reduced Neanderthals' reproductive fitness by 40 per cent compared with that of modern humans, which may have contributed to their eventual demise.[31] Nevertheless, the phenotypic differences between Neanderthals and modern humans must have been relatively minor because when members of these species met, they were able to have sex and produce viable offspring.

The timing and number of interbreeding events between

modern humans and Neanderthals remains unclear but these likely occurred on multiple occasions in western Asia and Europe between 65,000 and 40,000 years ago. Interbreeding between modern humans and Denisovans occurred in central and eastern Asia between 54,000 and 44,000 years ago.

Quite what these interbreeding events represent in terms of past behaviour remains unknown: rapes by an incoming population; one-night stands; occasional loving relationships; stable communities of modern humans and Neanderthals working and living together? My guess is a mixture, just as we find in the diversity of male–female relationships in the world today. Both species may have seen an advantage by interbreeding, and both may have gained more than a few of each other's genes. The most intriguing evidence was published in August 2023 involving re-analysis of skeletal remains excavated between 1949 and 1963 at Grotte du Renne, at Arcy-sur-Cure in central France. This identified a fragment of a hip bone likely coming from a modern human in the same layer as Neanderthal skeletal remains dating to 45,000–41,000 years ago. Whether the group occupying the cave had members from both species or whether the cave had been used on separate occasions by Neanderthals and modern humans with their bones becoming mixed into a single layer remains unclear.[32] My guess is the former.

Interbreeding led to the introgression of Neanderthal and Denisovan alleles into the modern human genome. We carry the legacy today because about 1–2 per cent of the DNA in present day non-Africans can be traced to Neanderthal introgression, while people in Southeast Asia may have a further 5 per cent from the Denisovans. The original extent of introgression is likely to have been higher because deleterious alleles have been gradually removed by natural selection – those modern humans carrying the deleterious Neanderthal alleles had less success in producing reproductively viable offspring for the next generation than did other modern humans.

By examining the Neanderthal DNA remaining within modern humans today, we find their alleles provided both advantageous and deleterious effects.[33] On the positive side, modern humans inherited alleles associated with skin pigment and immunity to disease. These are likely to have evolved within the Neanderthals as adaptations to life in northern latitudes.[34] Their inheritance by modern humans may have been critical to their eventual success in colonising such latitudes after dispersing from Africa. That came after one or more failed attempts, perhaps by populations who had yet to interbreed and gain the valuable Neanderthal alleles.[35]

Although the Neanderthals knew nothing about the genetic consequences of inbreeding, they may have been attracted to new reproductive partners in the form of incoming modern humans by the limited choice and availability within their own communities. There is, however, little evidence for modern human alleles within the Neanderthal genomes. This is a common outcome when a colonising population encounters a resident population, because introgressed alleles only attain sufficiently high frequencies to become fixed when a population expands – that was the case for the incoming modern human but not for the resident Neanderthals.[36]

The downside from interbreeding for modern humans was that the deleterious alleles accumulated within the Neanderthal genome spilled into the modern human population and then took a long time to be purged by selection.[37] Because of inbreeding, the Neanderthal reproductive fitness is likely to have been only 40 per cent lower than that of modern humans; by interbreeding, modern humans reduced their own fitness by an estimated 0.5 per cent. Further, several of the Neanderthal alleles that remain within modern humans are associated with ill effects, including depression, susceptibility to skin damage (actinic keratosis), and excessive clotting of the blood (hypercoagulation).

There are also 'introgression deserts', two regions of the modern human genome where Neanderthal alleles are strikingly rare: the testes and the brain.[38] In both cases, any introgressions have been heavily selected against and hence removed from the genome. Regarding the testes, this is explained by the Neanderthal alleles having caused a decrease in male fertility when placed within a modern human genetic context.[39]

The brain is of more interest to us.[40] Neanderthal alleles are rare within the brain-relevant areas of the modern human genome, although some exist, as demonstrated by Gunz and Fisher's study of the genetics behind globularisation. These areas of the modern human genome contain a higher frequency of recently mutated genes, indicating attempts to counteract the ill effects of Neanderthal introgressions before their alleles were removed by selection. Those which remain have often been down-regulated – made less active – notably in the neuron-rich region of the cerebellum and the basal ganglia. Both these areas are associated with language processing (among other things).

The rarity and down-regulation of brain-related introgressed Neanderthal alleles supports the genetic evidence that, following their divergence, the Neanderthal and modern human brains had evolved in different directions. During their evolutionary history in Africa, modern humans had acquired a larger cerebellum, enhanced neuron production and new patterns of connectivity. It is not unreasonable to suppose these affected their language capability, shifting it from whatever had existed within *Homo heidelbergensis* and appears to have remained little changed within the Neanderthal lineage.

. . . and possibly words

A final consideration is that modern humans and Neanderthals may have exchanged more than a few genes, especially if their interbreeding reflects prolonged periods of contact, as may have happened at Grotte du Renne. The incoming modern humans may have learned about the resources, seasonal rhythms, the pitfalls and the opportunities of life in high latitudes; Neanderthals may have learned about new ways to make tools and the use of symbols. Although such learning could have arisen by observation alone, it seems likely that they communicated with one another in whatever way they could make themselves understood, perhaps with some partial learning of each other's languages. By doing so, both modern humans and Neanderthals may have adopted some of the sounds, words, phrases and structures of each other's languages into their own, in the same manner that people speaking different languages do today. Just as Neanderthal alleles contribute to modern human genetic diversity, elements of long-lost Neanderthal languages may have had a lasting impact on linguistic diversity in the world today. With that intriguing thought, it is time to return to language and assemble the next fragment of the language puzzle: how words and the lexicon are constantly changing through time.

13

WORDS KEEP CHANGING

Language never stays still. Words are forever changing what they mean, how they sound and how they are used. Some words disappear altogether, others lose their independence and become parts of other words, while new words are invented – usually by combining those that already exist. Such processes of change cause languages to diverge into dialects and then become unintelligible to each other (Figure 5). We can hear this happening today – we all speak differently from our grandparents and our grandchildren.

Words have been changing ever since the first words were spoken, when they were few and before any grammatical rules had appeared. This has long been appreciated. Writing in the fourth and third centuries BC, Epicurus proposed that after people had got used to the iconic names given to things such as trees, conventional names began to be introduced, followed by the use of analogies, abstract terms, and new categories of words such as pronouns and prepositions.[1]

In this chapter we will look at the processes of word change in the recent past with a view towards projecting these into the distant past and considering their long-term consequences, following the example of nineteenth-century scientists such as Charles Lyell who sought to understand geological strata and Charles Darwin for the evolution of species. We will look at four dimensions of such change: how the meanings of words change; the invention of new words; how pronunciation shifts;

and how lexical words (those with content) become grammatical words (those which join lexical words together to create meaningful utterances). The examples in this chapter draw primarily on those from the English language. This is simply for my own expediency, reflecting my limited knowledge of other languages and the predominance of published research. There is nothing special or odd about English: the processes of change are universal to all known languages.

Words change their meanings

The meanings of words are constantly changing and multiplying – a process termed by linguists as semantic change.[2] We have a hint of this from our own lifetimes: is a *wicked mouse* an evil rodent or an excellent hand-held input device for a computer? The word table once only referred to a piece of furniture with a flat top, but has not only become a verb, but one with two opposing meanings. In the UK, 'to table' means the process of presenting a proposal, while in the US, 'to table' means postponing the consideration of a proposal. We can, however, only appreciate the scale and significance of word change when comparing historical sequences of written texts of the same language.

English is an ideal case study because we can compare word meanings in texts written in Old English between the seventh and eleventh centuries, such as *Beowulf*, with those of the seventeenth century by considering Shakespeare's plays, and then with their present-day usage. Even better, we can follow the same text, and hence words in their written context, through time by using different versions of Bible passages. These can be tracked from the original Latin to Ælfric of Eynsham's Old English translations of the tenth century, and John Wycliffe's Middle English translation of the Bible of 1382 to those within the Early Modern English King James Bible of 1611 and finally to their modern-day versions. We might be more ambitious, beginning with words before their adoption into Old English, such as in Latin and even the reconstructed Proto-Indo-European (PIE) of 4500 BC (6,500 years ago).

While historical texts are a critical resource for exploring semantic change, they lack the contextual evidence that is essential for inferring the nuances of meaning from the spoken word: the prosody, accompanying gestures, age, gender and status of the speaker, their surroundings, props and social relationships with whoever is listening. As we will see, these so-called pragmatics of language partly drive the process of semantic change. Moreover, written language that lacks any pragmatics is inherently conservative – it is slower to change and will lag behind the way people talk to each other.

Three examples of how the meaning of words have changed

The Old English word *sælig* meant 'blessed'. By *c.*1200 its meaning had changed to 'innocent'; by *c.*1300, it had further changed to 'harmless' and then 'weak'; by the 1570s it had come to mean 'feeble in mind and lacking in reason', which is close to the meaning of *sælig*'s modern spelling: *silly*.

Nescius in Latin means 'ignorant'. This word was carried into English via French in the early fourteenth century as the word *nice*, after which a new meaning emerged: 'a person or clothing that is luxurious'. *Nice* then began to mean 'a person who is finely dressed' or 'something that is precise or fussy'. By the late 1500s *nice* was used to describe something that was 'refined', from which a catch-all word for something which is 'pleasant' or 'agreeable' emerged – the meaning of our present-day word, *nice*.

A more recent example is the word *woke*.[3] This emerged within American Black slang in the 1930s as in the phrase 'best stay woke', meaning to stay alert to racial prejudice, especially from the police. It was first defined in print in 1962, with a subtle change in meaning to someone who was 'well informed and up to date'. But *stay woke* continued as a warning against racial discrimination and became pervasive in the Black Lives Matter movement following the police killing of teenager Michael Brown in St. Louis on 9 August 2014. With the hashtag *#stay-woke*, the word spread via the internet where its meaning began to change. *Woke* came to signify 'a progressive outlook on a host of issues as well as on race', becoming associated with people of varied races on social media rather than BLM activists. Its meaning continued to change, largely under the influence of the conservative right wing in the United States who found that targeting 'woke' as a byword for 'smug liberal entitlement', and then promoting the notion that it stood for an intolerant and moralising ideology, was an effective way to foment political division and garner support. The word is now used differently by different people, with all meanings still in play.

Further examples of semantic change can easily be found within an etymological dictionary, with some of the most surprising described in John McWhorter's excellent 2016 book *Words on the Move*. Virtually any word in any language appears to have a history of semantic change. The most frequently used words have the greatest turnover in meanings. There are some exceptions: *brother* is a frequently used word but appears not to have changed its meaning since its use in Proto-Indo-European 6,500 years ago. In their attempt to find order in what appears to be a suite of meandering if not random changes in semantics, linguists have identified three key axes of change and a pile of so-called 'dead metaphors'.

Axes of semantic change and dead metaphors

The first axis is that words can become either narrower or broader in their meaning. The Old English words *mete*, *déor*, *steorfan* and *gyrle* had broad meanings of *food*, *animal*, *die* and *young person* respectively. Their modern-day versions have much narrower meanings: *meat, deer, starve* and *girl*. Conversely, the present-day word *holiday* has a broader meaning than when used in Old English as *haligdæg*, which meant 'Holy day'. Some words have mirrored each other; as one narrowed, the other broadened: in Old English the generic word for *dog* was *hund*, while a *dogge* was a particularly large and fierce type of *hund*; today the *hound* is a type of *dog*.

A second axis of change is that the meaning of a word can become either more positive or more negative, referred to as amelioration and pejoration. *Nice* is an example of the former, changing from meaning ignorant to something that is pleasant. Pejoration is more common, illustrated by *silly* deriving from *blessed*, and the transformation of *woke* from a concern with injustice to a negative term for those who are intolerant of right-wing values. *Villains* are scoundrels or criminals, but the word derives from that for a fully respectable medieval serf; similarly, being *cunning* has overtones of trickery and deceitfulness, whereas the fourteenth-century meaning of *conning* was being *learned, knowledgeable and skilful*.

A third axis of change concerns the weakening or strengthening of meaning, with the former being more frequent. *Literally* was once used to indicate that one should interpret any following words according to their exact meaning, but we now use the word as a means of emphasis. The word *awesome* was once restricted to having emotions of great reverence or dread, but now means merely impressive.

As we noted, frequently used words are most prone to change. Words of greeting and farewell tend to become formulaic, both weakening and broadening their original meaning. We casually say *goodbye* without appreciating that this had once been *God be with you*. This is the same in all languages. It has been said that '*Hello, Good morning, How are you?, Bonjour, Ciao, Szervusz, Zdrávstvujte, Shalom, Goodbye, So long, Adieu, 10-4,*

Arrivederci, Do svidaniya, Güle güle and their kind, are a 'grave-yard of forgotten or neglected meanings'.[4]

Dead metaphors are words that were originally used to elaborate on the meaning of another word or phrase but lost their metaphorical use to become an ordinary word. While the word *field* maintains a similar meaning to its Old English counterpart of *feld* (a large tract of open country), we now use it to refer to a *field of study* or a *field of vison* without appreciating any metaphorical implications from the physical landscape. Similarly, we use the word *broadcaster* about a radio or TV presenter without needing to appreciate that this is ultimately a metaphorical use of the eighteenth-century word for scattering seed.

Changes in word meaning can also arise from metonymy – by simply being associated with the meaning of another word. A *crown* is a physical object worn by the monarch, but *the Crown* now means the monarchy and its associated processes of governance, and more recently a renowned Netflix series. Similarly, the *White House* is a building in Washington but these words can also mean the US president and his administration.

Word conversion

The changes of meaning so far considered have referred to words remaining within the same category, primarily nouns remaining as nouns. Another dimension of change is known as word conversion by which a word can take on all or some properties of another class of word, as we saw with *table*. The most common is when nouns become verbs, but adjectives can also become verbs, while verbs can become nouns, and nouns can become adjectives.

Verbification – a noun becoming a verb – is the most common, with the resulting verb being described by linguists as a denominal verb or a gerund. This is likely to have occurred throughout the history of every language and is a process that

English speakers are readily familiar with, even if a new verb has an infrequent use and soon disappears. Many remain as steadfast elements of the language. *To rain, to snow, to pocket, to father, to parrot, to mirror* and *to captain* are just a few examples we use frequently without appreciating their source; likewise, when we are *buttering* our bread, *lacing* our shoes and *elbowing* our way out of a crowd.

One of the striking features of denominal verbs is that they can be so easily invented and understood by others.[5] This also applies to the use of proper nouns as verbs. If I said to my wife that 'my friend Houdinied his way out of the mess he was in' she would understand, knowing that Harry Houdini was a famous escapologist of the early twentieth century. My grandson, however, never having heard of Houdini, would be perplexed. Similarly, if in the summer of 2022 I said to someone that I had been *totally Borised* they would readily understand this meant I had been lied to (in a jovial manner). Within a few years, however, someone might be as puzzled as my grandson because the reputation of Boris Johnson, the former British prime minster, may have faded away. As these examples illustrate, turning nouns into verbs in English is easy, and often accomplished simply by adding an *-ed* suffix.

Verbs can also become nouns: we can refer to someone as a *bore* or a *cheat*. As with nouns, verbs can be converted by adding a suffix, often an *-ance, -ment* or *-tion*, as in *appearance* from to appear, an *appointment* from to appoint, and *information* from to inform. When verbs with two syllables are converted to a noun, the first syllable becomes emphasised. For instance, when the word *suspect* is used as a verb, emphasis is placed on the *-pect*; we say *sus-PECT*. When used as a noun, however, we say *SUS-pect*. This sound change was noted by the linguist John McWhorter who calls it a backshift. As he explains, someone who *re-BELS* is a *REB-el*, whose crimes you can *re-CORD* and thereby leave them on *REC-ord* for all to see.[6]

Nouns can be easily changed to adjectives by adding a -y suffix, as in *icy* from *ice* and *bloody* from *blood*. Sometimes it is not clear which came first: is the noun *spy* derived from the verb *to spy* or the converse? Moreover, the process can operate in reverse. That the nouns *grot* and *greed* are derived from the adjectives *grotty* and *greedy* assumes that a suffix had once been applied to the noun; in fact, these adjectives arose by a different process.[7]

Processes of change

Identifying and categorising changes in word meaning is easier than understanding how and why they occur. That is because we lack the context in which the words were once used, both in the moments of utterance and their wider social and economic environments of use. What is readily apparent, however, is that words do not change their meanings abruptly: new meanings often co-exist with the conventional meaning(s) for a prolonged period until the latter become lost.[8] We are familiar with such polysemy today – the same word having two or more meanings: *skinny* (as in thin, naked, low fat), *wicked* (evil and marvellous) and *get* (procure, become, understand, etc). In fact, polysemy abounds within the English language, rarely causing any confusion because we always use context to infer the appropriate meaning.

New meanings can arise from a deliberate, innovative act by an individual speaker. More frequently, they appear to emerge without any intent and from multiple minor acts of conversation that cause one peripheral sense of a word, perhaps even a misunderstanding of its conventional meaning, to become emphasised and acquire salience.[9] That adjusted meaning might become prominent and used consciously by a particular group within a speech community, either because it is particularly useful for them or as a means of identity, before it becomes widely adopted and usurps the original, conventional meaning.

In rare circumstances, the date and identity of the person who changed, or at least supplemented, the meaning of a word is known. The term *cloud computing* was first used in an IT business plan in 1996, but gained prominence on 9 August 2006, when Google's then CEO Eric Schmidt introduced the term to an industry conference.[10] For most of us, it is only in the last few years that *the cloud* has become a store of digital information rather than a suspended mass of water droplets in an otherwise clear blue sky.

Although we can rarely identify when or precisely how a new meaning has arisen, we can explore how different meanings of the same word vary in their usage within living speech communities. This enables us to investigate how factors of age, gender and socioeconomic context influence semantic change. Consider, for instance, the various meanings of the word *skinny*, as documented by the linguist Justyna Robinson within a community in South Yorkshire in the early 2000s.[11]

Robinson interviewed seventy-two people ranging in age from eleven to ninety-four, equally divided between male and female, and of variable economic standing, to find out how they defined *skinny*.[12] Although 'thin' was the most salient meaning of *skinny* for all participants, there were significant age-related and socioeconomic-related differences in the use of other meanings. *Skinny* as in 'mean' was only used by the older generation (people over sixty years old) and was evidently disappearing from use. Moreover, it was primarily used by those whom Robinson categorised as working class. This was not unexpected because dialect words are usually retained for longest within closely knit communities who have limited geographical and social mobility.

The sense of *skinny* as 'showing skin' was also largely restricted to the older generation, but primarily used by those designated as middle class. Robinson suggested several reasons why this might be the case: because this sense emerged in North

America, the middle classes might have had more contact with American English via their types of employment or travel to the United States; they may have had more opportunities to holiday where skinny-dipping was au fait; rather more speculatively, Robinson notes that nude swimming had been made compulsory in certain public (i.e. independent) schools in the north of England during the 1970s, when her older generation would have been teenagers and likely attended such schools. Swimming facilities and lessons would have been absent from state schools.

Not surprisingly, those using *skinny* to refer to 'low fat' were primarily in the younger age groups. This meaning had also originated in North America and is mainly restricted to coffee in the UK, notably 'skinny latte'. As we would expect, Robinson found its use most prominent among women from the upper working class and the middle class who had opportunities to frequent coffee shops.

Although 'thin' was the most salient meaning of *skinny* for all speakers, there were differences in how this meaning was evaluated. When asked to give examples of its use, younger speakers and women tended to be more neutral and positive, as in 'looking good' and 'being fit'; older and male speakers were more negative, referring to unhealthy people and anorexia. Overall, Robinson felt there was an ongoing amelioration of thin as the most salient meaning of *skinny*. None of the participants below the age of eighteen cited any meaning other than thin. That was, however, two decades ago and I suspect that the low-fat meaning of skinny has now spread to that generation while skinny as being mean has disappeared.

This case study illustrates several generic points: a word can have several meanings in use at any one time; age, gender and socioeconomic position will influence which meanings will be encountered and adopted; semantic change is ongoing, occurring with different rates of change for different types of

meaning; young adults, rather than children or the older generation, are most likely to be the initial adopters of new meanings.[13]

I can see no reason why the same principles should not apply
to the words spoken by our ancestors, whether *Homo erectus*,
Homo neanderthalensis or early *Homo sapiens* – assuming for the
moment that they all had words to say. Individuals within their
communities also varied by age, gender and socioeconomic
position; they would have had different life experiences, pref-
erences and desires, and would have been keen to forge their
identities by the words they used.

Efficiency and effectiveness as drivers of change

An innate desire for efficient and effective verbal communication
sits behind all aspects of linguistic change through all recorded
time, and again can be reasonably extended into the recesses of
the Stone Age.[14] Speakers may find themselves needing to label a
brand-new entity – perhaps a new type of event, an animal never
previously encountered or an invention. Rather than making up
a completely new word, which would then need to be learned
by the speech community, it may be more efficient to borrow
an existing word that has some relevance. Using the word *mouse*
for a hand-held computer input device is readily understood by
others because of its visual resemblance. Similarly, it is efficient
to use a metaphor rather than invent a new word or engage in
a long description when talking about a complex idea, such as
an abstract concept. Those metaphors often become dead metaphors, leaving us with a word with two different meanings,
such as *cloud* and *cloud*. Likewise, borrowing a noun to become
a verb is quick and easy.

In all these and other cases about enhancing efficiency, such
as eroding the beginning or ending of words to save effort, there
is likely to be a tension between the speaker and the hearer.
Those making the mental and physical effort of speaking may

wish for an economy of expression that saves them time and energy. The priority of those listening, however, is with semantic transparency – they need to grasp immediately what is being said. Abbreviations and acronyms used in social messaging are a great time saver but can leave recipients rather puzzled. LOL and OMG are commonplace, but fewer people will know the meaning of IDK, NBD and GOAT.[15]

Adjustments to word meanings to maintain efficient communication may arise because of changes in wider society and culture. A nice example is how changing lifestyles in France led to changes in the meaning of the words for meals. In the sixteenth century, both *disner* and *desgeüner* referred to the main meal taken in the middle of the morning while *souper* was a lighter meal in the afternoon. Changes in lifestyle shifted the main meal to noon and required the introduction of a breakfast. That caused a semantic differentiation between *disner* and *desgeüner*. Now using their modern-day spelling, *déjeuner* became breakfast, *diner* the main meal and *souper* an evening meal. In the nineteenth century, the urban professional class preferred to have their main meal in the evening, which made *diner* acquire the sense of dinner, and *déjeuner*, lunch. A new word was part-invented for breakfast, *petit déjeuner*, and *souper* now designated a later evening meal.

More recently, we might allude to the impact of social media on the change of word meaning. Although the word *friend* has long been used as a verb – Shakespeare did so in *Henry V* and we have always been able to *befriend* someone – most people encountered this usage for the first time when they opened Facebook accounts and discovered they could *friend* someone.[16]

Efficient and effective communication means attention to the social context of word use. Speakers may be motived to deliberately innovate or casually adopt a new meaning of a word to maintain a positive social relationship when discussing subjects that might be taboo. *Passed away* has acquired the meaning of *died* because it seems a less upsetting phrase to use.

And rather than ask someone if they would like a low-fat milky drink, which might imply they are overweight, the word *skinny* provides a good euphemism – it is both veiling and explicit at the same time. New meanings might also be adopted to construct new identities. In North America during the 1930s the word *cool* acquired a new meaning of 'intensely good', this emphasised by the word being spoken in an especially casual manner. This meaning of *cool* arose in African American slang and flourished in the jazz culture, serving to identify its users as apart from mainstream society.[17]

New words from old words

While words can be repurposed, new words are sometimes required to talk about new things, whether ideas, events or inventions. This necessity is at the heart of this book: new words were needed to talk about the new cultural innovations that enabled farming to arise, which ultimately took us out of the Stone Age. Because new words are primarily constructed from existing words, this topic is a natural continuation from semantic change. To start, we need to make a four-way distinction between word creation, word conversion, word borrowing and word formation, with the last of these being the focus of this puzzle piece.

Word creation, sometimes known as coinage, involves inventing completely new words, known as nonces. These are surprisingly rare, with the same few examples being frequently re-quoted, notably *Jabberwocky* (by Lewis Carroll), *kodak* and – less convincingly – *google*.[18]

We have already considered some types of word conversion such as how nouns become verbs (*rain, friend, parrot*) and verbs become nouns (*bore, cheat*). Proper names are converted into words known as eponyms. *Boycott*, for instance, derives from Captain Charles Boycott who was a land agent in Ireland during the 1880s who tried to evict tenants from their farms. Rather than rising against him violently, the tenants stopped working and local businesses refused to deal with him; the captain had to recant and the word became associated with this mode of action. Both word and action were rapidly adopted elsewhere. Other eponyms include: *wellington*, derived from the Duke of Wellington, victor of the Battle of Waterloo (1815); *Alzheimer's*, from Dr Alois Alzheimer, who first identified this form of mental illness in 1906; *atlas*, from the Greek god who supported the heavens on his shoulders; and *sandwich*, from John Montagu (1718–92), the 4th Earl of Sandwich, who is said to have spent twenty-four hours at a gambling table with nothing but slices of beef placed between slices of bread for refreshment. Similarly for words such as *Fahrenheit, Caesar salad, America* and *nicotine*.[19]

There are two types of word borrowing. A word from one context can be used in another, often because of an analogical association. The word *crane* when used for a long-armed lifting machine is borrowed from that for a long-necked water bird. A more frequent type of borrowing is when words are taken from another language, often with minimal phonetic and morphological change, although sometimes with an adjustment of meaning. Many, perhaps all, languages do this, with the practice particularly rife in English: *tofu* is a loan word from Japanese, which had borrowed it from Mandarin Chinese; *cigar* comes from the Spanish *cigarro*, which had been taken from the Mayan word *sicar*; *café* comes from French, with a semantic change from coffee to an eating and/or drinking place. In the previous chapter, I speculated that during the occasions that led to interbreeding, modern humans might have borrowed words from Neanderthal languages and vice versa.

Word formation is the process of making new words from existing building blocks, which are the morphemes within current words. There are seven ways that existing morphemes can be recombined to make new words.[20]

Compounding. Forming a word or phrase out of two free morphemes, which may have been independent words. The morphemes/words are most frequently two nouns, but can be other combinations, such as adjective-noun (*blackbird*) or verb-noun (*pickpocket*). Most compounds have two components, such as *fiddlesticks, claptrap, bailout, daydream, awe-inspiring* and *environmentally friendly*. They can also be made from more than two components, such as *pick-up truck*. One interesting class of compounds are those that have rhyming elements, as in *lovey-dovey*.

Making compounds often involves a change in pronunciation by placing emphasis on the first syllable, just as happens when verbs become nouns. While we might see a black bird, the compounded word is pronounced *BLACK*bird, just as we say *BLACK*board to refer to an object rather than to a board that happens to be black. Similarly for *ICE cream* and *SUPERmarket*.[21]

Derivation. Adding either a prefix or a suffix (or both) to an existing word. The word *democratise* was invented in 1798, followed by *detonator* in 1822, *preteen* in 1926, and *hyperlink* in 1987.

Backformation. This is the deletion of an affix or part of a word. *Sleaze* was backformed from sleazy in about 1967. A similar process brought about *pea* (from pease); *liaise* (from liaison); *enthuse* (from enthusiasm); *aggress* (from aggressive); *donate* (from donation). The term backformation is used because it reverses the process of adding a suffix to transform a word – even if that had not ever happened for the word being backformed. The word *pease*, for instance, was always a singular form but because adding an -s is a frequent way of creating a plural, it was misinterpreted as a plural and backformed into *pea*. Similarly, the word *greed* was backformed from the adjective *greedy* because it was assumed that a suffix had once been applied to a noun – which had not been the case.[22]

Reduplication. This is the repetition of a word or word-like element either unchanged or changed with a different vowel or a different consonant. For instance, *hip-hop*, *boogie-woogie*, *tick-tock* and *pitter-patter*.

Blending. This combines at least two words and either shortens one or both. Unlike compounding, this disregards the original morphemes and often places emphasis on sound structure. Examples include *sitcom*, *paratroops*, *internet*, *sexting*, *brunch*, *mocktail* and *stagflation*.

Clipping. This is the deletion of the initial or final portion of a word, while preserving the original meaning. Examples include *pram* from perambulator, *taxi* and *cab*, both from taximeter cabriolet, and *curio* from curiosity.

Acronym. This deletes everything except for the initial letters or syllables of a two-word or longer phrase to make a word that preserves the original meaning. Examples include *laser*, from 'light amplification by stimulated emission of radiation', *scuba* from 'self-contained underwater breathing apparatus', and *Gestapo* from 'Geheime Staatspolizei'.

How does a speaker invent a word?

The study of word formation is a large and complex area of linguistics.[23] It has adopted two complementary approaches since its emergence during the early twentieth century. The most prominent has been semasiological, this term coming from the Greek words *semasia*, meaning, and *logos*, study. This approach begins with words and aims to understand how specific meanings, or concepts, are attached to them. Words are analysed to establish the rules by which they are formed, seeking to identify those that apply across all languages. It has been commented that the semasiological approach seems to forget that words are created, spoken and understood by people, who appear to have a minimal if any presence in the vast tomes on issues such as affix combinability. That is, perhaps, unfair, because hearers need to grasp the meaning of a word and must proceed semasiologically, especially when hearing a word that has never been uttered before.

The complementary approach is more relevant to our needs and is referred to as onomasiology, the term coming from the Greek *ónama*, name, and *logos*, study. This begins with the concept that needs to be designated by a word, which can either be a new object in the world, a new action, or a new idea within the mind. How does a speaker invent the new word?

Onomasiology developed in the early twentieth century but only gained prominence from the 1960s onward, and more particularly from the 2000s following seminal works by the Czech linguist Pavol Štekauer.[24] He built on the work of other eastern European linguists who had thought the current hard-and-fast semasiological rules for word formation were too detached from the human and social component in language.[25]

The aim of the onomasiological approach is to provide an accurate portrayal of the naming act. A recent model for this has been proposed by the German linguist Joachim Grzega, who combined the key elements of models previously proposed by Štekauer and his predecessors into what he calls CoSMOS – the Cognitive and Social Model for Onomasiological Studies. Grzega illustrates his model by going through the stages that might be involved in the naming of an animal as the Bengal tiger.[26] I will follow his stages but use them for a different type of animal, while also avoiding much of the arcane terminology in the model.

The starting point is either a thought or a referent – something perceived within the non-linguistic world that requires to be designated by a word. The word devised will depend on the nature of the speaker: their gender, age, education and experience. It will also depend on their situation and to whom they will say the word: it might be to themselves, because speaking to oneself is often important especially for young children. Conversely, the speaker might be addressing a vast number of people, such as via a television broadcast. Who is listening, the communicative goal and the ongoing utterance into which the new word will be placed will influence the word that the speaker will devise. An example of a referent requiring a name might be a previously unseen type of animal.

The speaker will classify the referent by processing its observed traits and seek to place it into an existing conceptual category. Some of these traits will be 'global', in my case noting that it is a living thing that moves. This places it into the superordinate category of 'animal'. Other traits will be 'local', which may place it into an existing base or subordinate category. In my example, the speaker is visiting wetlands in Sussex and has seen an animal in the reeds at the edge of a pond. It is about the size of her hand, green with brown spots, with protruding eyes. It made a loud, deep, vibrating noise, before plopping into the water and swimming away.[27]

If the animal is sufficiently like the prototypical example of a category, the speaker may decide to use the existing concept/word for that prototype; if it is perceived as significantly different, the speaker may decide to create a new word, sometimes more and sometimes less consciously. In our example, the speaker recognises that the observed animal is a type of frog, but one unlike they have seen before. It is much larger and louder than the prototypical frog.

Grzega's next stage is that the speaker will perform an unconscious cost–benefit analysis. The speaker considers what they are attempting to achieve by their forthcoming utterance that will include the new word: do I want to sound like the person I will be speaking to, or do I want to sound different? Do I want to sound intelligent, expressive or polite? Do I simply want to take the first word that comes to mind? If the speaker wants to coin a new word, they will go through several steps of a name-giving process.

The speaker will return to the referent, with a focus on its local traits and its context. In our example, the speaker recognises it as a frog and selects the features of its large size and loud croak as being significant. This is referred to as the onomasiological stage and is concerned with identifying the semantic qualities of the referent that will provide the basis for its linguistic representation. Next, the speaker searches for the morphemes that represent the identified semantic categories, referred to as the onomatological stage. The speaker may select one or more free morphemes – existing words – and/or bound morphemes (such as affixes on existing words), along with a process by which they can be modified and/or combined, such as compounding, derivation, blending and clipping.

In my example, the free morphemes *frog* and *bull* are selected, the first taking priority by placing the animal into the relevant category and the second reflecting how it differs from the prototype by its size and sound. *Bull* is selected because a bull is a particularly large animal that makes a deep bellowing noise, having the same relationship regarding size to a cow as this large frog does to the prototypical frog. Regarding the word-formation process, our speaker chooses compounding and derives the word *bullfrog*.

Compounding is more likely than another process such as blending. That would have reduced one or both morphemes creating words such as *bullog*, *bulf* or *froll*. These are unlikely because our speaker needs to tell someone what she has seen in an informative way, although they will not want to use more effort than necessary when doing so. A listener will readily appreciate the size and loudness of a *bullfrog* as soon as its name is stated but would remain rather mystified by any of the blended words. In general, words created using complete morphemes by compounding or derivation are more easily identifiable and far more common than those arising from blending, clipping or the use of acronyms – *bullfrog* has what is termed 'semantic transparency', a quality lacking in *bulf* and *froll*. Semantic transparency can, however, be costly for the speaker. Had the speaker coined the word *bullploppyfrog*, she would have invested too much effort for too little gain in the additional information he provided.

Pavol Štekauer has emphasised the role of individual creativity in word formation. Since everyone has their own experiences, general knowledge, intellectual capacity, imagination, education, professional interests and existing lexicon, they will vary in the naming task, differing as to which features of the referent they wish to denote, the morphemes they select and the method by which these are combined.[28]

How are new words adopted within a language community?

The use of a new word by a speaker within a speech community is just part of the process of linguistic change. Why do some words become adopted and diffuse through the community, while others are largely ignored and soon forgotten? How does the process of diffusion take place?

For these questions we can invoke the same type of answers we found when looking at how semantic and sound change diffuse through speech communities. New words will be adopted because they make communication more efficient and effective. Those that do not, because they either involve too much effort or lack semantic transparency, will usually disappear from the lexicon. This will not always be the case because new words may be adopted as a social strategy, even if they have a negative impact on communication. A speaker may wish to align themselves with someone who is perceived of higher status or has other appealing traits that they wish to imitate. In my experience, young academics often adopt words that have been invented by their professors or other leaders in their discipline whom they wish to impress or imitate. Unfortunately, they are often used unnecessarily and incorrectly, making their lectures and publications too long and difficult to understand – the converse of efficient and effective communication.

Brexit is an interesting example.[29] This is a blended word arising from *Britain* and *exit*, originally coined in a blog of 2012 to refer to 'the withdrawal of the United Kingdom from the European Union and the political process associated with it'.[30] *Brexit* is certainly more economical than saying that phrase every time this matter is discussed, and this helps explain why this word spread so rapidly before and following the 2016 referendum. While the word has high semantic transparency at one level – exit from the European Union – the implications of that for the British people remained entirely unclear. People were free to load additional meanings onto the term which they often felt passionately about, such as 'the return of national sovereignty', 'an end to immigration', 'inevitable economic decline' or 'loss of freedom to work and live in Europe'. All of these may have been true to a greater or lesser extent but no one knew for sure, enabling the word to be widely adopted, with approbation by leavers and pejoratively by remainers. Theresa May, the soon-to-be British prime minister was only able to say that 'Brexit means Brexit'.[31] *Woke* appears to have a similar attraction to those of opposing views: some using it as a positive expression of being alert to injustice and discrimination, others as a negative term for those who are intolerant of right-wing values.

As with sound and semantic change, factors of age, gender and socioeconomic class will play a role in how new words diffuse. As does the wider cultural context. During the Covid-19 pandemic, several long-lost words in the English language returned, filling a need that had arisen. To many people the word *furlough* was a new word, despite it having been used in English from at least the eighteenth century, having been borrowed from the Dutch word *verlof* (leave of absence) and with roots in Proto-Indo-European. With the pandemic over, the word has virtually disappeared again.

Changing cultural attitudes to gender are becoming expressed in language, partly by deliberate acts of innovation. In English, the word *they* has become more widely used as a singular, gender-neutral pronoun, while in Swedish, the non-gendered pronoun *hen* has been added to the existing masculine *han* and the feminine *hon*. In France, *iel* has been proposed, a word that blends together the masculine il and the feminine *elle*, but this has gained little traction and has not been officially accepted by the Académie Française.[32]

Sound change

Just as the meanings of words are constantly changing, so too are the way they sound. You may have a sense of this from your own experience, noting how your pronunciation of certain words differs to how your parents had spoken, and how there are further changes by successive generations. We are readily aware that word pronunciation varies both between dialects of the same language and between individuals – some people speak louder, have a more precise articulation of syllables, and more variations in tone and rhythm than others. The listener also plays a key role in how pronunciation is perceived and then repeated; some people are more sensitive to variations in pitch, timbre, loudness and so forth, and better at repeating the sounds they have heard.

Sound change is both one of the oldest and most debated areas of study within linguistics.[33] Perhaps that is because of its inherent difficulty – there are no recordings of speech before the mid-nineteenth century so long-term change in pronunciation is difficult to prove. Fortunately, sound change is an ongoing process enabling short-term change to be studied in any existing language.

Within our speech communities, and those we may visit on our travels, there is always a pool of variability in how words are pronounced. Whether as children, adolescents or adults, our own pronunciation will be influenced by that of others.[34] Some of the variants will catch on and spread, notably those that save effort in speaking without compromising the ability of others to understand what we are saying, and those perceived to be of social value – helping to build the social relationships and projecting the identity we desire.[35] Some examples of sound change will be helpful.

We must first recall that the human vocal tract enables an immense diversity of sounds to be made, these providing the phonemes used as syllables to form words. Within known languages, the number of phonemes vary from as few as eleven to almost 150. They arise from how we manipulate our vocal folds, position our tongue in relation to our palates and teeth, and how we use our lips, to control the expression of air from our mouth. As we discovered in Chapter 5, a similarly flexible vocal tract was present 500,000 years ago in *Homo heidelbergensis*. It is likely that even earlier ancestors had the ability to make and use a diversity of sounds that we would recognise as vowels and consonants.

The way phonemes are pronounced is one aspect of sound variation and change that has been recorded in recent times. Another is how strings of syllables are pronounced within words, and the strings of words themselves – variation in where the stress is placed, in the pitch, tone and rhythm of the utterance. One noticeable change in Britain of the last two decades is the spread of upspeak (sometimes called uptalk). This is a rising pitch at the end of an utterance, which has long been used in Australia, New Zealand and California as a means of asking a question. It is now also used for making a statement, notably among younger people and especially women.

Sound shifts from becoming lazy

If upspeak is a recent development, sound change must have been occurring ever since words were first spoken. Its recognition came in the early nineteenth century by noting that the emergence of Germanic from Proto-Indo-European (PIE) involved changes in the initial consonants of many words. Because PIE is no longer spoken, its words are reconstructed by examining cognate words in its daughter languages, including German, Danish and English (from the Germanic branch), and French, Italian and Spanish (from the Romance branch that emerged via Latin). Reconstructing the PIE words is done by taking the most common form of the word in the surviving languages.[36]

This process of reconstruction identified that a widespread change had occurred in the shift from PIE to Germanic: the consonants at the start of words had changed. Several plosives were weakened to become fricatives: /p/ had become /f/, /t/ had become /θ/ = *th*, and /k/ had become /tʃ/ = *ch*. Moreover, several voiced consonants became unvoiced: /g/ had become /k/, /d/ had become /t/ and /b/ had become /p/. For instance, **pods* became **fot* in Germanic, which is now *foot* in English, while **grno* became **kurn*, with *corn* as the modern English equivalent. Because this sound change did not happen in Latin and the later Romance languages, English has systematic differences from French, Italian and Spanish. For instance, words spelt with an 'f' in English are frequently cognate to those spelt with a 'p' in French and Italian, such as *father* (père and padre), *fish* (poisson and pesce) and *few* (peu and poco).[37]

Why this sound shift occurred in the Germanic branch of Indo-European but not in the Romance languages remains unclear, as is precisely when it occurred – the best guess is around

500 BC. There is an explanation, however, for the cause of the change, one that we also noted when considering why words change their meaning: least effort.[38] It takes more time and energy to momentarily close the mouth to make a plosive than to narrow the airway to form a fricative; similarly, more energy is expended to vibrate the vocal folds to make a voiced consonant than one that is unvoiced. The savings in time and energy might appear marginal, but when we are speaking 16,000 words a day, and up to 200 words a minute, the tiny savings soon accumulate.

Another way to save time and effort is to shorten or entirely skip the ends of words – what would be the point if one is confident that the listener has already grasped the meaning of the word being spoken? The past tense of English verbs is often formed by adding an *-ed* at their end, as in *jumped, walked* and *bathed*. In the sixteenth century, the *-ed* was fully pronounced as in *jump-ed*, whereas today we have lost the vowel and say *jumpd*, maintaining the final /d/ to indicate that the word is past tense. In other cases, this 'sound erosion' leaves the potential for confusion. French verbs are a good example.[39] *Parler* means 'to speak'. When written down using the present tense, the word has five different endings reflecting who is speaking: *je parle, tu parles, il/elle parle, nous parlons, vous parlez* and *ils/elles parlent*. Today, there are only three different pronunciations for the five different endings, with 'parl' used for *parle, parles* and *parlent*. While possible confusion is avoided by use of the personal pronouns, that is not always the case in other languages. In Spanish, for instance, personal pronouns are not required and verb forms with different meanings are spelt and pronounced the same.[40]

An important type of sound change is known as the umlaut effect: the first vowel becomes more like the second. This is another example of the least-effort principle because the umlaut effect reduces the time and energy when saying a word with two vowels. Note that the word *umlaut* is also used for an accent over a vowel that has undergone such change. The umlaut effect explains why the plural of mouse is not mouses but mice.[41] In Germanic the word for *mouse* was **mu:s*, with the asterisk indicating this is a reconstructed word and the two triangular dots indicating the back vowel /u:/ is rounded and pronounced long, as 'oo'. To make the plural in Germanic, the suffix *-iz* was added, making **mu:siz*. The final *z* became lost by sound erosion, and then the umlaut effect caused the first vowel to become more like the second, creating **my:si*. Next, the final vowel was lost by sound erosion, leaving **my:s*, for which the modern English equivalent is *mice*. In this case the second vowel of **mu:siz* was a front vowel, /i/, and we refer to the second vowel, the /u:/, as having been fronted. In other cases, the first vowels can be raised, lowered or rounded, all depending on the following vowel, and all making pronunciation of the entire word a little easier and quicker. Similarly for *louse* and *lice*, *foot* and *feet*, *goose* and *geese*.

Vowel shifts as social tactics

Saving effort cannot, however, be the only reason to change the pronunciation of words. Language is a social construct: we modify the way we speak to build relationships and forge identities, even if that involves a little more effort when we speak.

There are periods and places in history which experienced an overall shift in the way people pronounced their vowels because of contact between speech communities of different dialects that stimulated such social strategies. The most renowned but least understood is the Great Vowel Shift (GVS) that occurred in England between 1400 and 1700, probably originating in London.[42] Although this involved some consonants becoming silent, as in *knight* and *enough*, its major impact was on long vowels. These became heightened – pronounced with the tongue being placed higher in the mouth; when the tongue was already as high as it could go, as with the back vowel /u:/ and front vowel /i:/, the vowels became diphthongs, /aʊ/ and /aɪ/ respectively.

Some examples. The word *bite* had been pronounced as 'beet' by having the long vowel /iː/; the GVS caused the /iː/ to become /aɪ/, creating its modern-day pronunciation. Similarly, the first vowel in mate had been hard, /aː/, and the final vowel long, /eɪ/ (as in 'ay'), causing the word to be pronounced as mah-tay. The GVS caused the first vowel to become long, /eɪ/, and sound erosion removed the final 'ay', leaving the pronunciation of mate to rhyme with eight. All these changes are related and can be described as a chain shift: as one vowel was heightened, it left a space for the one below to move into.

The Great Vowel Shift is often presented as a turning point in the development of the English language. It explains why today's pronunciation is so out of kilter with its spelling and why pre-sixteenth-century texts such as those of Chaucer are so difficult to read. It is curious, therefore, that there is little consensus on its cause.[43] Some argue that the GVS arose from an influx of migrants into southern England, having been displaced by plague, warfare and other turmoil of the fourteenth century. This led to a panoply of accents and dialects – a vast pool of variation in how words were spoken. The Londoners may have changed their pronunciation to differentiate themselves from the incomers, who in turn copied them causing an overall linguistic change. Others emphasise the role of French. This was a period when numerous French words entered the English language and French pronunciation may have been prestigious – the aristocracy were only just switching from speaking French to English. Consequently, the vowel shift arose as people wished to sound more like the French.

The Great Vowel Shift occurred between the fifteenth and seventeenth centuries but such vowel shifts are always occurring.[44] There are indeed many examples of similar vowel shifts, some underway in the present day. One of these is the Northern Cities Vowel Shift, or simply Northern Cities Shift (NCS). This has been ongoing across a swathe of the United States from New

York to Wisconsin since the start of the twentieth century. It gained pace and became subject to study from the 1960s, notably by William Labov who has contributed several seminal works on sound change through his long career.[45] The NCS began in cities such as Chicago and Detroit, with the heightening of the short vowel /ae/ and turning it into a diphthong so that words such as *cat* were pronounced as 'kay-it' and *trap* as 'tray-up'. The new pronunciations seem to require more rather than less effort to make. The vowel change led to knock-on effects on other vowels, so that *gosh* and *lock* turned into *gash* and *lack*.

Explanations for the NCS have similarities with those for the GVS. William Labov argues that the industrialisation of the Great Lakes region in the early nineteenth century attracted migrants from across the United States and Europe, resulting in dialect mixing.[46] The raising of the /ae/ vowel may have been to create a higher degree of similarity between the accents within the dialect mix, enhancing social cohesion. Others have emphasised the role of native German speakers in the Great Lakes region because their pronunciations already had aspects of the vowel shifts that later occurred. Another view is that white northerners may have been prominent in making the vowel shifts to differentiate themselves from the Black population.[47]

An ongoing and inevitable process

Being able to monitor sound change in real time enables insights into how new pronunciations spread through a speech community. As with changes in the meanings of words, factors of gender, age and socioeconomic class are relevant to sound change. Studies of the NCS have shown that sound changes are predominantly initiated in urban areas, primarily by the working class and especially by adolescents; young women often lead the way.[48] Speakers are unaware they are making such changes until they become established.

A study of working-class and middle-class adolescents in the same Detroit high school during the 1980s showed how the former were more ready to adopt the new vowel sounds.[49] This was partly because the social networks of the working-class students had greater engagement with the metropolitan areas of the city where the changes arose and where they hoped to gain employment. In contrast, the middle-class students aspired to transcend their local community, anticipating a move to a college for further study. The hostility between these two groups of students caused them to further highlight their differences by differentiating their word pronunciation.

Reference to the prominent role of women and adolescents in sound change returns us to the phenomenon of upspeak. This is rising intonation at the end of an utterance. It had once been restricted to asking a question but is now often used when making statements. Studies have confirmed that this is predominantly used by young women and is abandoned as they age. There are several competing explanations for its use including that it conveys uncertainty, is a means to indicate that the speaker has not finished, and is an invitation for the listener to intervene.[50]

Although I have provided only a few examples of sound change, these demonstrate it is a normal and constant feature of spoken language. Sound change may be more intense in some periods and regions than others, but that will be a consequence of socioeconomic factors rather than any inherent stability in the way we speak. Just like word meanings, the sounds of language change because of the nature of language itself, including the process of transmission between generations, the variations in how people speak and hear, and the social strategies they might employ.

That is the case for today and there is no reason to expect otherwise for the duration of the Stone Age. According to the linguist John McWhorter:

even if a language were spoken by a community myste-
riously condemned to live for millennia in a cave, under
staunchly unchanging conditions, after three thousand
years the language of that community would be vastly different from the one spoken when they were first herded
into the cave, and outsiders would most likely hear it as a
different language entirely.[51]

Words change their class and type

Having described semantic and sound change, we now need to
grasp the basic principles of a third process of language change
known as grammaticalisation. I stress 'basic principles' because
this process and the studies surrounding it are complex.

Grammaticalisation is the process by which lexical words
become either grammatical words or morphemes attached to
words that provide a grammatical function, known as grams. To
recap, words fall into two categories: lexical and grammatical.
The first are nouns, verbs, adjectives and adverbs. These word
types have content. Grammatical words lack content but help
define the meaning of lexical words and the sequences of such
words in spoken and written phrases. These are the articles, pronouns, prepositions, demonstratives, conjunctions and so forth
of language, the the-s, him-s, her-s, they-s and them-s, the in-s, at-s
and on-s, the this-es and that-s, the but-s, and-s, for-s and when-s.
The lexical–grammatical word distinction is not entirely strict:
some words, such as auxiliary verbs, fall midway between the
two, having their own meaning while also helping that of the
main verb.

Early nineteenth-century studies of language change came
to appreciate that grammatical words and structures arose from
lexical words, leading to the idea that the very first forms of language contained only words for objects and actions, the nouns
and the verbs. The term grammaticalisation was coined in the

early twentieth century and has been a major area of historical linguistics since the 1970s.[52] The following primarily draws on the 2007 study by Bernd Heine and Tania Kuteva called *The Genesis of Grammar*. I chose this because the authors have a keen interest in the earliest stages of the language evolution and propose a reconstruction by drawing on grammaticalisation as observed within recent language change. I also like the many examples they use from non-European and especially African languages.[53]

Processes of change

Heine and Kuteva have a broad concept of grammaticalisation, one that encompasses the conversion of lexical words from one class to another, such as from nouns to adjectives and to verbs. They identify four processes of grammaticalisation: *extension*, the rise of new meanings when words are extended to new contexts; *desemanticisation*, the loss of meaning, otherwise known as semantic bleaching; *decategorialisation*, the loss of properties that characterise lexical words, such as inflections for number, gender, case and tense; and *erosion*, the loss of sound segments and status as independent words. Overall, grammaticalisation has a direction of change, creating phrases with reduced lexical content but increased morphology.

These processes can be illustrated with a few examples. We have considered how nouns can become verbs when examining semantic change, so now we can see how they can also become adjectives. *Orange* and *bronze* are nouns referring to types of fruit and metal respectively but have become adjectives. In this role, they are used in different contexts (extension), lose aspects of their meaning (those unrelated to colour, desemanticisation), and can no longer be made plurals as they could when nouns (decategorialisation). In many languages, such as Swahili, the words for *father* and *mother* have similarly been grammaticalised to refer to male and female in general, being used as adjectives rather than nouns.

The transition of one noun, the Old English word *lic* that meant shape or body, has created a whole class of adverbs. *Lic* lost its lexical status but was retained as a suffix *-lice*, which through sound change became *-ly*. We now make adverbs by attaching *-ly* to adjectives, as in *quickly* and *loudly*. The ending *-lice* also gave rise to the modern word *like*, which we retain as a verb but also combine with nouns to make adjectives, such as *lifelike*.

Nouns have also given rise to adpositions – words used to express spatial or temporal relations, such as in *in, under, towards* and *before*. In English these usually come before the noun and are called prepositions; postpositions come after the noun. In many languages, nouns for parts of the body have become grammaticalised into prepositions. In English we use *to the back of . . .* and *in front of . . .* to refer to spatial locations rather than our body parts, these further illustrating the extension of context and loss of original meaning. Languages as diverse as Icelandic, Welsh, Kpelle (West Africa) and Imonda (Papua New Guinea) have independently developed their words for *back* (of the body) into adpositions that denote *behind*. Other body-part words have been grammaticalised in a similar manner. Words for *bowels, breasts, buttocks, faces, eyes, hearts, heads, mouths* and *necks* have all given rise to adpositions within a variety of languages.[54] A further development is a metaphorical extension of such words from the spatial to the temporal domain. In English, for instance, we refer to *back in time*.

The word *lot* is of interest because we use this as a noun, a pronoun and an adverb. The Old English word was *hlot*. This was eroded to *lot* and meant a portion of something assigned to someone. That remains one of its meanings, as in *a lot of land is for sale* and *that's your lot!* The word has become grammaticalised into a pronoun, so one might say *lots of land is for sale*, and also an adverb (known as a degree modifier), as in *a lot busier*.

These few examples illustrate how grammatical words have arisen from nouns, either in the recent or more distant past. Heine and Kuteva explain how nouns have given rise to not only adjectives, adverbs and prepositions, but also case markers (which assign properties to nouns, such as whether they are the subject or object of a phrase). Such case markers are usually affixes or clitics (words that have merged with an adjacent word, as in *we've*). Similarly, complementisers (words that introduce a complement clause, such as *if* in *I wonder if she will visit*) and subordinators (words that indicate the subordinate nature of a following clause, such as *although*) have also derived from nouns.

Grammaticalisation beyond nouns

Verbs, adjectives and adverbs also become grammaticalised. That of the verb *to go* is another illustration of a shift from space to time. *To go* refers to movement but can also be used as a marker for the future tense, as in *I am going to write this book* – a shift from a reference to space to that of time. In this context, *going to* is sometimes eroded into *gonna*. Similarly, the Old English verb *willan* that meant *to want* gave rise to the auxiliary verb *will*, which we also use to express future tense, as in *I will write this book*. Verbs meaning *want*, *come to* and *go to* have turned into future markers in many languages throughout the world.

The grammaticalisation of verbs has created demonstratives (words that point to and can replace a noun, as in *this*, *that*, *these* and *those* in English); negations (words that reverse the meaning of a following word or phrase, as in *no*, *not* and *never*); tense markers (as in *going to*); and aspects (words or affixes used to express the temporal nature of an event or state of the supported verb). The verb *keep*, for instance, has been grammaticalised for a durative aspect as in *he kept complaining*.

Once grammaticalisation has created demonstratives,

prepositions and so forth, these words themselves become subject to further grammaticalisation. Demonstratives, pointing words, have given rise to pronouns, such as *me, you* and *him*. Latin has three principal demonstratives. When referring to male nouns these are *hic* (*this*), *iste* (*that*) and *ille* (for something in the distance, lacking an equivalent word in English). These map onto the pronouns *-me* (pointing to myself, this) *-you* (pointing to another person present, that) and *-him* (pointing to a third person not present).[55] The Latin *ille* (masculine) and *illa* (feminine) provided the source of two of the third person singular pronouns in French, *il* and *elle*.

Demonstratives have also led to definite articles. The Latin word *ille*, and its feminine form *illa*, were semantically bleached (removing the reference to location) and eroded by losing its beginning to form the French words *le* and *la*. Similarly, in English, *the* is an eroded and bleached form of *that*. Such creation of definite articles from demonstratives is common in many languages, either creating an independent word used before the noun, as in English, or creating an affix for the noun to indicate its definiteness.

Phrases rather than lone words can become grammaticalised. The preposition *beside* is derived from the grammaticalisation of the phrase *by the side of* as used in Middle English (*c.*1150–1450). This also involved a semantic change with *beside* coming to mean *outside*. Hence, someone who is *beside themselves* is *out of their mind*. Similarly, *because* is derived from *by the cause of*, and is sometimes further eroded to *'cos*.

It is not only words and phrases that evolve incrementally through time via the repeated transmission from one generation to the next, but also the way they are ordered, known as syntax.[56] As described in Chapter 8, this has been demonstrated by the automatic changes that arise in artificial languages when they are transmitted from generation to generation in a computer simulation.[57] It has also been observed in real life. When the development of new sign languages within deaf communities has been monitored, syntax emerges incrementally through time.[58]

Such incremental development has been proposed for spoken language, notably by the distinguished linguists Jim Hurford and Ray Jackendoff, who have both written extensively about the origin and development of syntax and grammar.[59] They propose that syntax develops incrementally, with each addition enhancing the communicative efficiency of language. To do so, syntax would need to be evolving hand in hand with the lexicon that provides the grammatical words and inflections that help to join phrases together. It was once assumed that incremental evolution of syntax was the outcome of a biological process as natural selection constructed specialised linguistic processes within the brain that support syntactical structures. We now recognise that syntax is likely to be no more than the outcome of cultural evolution – the transmission of language from one generation and one person to another, entirely reliant on general-purpose learning processes.[60]

A slow and gradual process

None of this happens overnight: grammaticalisation of words and syntax is a slow and gradual process. It occurs as chains of stepwise developments, often referred to as clines, that are common across languages. A typical cline consists of a lexical word, which becomes a grammatical word, then becomes a clitic and finally no more than an affix.[61] For instance, the Old English verb *willan* (*to want*) became a modern English auxiliary verb (as in *I will see you later*) and then became a clitic (as in *I'll see you later*). Hypothetically the next stage is for the clitic *'ll* to become a fused suffix (*I seell you later*), as has occurred within the same cline in other languages.[62] Some linguists have suggested a further step: that the gram (the clitic or affix) will disappear altogether. Ancient Egyptian, for instance, appears to have changed between relying heavily on morphology to a reliance on word order and then back to morphology – a linguistic cycle.[63]

How long does each step, the cline or a whole linguistic cycle take? Frustratingly, this is one of the least researched areas in linguistics. Studies of pidgins (relatively simple languages developed between people who do not speak the same language) and creoles (a fully developed language arising from the mixing of two or more languages) suggest that new grammatical elements can arise within less than a century. That is swift compared to Guy Deutscher's discovery that it took nearly 2,000 years for a quotative marker, a grammatical device to indicate quoted speech (such as *he said* "I love you"), to arise from a multi-word clause that had introduced such speech in the ancient Akkadian language (spoken between c.3000 and 700 BC). In this case we must be cautious because Deutscher inevitably relied on written texts which are known to be slower to change than spoken language. Once grammatical elements have arisen, their adoption can be swift. A study of spoken English among Canadians in their twenties found that the use of the quotative *be like*, as in 'I'm like, "I don't know"', increased from 13% of all types of quotatives in 1995 to 58% of them by 2002.[64]

Another under-researched question is how grammaticalisation occurs within a speech community. The changes in Canadian English were led by adolescent women; men also played an active role but did so at a slower rate. Studies of recent grammatical change in France have focused on how negation is made – called the *ne* deletion.[65] While the traditional French way of making a negative is to surround the verb with *ne . . . pas*, this is becoming croded by removing the *ne*. A study of Parisian speakers in 1975 showed this was led by younger women, with the highest deletion rate among those defined as lower middle class, and the lowest among those of the upper middle class. By 1990, those under fourteen years old in Paris did not use *ne* at all, indicating that this process of grammaticalisation had run its course in the capital. The *ne* deletion will likely become universal within spoken French, and written French will catch up some decades later.

Socioeconomic class has influenced the adoption of the *-ly* suffix to create adverbs. This spread among the literate social ranks in England between the fifteenth and seventeenth centuries. The word *extremely* came into use in the seventeenth century and was adopted primarily by the higher social class, and by women rather than by men; merchants and other social ranks below the gentry used extremely only rarely. Similarly, a study of Scottish English found that *-ly* adverbs were used by middle-class speakers much more frequently than by those of the lower class.

Drawing these few strands of evidence together, we can conclude that factors of age, gender and social class influence how grammaticalisations arise and diffuse through speech communities. Moreover, it is reasonable to assume that the motivations for grammaticalisation are the same as those for semantic and sound change: to communicate efficiently and effectively, reducing the effort in speaking while ensuring listeners understand what is being said.[66] Repetition within the speech community will cause new innovations to spread, become fixed and potentially obligatory.

Grammaticalisation in evolutionary time

The quest for efficient and effective communication provides a continuous driver for new grammatical forms. Heine and Kuteva propose that there are up to six layers of grammaticalisation with the process ongoing in all languages. At the base are nouns (layer 1); these lead to verbs (layer 2) and then adjectives and adverbs (layer 3). At layer 4, demonstratives, prepositions, aspects and negations arise; these are joined by pronouns, definite articles, case markers, complementisers, tense markers and relative clause markers (level 5), and finally by further grammatical functions that allow construction of hierarchical phrase structures and recursion (level 6).

Syntactical structures would have to be evolving hand in hand with the changes in word form. The linguists Hurford and Jackendoff agree with Heine and Kuteva that the first words would have been nouns and verbs. Such words would have been used in isolation from each other with their meanings dependent on their individual sounds alone. As such, there was no syntax. A second stage, encompassed by Heine and Kuteva's levels 3 and 4, would be the stringing of two or more words together, with prosody indicating they comprise a single phrase. Their combined meaning would then be reliant on the emergence of simple syntactical rules such as Jackendoff's suggestions of 'Agent first' or 'Focus last'. These are the types of rules that can automatically arise with computer models of language transmission. The third stage fits with Heine and Kuteva's levels 5 and 6. This involves the combination of phrases into a hierarchical structure, enabled by the emergence of grammatical words.

Although the successive layers of grammaticalisation have been inferred from the recent history and ongoing change of languages throughout the world, Heine and Kuteva suggest it may provide an evolutionary sequence for the emergence of language itself. They propose that the very first language – entirely unspecified in time, place or hominin species – consisted of nothing more than words referring to objects, the utterance of a single noun or a string of such words. At a later, equally unspecified date, nouns were joined by verbs, and then by adjectives and adverbs, and finally by all the grammatical forms of layers 4, 5 and 6 of their scheme – recognising that no language need to have gone through all six layers.

Whether or not one agrees with these six layers of grammaticalisation, the notion that the first language consisted of words with limited, if any, grammatical structures is widespread. It comes in several different forms. When concluding his 2001 overview of linguistic history *The Power of Babel,* John McWhorter suggests that the first language had words that could be used in different word classes – employed as either nouns, verbs or adjectives; he doubts there were any grammatical markers such as affixes and suggests there was limited if any marking of tense, with context taking care of most needs.

Guy Deutscher similarly concludes his 2005 book *The Unfolding of Language* by speculating about the first language. He suggests this likely had a clear divide between words for things and actions (avoiding calling these nouns and verbs). He refers to this as a 'Me Tarzan' stage of language evolution – an unfortunate phrase to use in an otherwise serious study of language change. Putting that gripe aside, his early phase of language is one in which thing words and action words are simply strung together – Heine and Kuteva's level 2 of grammaticalisation. He suggests a story that might have been recounted at some unspecified date in the past (also exposing outdated stereotypic views of the past):

> *Girl fruit pick . . . turn . . . mammoth see*
> *Girl run . . . tree reach . . . climb . . . mammoth tree shake*
> *Girl yell yell . . . father run . . . spear throw*
> *Mammoth roar . . . fall*
> *Father stone take . . . meat cut . . . girl give*
> *Girl eat . . . finish . . . sleep*

We can understand this story, but it is not expressed in the manner we would today: it lacks detail, description and eloquence; it is open to various interpretations; it is neither efficient nor effective communication. Deutscher provides a modern version:

> A girl who was picking fruit one day suddenly heard some movement behind her. She turned around and saw a huge

mammoth charging straight at her. She ran to the nearest tree and climbed up it, but the mammoth shook the tree so roughly that the terrified girl started yelling hysterically. Her father, who heard the loud screams coming from the forest, realized that his daughter must be in danger, so he grabbed his spear and ran towards her. He threw his spear straight at the mammoth, which led out a blood-curdling roar and fell to the ground. With a sharp stone he cut some pieces of meat for the girl, who ate them up before falling asleep.

This version has the grammatical words and functions that arise in levels 3 to 6 of Heine and Kuteva's hierarchy: adjectives, adverbs, pronouns, prepositions, demonstratives, aspects and tense markers, along with subordinate clauses. Deutscher shows how that version could have arisen by the gradual, step-by-step, piecemeal processes of grammaticalisation, that he refers to as the 'forces of creation', as the same and similar stories were retold across many generations. Quite how many generations and how long this transition from Heine and Kuteva's level 3 to level 6 would have taken is assiduously avoided. As is another key question: where did those original words, those for things and actions, come from?

Deutscher offers no suggestions. Indeed, he is dismissive of ever being able to answer this question. He says we must take it as given that language already used arbitrary signs to refer to objects and actions; that these were conveyed vocally by using vowels and consonants, and that meanings were shared within a speech community.

Don't despair

That bleak outlook arises because Deutscher hadn't spotted the other pieces and fragments of the language puzzle, some lying elsewhere in linguistics and some in other disciplines – archaeology, psychology, genetics and so forth. Throughout this book I've been collecting them, chapter by chapter, and the table is now getting crowded with evidence. This chapter has provided a particularly large and clearly defined fragment of the puzzle: that the meanings of words, how they sound and how they are used are constantly changing. I have documented this primarily in English and with examples spoken in recent times. There is no reason to doubt that the principles we have found, notably the drive for efficient and effective communication and the use of words as a social strategy to define individual and group identity, extend into the distant past, perhaps from the time that words were first spoken.

We now have just two more fragments of the language puzzle to assemble before we can fit them together and discover what they reveal about the evolution of language. The next asks a fundamental question about language, challenging an assumption that has so far pervaded this book: is language used only for communication?

14

LANGUAGE, PERCEPTION AND THOUGHT

Words and language provide a means of communication. They also influence the way we perceive and think about the world. The extent of that influence is heavily contested: to some academics it is self-evident that speakers of English, Mandarin and Hindi experience the world in different ways, even before they start thinking about it. That is what bilingual speakers express, tending to say they feel like different people when using their different languages.[1] To others, language is merely a veneer over universal processes of perception and thought. A third group – to which I am aligned – is sympathetic to a cognitive role for words but requires empirical evidence. Experimental tests undertaken throughout the twentieth century to find such evidence had inconclusive and sometimes contradictory results. During the last two decades, however, a new level of experimental sophistication has provided convincing evidence for the impact of words on perception, while the case for words as a tool for thought is compelling. As such, the nexus between language, perception and thought provides a critical fragment of the language puzzle.

The notion that language influences thought, perception and one's world view in the most general sense reaches back to at least the eighteenth century, notably in the works of Johann Gottfried Herder (1744–1803) and Wilhelm von Humboldt (1767–1835).[2] Herder and Humboldt influenced the anthropologist Franz Boas (1858–1942), a German-born American who pioneered modern anthropology. One of his many interests was the languages of Native Americans and how these influenced their culture and thought. He proposed that 'the categories of language compel us to see the world in certain definite conceptual groups which, on account of our lack of knowledge of linguistic processes, are taken as objective categories, and which, therefore, impose themselves upon the form of our thoughts'.[3] As such, language provides the foundation for cultural and cognitive diversity.

By 'conceptual groups' – or simply concepts as we would call them today – Boas was referring to bounded units of knowledge stored in long-term memory that provide the building blocks for thought. Their significance for learning language and their association with words was considered in Chapter 9. Let's remind ourselves of their core features.

Concepts include the following: physical entities in the world, such as dogs, tables and chairs; feelings, such as sadness and anger; relationships, such as friends and enemies; and abstract ideas, such as freedom and betrayal. Our words are labels for concepts held within our minds, while concepts are the mental representations of categories, the members of which share several essential features. Without such categories, our minds would be overwhelmed and incapacitated by having to recall facts and figures about every single item in the world. Categories are hierarchically organised: the base category will contain several subordinate categories and may be an element of a more general superordinate category. The category of 'dogs', for instance, contains terriers, working dogs and toy dogs, while being part of mammals and, at a higher level, living things.

Categories tend to have prototypes, which are their most typical examples, such as a robin for the category of bird.[4] Our concept of dog is typically formulated around a prototype of a hairy animal with four legs, a tail and a bark, even though individual members of that category might lack one or more of these features. Concepts often have fuzzy boundaries: is a tomato a type of vegetable or fruit?

Some concepts are universal, possessed by all humans, while others are culturally specific. The former relates to aspects of the world that are of such long-standing importance that they have become embedded into our DNA, saving us the bother of learning about them, although we may need to give them names. Knowing that things fall when dropped, for instance, appears instinctive – we are a born with the concept of gravity, although without a word for it.[5] Others will relate to recurrent features of the world. Most people in the world today will have a concept of a telephone, although there is nothing in our DNA to automatically place that concept into our minds.

Regarding the biologically based concepts, we are born with an intuitive understanding of the physical world, with our minds already possessing concepts of gravity, inertia and solidity, although we have no words for such phenomena.[6] Similarly, we have an intuitive understanding of the difference between living things and inanimate objects.[7] Young children evidently have ready-made concepts about living things because they are compelled to attribute each with an essence – a horse remains a horse even if it is put into striped pyjamas to look like a zebra.[8] Children attribute mental states to other people when seeking to explain their actions, indicating they have concepts of beliefs and desires.[9] Initially, these may be their own mental states, but after the age of four, they understand that people can have beliefs and desires that differ from their own.

These are the categories of domain-specific knowledge for which the brain has functionally specific regions connected by long-distance neural networks, as we considered in Chapter 11. The innate concepts they provide enable us to function in the world and begin the process of learning by which further concepts are acquired. Franz Boas suggested this occurs via the words we learn. But what did he mean by writing that 'the categories of language compel us to see the world in certain definite conceptual groups'?

Consider the words for caribou in Inuktitut, the language spoken by Inuit caribou hunters of the Qikiqtani Region in the Canadian Arctic and those used by the English-speaking biologists who co-manage the herds.[10] Both types of speakers have a concept of caribou as a distinct type of animal, this known as *tuktuit* in the Inuktitut language. At the next level, however, the Inuit and English speakers differ in their subordinate caribou categories / concepts. The Inuit distinguish between *iluiliup tuktuit* (inland caribou), *kingailup tuktuit* (island caribou), *qungniit* (reindeer) and a mixture of *iluilip* and *kingailup tuktuit*. These names reflect the places and times that they hunt (or once hunted) caribou, emphasising the human relationship with the animal. In contrast, the biologists name specific herds according to where they calve, such as the Wager Bay herd, the Melville Island herd and the Bathurst herd.

These contrasting ways of categorising caribou led to challenges in developing management plans because they provided different views of how the world is structured. The challenges were overcome only when the native Inuit and the biologists learned a little of each other's language, engaging in what Boas called 'knowledge of linguistic processes'. By doing so, the Inuit and the biologists came to appreciate how they were categorising the caribou in different ways, with the Inuit becoming more conversant at switching between the different schemes than the English-speaking biologists. Had the two groups never met each other and remained unaware of each other's language, they might never have appreciated the different ways in which caribou can be categorised. Both may have taken their words for caribou-types as objective categories / concepts.

Linguistic relativism

Boas' ideas were further developed by his student Edward Sapir (1884–1939). Writing in 1929, Sapir stated that 'we see and hear and otherwise experience very largely as we do because the language habits of our community predispose certain choices of interpretation'.[11] Benjamin Lee Whorf (1897–1941), an engineer with an interest in linguistics and an acolyte of Sapir, further developed such ideas, again working primarily with the languages of Native Americans. Their views have been combined into what is now known as the Sapir–Whorf theory of linguistic relativism. This is summarised in a famous passage written by Whorf that is useful to quote at length (including his use of capitals to stress a phrase):

> We dissect nature along lines laid down by our native language. The categories and types that we isolate from the world of phenomena we do not find there because they stare every observer in the face; on the contrary, the world is presented in a kaleidoscope flux of impressions which has to be organized by our minds – and this means largely by the linguistic systems of our minds. We cut nature up, organize it into concepts, and ascribe significances as we do, largely because we are parties to an agreement to organize it in this way – an agreement that holds throughout our speech community and is codified in the patterns of our language. The agreement is, of course, an implicit and unstated one, BUT ITS TERMS ARE ABSOLUTELY OBLIGATORY; we cannot talk at all except by subscribing to the organization and classification of data that the agreement decrees . . . We are thus introduced to a new principle of relativity, which holds that all observers are not led by the same physical evidence to the same picture of the universe, unless their linguistic backgrounds are similar, or can in some way be calibrated.[12]

And so began one of the longest-running debates between linguists, anthropologists and philosophers – the idea that the language we speak influences the way we perceive and think about the world.

Following initial enthusiasm, by the 1970s this idea had become a bête noire of the academic establishment.[13] Despite being softened from linguistic determinism to linguistic relativism, Whorf's view had its obituary written by Steven Pinker in his 1994 book *The Language Instinct*, who simply stated that 'it is wrong, all wrong'.[14]

Pinker certainly found valid flaws in Whorf's work, criticising him for making naïve written translations of Native American spoken phrases that couldn't but help imply different ways of thinking from those of an English speaker. Little credence can be given to Whorf's ideas that the structure of language, its word classes and grammatical constructions, influences the way one thinks.[15] That does not, however, mean that it is 'all wrong', and linguistic relativism has undergone a renaissance during the last decade.[16] There has been a host of ingenious experiments that demonstrate – conclusively to my mind – that the words we use influence the way we perceive the world.[17]

Experiencing space, time and gender

One set of experiments explored the impact of the way people talk about space. Some languages use phrases that employ relative terms such as 'in front of' and 'behind' whereas others predominantly use absolute terms such as 'to the north of', or refer to fixed objects, such as a notable mountain. In the Australian language of Guugu Yimithirr, for instance, people sitting in a room would be described to the east or west of each other rather than being in front or behind.[18] Cleverly designed experiments involving non-linguistic tasks have shown that these different ways of talking about space influence the way people think and act in the world.[19] One of these compared spatial cognition between a group of Dutch-speaking children, who predominantly use a relative frame of reference for talking about space, and a group of Khoekhoe-speaking children from a hunter-gatherer community in Namibia, who predominantly use an absolute frame of reference. The experiment involved placing objects on a table and asking the participants to remember how they are arranged. The objects were removed and the participants rotated and/or moved in space before being asked to reconstruct the arrangement of objects. Their linguistic frame of reference for talking about space was found to influence how they rearranged the objects on the table.

Other experiments have explored time.[20] English speakers talk about time using spatial metaphors, employing terms such the 'good times that are *ahead* of us', referring to 'events that happened *before* the present', and noting how 'time is *rushing by*'. Speakers of Mandarin do likewise but also use vertical metaphors – earlier events are said to be *shàng*, or 'up', and later events to be *xià*, or 'down'. An experiment undertaken by the psychologist Lera Boroditsky primed the thinking of a sample of speakers of both languages by providing them with images of spatial relationships between objects. Some of the images had objects arranged horizontally while in others they were placed vertically. The experiment found that when both English and Mandarin speakers were asked questions about time using 'before/after' words, they could respond faster if they had been primed with images of horizontally rather than vertically arranged objects, reflecting the use of horizontal metaphors in the way they speak about time. When asked questions using 'earlier/later' words, Mandarin speakers were faster if they had been primed with images of vertically arranged objects, but this had no beneficial impact on English speakers – in fact, it slowed them down.

The impact of gender labels has also been explored: do the

arbitrary genders attributed to inanimate objects in languages such as French, German and Spanish influence the way in which their speakers perceive, recall and think about those objects?[21] Boroditsky and her colleagues compared how German and Spanish speakers think about bridges, which are feminine in German, *die Brücke*, but masculine in Spanish, *el puente*. When asked to describe bridges, they found that German speakers tended to select words that are stereotypically associated with women, such as beautiful, slender, elegant and pretty, while Spanish speakers chose words we often associate with men, such as big, strong and dangerous. In a further study Boroditsky and her colleagues presented the speakers with several objects, each of which was given a boy's or a girl's name. The participants were given time to memorise the names of each object and then tested on their recall. They found that when the name agreed with the gender of the object, the name was easier to recall than when it differed. This is a particularly interesting study because it did not require the participants to say the type of object that was being named and hence cite its linguistically defined gender when doing so. This suggests that male or female associations sit within the mind even when the names of objects are not being spoken.

The extent to which these experiments lend credence to Whorf's statement that 'we dissect nature along lines laid down by our native language' may remain contentious. Space and time are abstract concepts, while gender-varying inanimate objects such as bridges are part of culture rather than nature. Would we find the same when dealing with matters closer to nature, those in which our human physiology and sensory experience are involved, such as our emotions and the way we move our bodies?

Feeling emotion and talking about locomotion

We experience emotions as different bodily sensations, such as a beating heart, sweaty palms, a knot in the stomach, euphoria and so forth; we recognise emotions in others by their facial expressions, posture and behaviour. A prominent idea, championed by the psychologist Paul Ekman, is that we are born with a fixed set of basic emotions that are universal within our species, notably happiness, sadness, fear, surprise, disgust and anger.[22] Just as we attach the word gravity to our intuitive understanding about how objects move through space, we simply attach words to each of these innate and universal emotions once those words become available – there is no role for learning, individual experience or cultural variation.

An alternative view is that we make sense of the sensations we feel and the facial expressions we see only when we attach words to them – we develop rather than inherit our emotional concepts.[23] Key evidence is that children are unable to categorise facial expressions as representing different emotions until they have acquired a lexicon of words for emotions. Before having such words, faces that we might view as angry, sad or fearful are all categorised together as 'unpleasant' – a concept that is likely to be innate. By acquiring the words for different types of emotions while experiencing sensations or observing their expression in others, we develop a set of concepts into which those feelings can be placed.

Children will differ in the number and range of emotion words they learn, depending on the emotional intelligence of their parents and others from whom they learn language. Rather than just using the single word *happy* to describe their own and their child's emotional states, a parent might use *cheerful, joyful, jovial, euphoric* and so forth – all with subtly different types of meanings. The child acquiring such words will have a wider range of concepts into which their own sensations can be placed than will a child of a less emotionally articulate parent, especially one suffering from alexithymia – a difficulty in identifying, understanding and expressing feelings. Not surprisingly, alexithymia appears to be transmitted across generations: a child who fails to develop a multi-concept emotional understanding of their own sensations will lack the words to facilitate such understanding in their own offspring.

The significance of words for shaping our emotion concepts becomes further apparent when we appreciate how languages differ in their words for emotions.[24] Polish, for instance, does not have a word that perfectly corresponds to the English word *disgust*, and English lacks a word that fits to *tęsknić* in Polish, which can roughly be translated as being homesick – but not in the way that an English speaker would think of it. Russian separates *anger* into two types: *serdit'sia*, anger at a person, and *zlit'sia*, anger for more abstract reasons, such as about climate change; English speakers use the same word for both. The Australian language Gidjingali has a single word that refers to both fear and shame, emotions that are kept quite distinct in the English language.

While children who grow up speaking English, Polish, Russian and Gidjingali are likely to experience the same range of sensations, these will be dissected and understood in different ways because their languages have different words that shape different – although overlapping – concepts in their minds. As Franz Boas explained, without knowledge of other languages these concepts would be taken as objective, universal categories. That seems to be what Paul Ekman assumed, not questioning why his universal emotions should fit so perfectly onto the words used in the English language.

Words for complex emotions such as shame, embarrassment, remorse and jealously tend to be language specific; such emotions also lack distinctive sensations. Consider shame.[25] Can I feel shame without having a word for shame? Yes, but it would be ill defined – an unpleasant feeling that I would struggle to place within any of my existing emotion categories and hence reflect on and communicate to others. Could I distinguish between my feelings of shame and those of embarrassment, humiliation and guilt without having separate words for them? Could I identify these in others? Probably not. While shame has been described as a universal emotion, there is considerable cultural variation in its meaning, expression, prevalence and salience. Mandarin, for instance, has more than 100 words for different types of shame – those words providing a fine-grained partition of this otherwise ill-defined concept.[26]

We need to ask a fundamental question: does having so many words for shame, or a different set of words for emotions in general, really make a difference to perception, thought and behaviour? Or is linguistic diversity no more than a superficial cultural veneer, one that masks a higher degree of unity in conceptual knowledge than variation in words implies? After all, we use more than words to learn about the world: we directly interact with it by sight, touch, smell, hearing and taste. Those senses have taken hundreds of millions of years to evolve. Is it likely that words, 6 million years old at most and probably much younger, could so easily override these other methods of learning?

Probably not, according to an experiment that found that concepts and words about locomotion can be overridden by the actual experience of watching movement. This study considered words for different types of human locomotion as used in four languages, English, Dutch, Spanish and Japanese.[27] Each language dissected human locomotion in different ways. English had the largest number of terms (fourteen, including *walk*, *run*, *jump*, *stomp* and *shuffle*) and Spanish the least (five: *caminar*, *correr*, *marchar*, *saltar* and *trotar*), with no precise correspondence of words between the languages. By asking the speakers of these languages to make similarity judgements between video clips of different types of locomotion, the researchers found the participants' understanding of locomotion had a higher degree of overlap than their varied word inventories would imply. Such understanding – their concepts – seemed to be at least partially tied to their visual and biomechanical experience of locomotion, quite independent from the words they used.

Seeing the world in colour, and being half right

The view that words are just one of several influences on perception of the world matches a recent finding from the most explored and contested aspect of linguistic diversity: colour terms. Languages vary in their number of words for colours and how they map onto the colour spectrum as defined by physics. English has eleven basic colours: black, white, red, green, yellow, blue, pink, grey, brown, orange and purple – and then a myriad of infrequently used other terms such as vermillion, ochre and azure. The Papua New Guinean language Berinmo has only five colour terms and the Amazonian language Tsimané only three, these corresponding to black, white and red, while the Pirahã language only categorises colours as 'light' and 'dark'. Many languages, including Japanese, Korean and Old Welsh, collapse blue and green into a single colour, named as *ao*, *peruda* and *glas* respectively, these words translating literally as 'grue' in English.

There have been decades of ingenious experiments seeking to decide whether the colour one perceives is influenced by the colour words within one's language.[28] The principal experimental design is known as 'categorical perception' – the ability to discriminate between two colours. If language influences perception, one should be quicker at distinguishing between colours that fall either side of a linguistically defined colour boundary, such as between blue and green in English, than between colours within a category, such as different shades of red. A Japanese speaker should be slower than an English speaker at distinguishing between blues and greens because there is a single colour category word for these in Japanese.

Experiments built on this basic design have come to diametrically opposed conclusions, providing support to both universalists, such as Paul Ekman and Steven Pinker, and relativists who follow Whorf's view about the significance of words. The most recent research, however, has found that both are 50 per cent correct.[29] These experiments have differentiated between colour perception in the right field of vision and that in the left, their design relying on neural networks for language being predominantly found in the left hemisphere of the brain, which also takes control of the right eye. Consequently, if our words influence colour perception this would be more evident in the right field of vision than in the left. A range of experiments, some monitoring what is happening in the brain while subjects undertook discrimination tasks, have shown this is the case. A review concluded: 'It now appears uncontroversial that once language is learned, its categories shape perceptual discrimination primarily in the left hemisphere/right visual field and less so, if at all, in the right hemisphere/left visual field – a specific and unexpected sense in which Whorf was half right.'[30]

'Half right' seems a reasonable conclusion about the overall idea of linguistic relativism. As evolved biological organisms it is reasonable to conclude that we have many concepts embedded into our DNA that become manifest as we mature and to which we attach words. The words might tweak the content of those concepts but are unlikely to override the biological imperative to have a close match between our concepts and the nature of external reality. An English and a Gidjingali speaker may have a different concept of shame, but their sensations and resulting behaviour may have greater similarities than their words imply. Similarly, any concepts we build from words will be tempered by other sources of knowledge and are unlikely to cause us severe disadvantage.

If words gave us concepts that compromised our ability to operate in the world, those words are unlikely to become widely adopted and remain within a language. A hunter-gatherer reliant on acquiring plant foods to eat and who only has words for light and dark, must nevertheless be able to identify the redness of a ripe fruit and that of a poisonous fungus – perhaps by using their left eye only! Although the Dani people of the Central Highlands of Western New Guinea have only two colour terms in their language, they had no trouble learning the English set of colour categories.[31] Moreover, if words are leading us astray, we have strategies to bring our concepts back into line with the world: we either adopt new words, as did the Inuit caribou hunters, or change the meanings of those we possess.

An implication of this view is that words will differ regarding their impact on our perception and thinking. Those concerning abstract concepts which have no physical representation in the world and hence can only be learned from other words, such as *democracy*, *freedom* and *justice*, will have most impact on cognition.[32] Imposing the word *democracy* onto a non-English speech community that lacks other words to define its meaning will fail to convey the concept of the word. Because abstract words are speech-community specific, they make a substantial contribution to cultural diversity in the world – those speaking different languages will have different ways of perceiving and thinking about the world. In contrast, concepts that can be learned by multiple senses, such as types of weather – one can hear, feel, see and smell rainfall – will be least influenced by the words being used. Different languages may have different words for rainfall but it is unlikely they induce a different way of perceiving rain.

Words are tools for thought

Consider why we talk to ourselves, either out loud or inside our heads with what is known as inner speech. Rather than using words to communicate our thoughts to someone else we are using private speech to formulate those thoughts in our minds, thoughts we might never wish to tell someone else about. We also use private speech to shape our actions by talking ourselves through a complex task.[33] A well-rehearsed example is learning to tie shoelaces. This often begins by an adult providing verbal instruction; when the adult is absent, the child may repeat the instructions to themselves, either silently in their head or vocally, as means to remember what to do, focus attention, guard against errors, and complete the task.

The use of private speech by children is predominantly directed to their own actions and increases in frequency when they are alone and trying to master a complicated task. Those who make the most self-directed comments are the best at mastering the task. Private speech is not restricted to children. I use it all the time, both out loud and inside my head, to guide me through tasks I find difficult, such as composing this sentence.

It is a matter of philosophical debate as to whether our streams of inner speech constitute thinking itself. Our silently spoken words might bring our concepts to consciousness, such that inner speech itself can be considered as a type of thought.[34] That would be a supplement to most of our thinking that is undertaken without any awareness and without the use of words, referred to by psychologists as mentalese.

Whether or not we think in words, they certainly augment our thought.[35] While I might feel uncomfortable sensations after I have done something wrong, by attaching the word *shame* to those feelings, I can reflect on them, recalling what they were like, which may deter me from repeating such behaviour. Moreover, I can talk about those feelings to others, which is itself a form of thinking because it helps me to understand the cause and nature of those feelings.

More generally, simply the act of labelling items in the world, whether they are sensations, material objects, actions or abstract ideas, makes them salient and concrete.[36] By attaching words, we enhance our ability to remember them and target our thoughts on them; we can compare one item with another, by which we begin to establish relationships, and learn about the world.[37] As such, the language we speak influences how we remember the past.[38] Unlike English, Spanish has words for inside corners and outside corners, and consequently speakers of Spanish have a better memory for where items are present in a display. When speaking about accidental events, those using Spanish are less likely than English speakers to mention who caused the accident. A phrase such as *Se rompió el florero*, literally *the vase broke*, is an entirely natural construction – the person responsible does not need to be mentioned. A consequence is that when remembering past accidents, they are less likely to recall the person(s) involved than an English speaker. Not surprisingly, languages label those items which are most significant for their speech community's ways of life, those which their members need to remember and think about.[39]

Most importantly, we can chunk labels together, into a single new label/concept which enables us to use our powers of thought to their greatest effect. The linguist Stephen Levinson nicely summarised this: 'We don't have to think about a *hundred* as "ten tens" when doing mental arithmetic, or *aunt* as "mother's sister, or father's sister, or father's brother's wife, or mother's brother's wife" when greeting Aunt Mathilda.'[40] The chunking of labels, and then the chunking of those chunked labels, is a key means for complex concepts to develop. A long chain of chunking has taken us from thinking about stones and bones to thinking about black holes and human evolution.

We can also use words to invent concepts that have no physical manifestation in the world – concepts that can only be defined by using other words. These are the words/concepts that we describe as abstract in contrast to those that are concrete words/concepts, referring to objects and actions that can be experienced via our senses, by sight, touch or smell. How can ideas about morality and infinity ever arise unless we have words to build, define and describe them?[41]

We cannot, of course, develop such ideas by ourselves. Words augment thought by making thought a collaborative process. While I might have an idea, such as how to redesign my garden, by sharing those ideas with someone who knows about flowers, someone who has expertise on drainage and another who recently purchased garden furniture, I can develop my own thoughts by drawing on those of others. Words connect minds together just as the internet connects computers to create systems with immense memory capacity and computational power. Talking with others helps us to interrogate and develop our own thoughts. This is especially important for abstract concepts when we may struggle to understand what they mean. What is 'justice'? Nobody has a complete understanding and hence we share our ideas by talking to develop our own understanding.[42]

Critical to this process of thinking through words is the invention of words. New words can help establish a new concept within our minds, the word acting as a cognitive anchor and helping to spread the existence of this new knowledge into other minds by talking about it. The invention of the word 'dinosaur' by Richard Owen in 1842 enabled a set of heterogenous fossils discovered since the early 1800s to be placed together into a single new category. We now invent words with such frequency that we miss their significance for not just labelling new items and ideas in the world, but for also changing what we can think about.

Number words

Number words are a good example. As described in Chapter 3, languages vary in their number systems. Some, like English, can generate labels for any number of precise quantities while others, such as Mundurukú, only have words for inexact numbers and/ or small quantities.

It is not uncommon for number words in restricted counting systems to be based on body parts. Some of the many languages in New Guinea use a tally system rather than having a number concept, in which counting starts with the fingers on one hand and then goes up one side of the body and down the other. Others have words for numbers that use parts of the body. The word for 2 is sometimes that for *eyes* (because we have a pair of them); that for 5 can be *hand*; for 10, *two hands*; for 15, *three hands* or *two hands and a foot*; and for 20, *one man*.[43]

The reason for such variation has been much discussed and debated. Having a large set of number words is not necessarily an advantage: it involves a cognitive cost in learning and storing the words, which may have no value for the lifestyle of the language community.[44] Moreover, even without words for quantities larger than 1 or 2 it may still be possible to calculate larger quantities, although with difficulty.

A common assumption is that there is an evolutionary sequence from approximate and restricted number systems to large, abstract and recursive systems, this reflecting need as human communities increased in size and contact, perhaps engaging in trade that required systems for accountancy.[45] What is intuitively evident, however, is that by having words that specify precise quantities of any amount, our minds are opened up to the world of mental arithmetic. Our brains did not evolve to undertake multiplication, geometry and calculus. By using words for numbers, however, and then using further words for new concepts based on those numbers, such as integers and primes, we are able to recruit existing or build neural networks to undertake mathematics.[46]

Metaphor

This use of body words introduces the most powerful way in which words augment thought – the use of metaphor.[47] It is easy to think that metaphor provides no more than an embellishment of our phrases, one used by poets and politicians. In fact, metaphor pervades virtually every phrase we use in everyday life and is how we manage to entertain abstract ideas and think about complex issues. This was made evident in a ground-breaking book of 1980 by George Lakoff and Mark Johnson called *Metaphors We Live By*.[48]

Metaphor is the use of knowledge about one conceptual domain, the source, to inform that of another, the target. Some examples: ideas can either be 'lightbulbs' that come on in a flash, or 'seeds' that grow slowly and need nurturing before they flourish; love is a 'journey' that will have hills to climb and rocky patches to pass through; arguing is a form of warfare because views can be 'attacked' and claims can be 'indefensible' while criticisms can be 'right on target'; presidents can be 'lame ducks'.

The key insight of Lakoff and Johnson is that we use our knowledge about the physical word around us, such as our bodies and how they move, plants and animals, food and eating, warfare and so forth, to reason about unfamiliar, abstract or complex concepts: when used as metaphors, words enable us to move from the concrete to the abstract. The extent to which metaphors pervade our thinking is illustrated by how we use space to think about time. English and many other languages use the words that describe spatial relationships to also describe time. By using such words, we think of time as a line in space with the past behind us and the future stretching out in front. Could we possibly think about time without such spatial metaphors?

As the linguist Guy Deutscher has noted, the link between space and time has become so entrenched in our thinking that we struggle to understand that time cannot literally be long or short, that it can pass, that we can look forwards or backwards in time.[49] Moreover, we often fail to appreciate that we are using metaphor at all when talking and thinking about time. That is quite common – we can take almost any passage of speech or written text and find it laden with metaphors that pass by unnoticed. Deutscher gives a passage that begins:

> At the cabinet meeting, ground-breaking plans were put forward by the minister for tough new legislation to curb the power of the unions. It was clear that the unions would never go along with these suggestions, and the conflict erupted as soon as news of the plan was leaked to the press.

This short passage contains at least seven metaphors all deriving from the concrete world of physical actions (ground-breaking, put forward, go along), materials (tough, curb [a piece of metal placed in a horse's mouth]) and processes (erupted, leaked) to express the nature of human intentions and social interaction.

When metaphors become so frequently used and familiar, they lose their evocative power, becoming 'dead metaphors' – no more than content words for the target concept itself. Phrases such as 'pushing the meeting back in time' may have become an idiom, its meaning now understood without recourse to concepts of space. Because metaphors fade away and die, new ones are required to create an impact. In his 2016 presidential campaign, Donald Trump demanded that corruption is rooted out (to use another metaphor) by 'draining the swamp' – imposing the idea of a murky, insect-infested, disease-ridden location with slimy creatures onto Washington.[50] Although he was not the first to use this phrase, it was new to many Americans and evidently struck a chord (to use another metaphor). Draining the swamp influenced how many of the electorate reasoned about the state of US politics – not only that it is corrupt but that it can be cleansed – and helped Trump win the election. Such is the power of metaphor.

Metaphors play a critical role in science. They enable us to think about difficult ideas, to develop new concepts and progress our understanding of the world by drawing on what is familiar – to use the concrete to think about the abstract.[51] Many of the key breakthroughs in scientific understanding have been reliant on metaphors. In 1605 Johannes Kepler developed his concept of planetary motion by comparison with a clock, followed by Robert Hooke in 1665 who adopted the word *cell* to describe what he saw when he placed a piece of cork under his microscope, having been reminded of the small rooms, or cells, occupied by monks. We are familiar with the atom as a miniature solar system and with waves of light: how could we possibly think about these entities without such metaphors? They provide us with words to anchor, communicate and, for some, develop more advanced concepts. The word *string* in string theory provides non-scientists with an inkling about the idea that reality is made up of infinitesimal vibrating strings, smaller than atoms, electrons or quarks. How could anyone, scientist or otherwise, think about a region of spacetime where gravity is so strong that neither particles nor electromagnetic radiation such as light can escape from it without using the term 'black hole'? Equally, how can we think about and debate the working of the brain without comparing the activation of neurons with either ripples across a pond or a telecommunications network?

For me personally, I struggled for years to think and write about the evolution of language because there were so many different strands of evidence. Only when I imagined the challenge as a jigsaw puzzle, needing to find the pieces and assemble the frame and fragments, could I make progress.

The benefits of bilingualism and multilingualism

With language providing such a powerful tool for thought, it should not be surprising that bilingual and multilingual people appear to have a cognitive advantage. This has been promoted by Viorica Marian in her recent book *The Power of Language* to the extent that I, as a monolingual person, began to feel quite inadequate. She explains that because the brain keeps all its known languages co-activated, those with multiple languages will have multiple concepts stimulated when hearing a single word – not only its meaning in the language being spoken, but also the meaning of words that sound the same in other languages known to that person. Although those additional meanings remain within the unconscious, not interfering with the ongoing discourse, they will facilitate connections between seemingly unrelated things should the need arise.

Marian provides an example. When an English speaker sees a fly, they will not only have the word/concept for *fly* activated in their brain, but also concepts with words that sound similar, such as *flag*. When a Spanish speaker sees a fly, they will have the Spanish word for *fly* activated, *mosca*, and similar-sounding words, such as *molino*, which means windmill. When a bilingual English-Spanish speaker sees a fly, they will have all three concepts of *fly*, *flag* and *windmill* activated in their brain, irrespective of the language they are currently using. This may explain why multilingual people consistently perform better on creativity tasks. Earlier I noted how bilingual children acquire mind-reading skills at an earlier age than monolinguals. This persists into adulthood: bilinguals are better at detecting when they are being verbally manipulated, such as by politicians, advertisers or in any type of daily discourse.[52]

If language influences how the world is perceived, then multilinguals will experience the world differently depending on the language they are currently engaged with.[53] Formal tests have demonstrated that an individual's personality and cognitive skills change depending on whether they are speaking their native language or a secondary language. Native languages tend to elicit stronger emotional responses, most likely because they are acquired in an emotionally rich context rather than a classroom. Conversely, secondary languages lead to more logical and rational decision making. Chinese-English bilinguals emphasise their group rather than individual identity and express lower levels of self-esteem when speaking in Mandarin than in English. The same individual will have different memories about the past depending on which language they are using because memories are most accessible in the language present when they were encoded. By having more than one set of words for emotions, colours, numbers and metaphors, bilinguals and multilinguals will have different ways of categorising and perceiving the world. Viorica Marian sums this up by claiming that 'multilinguals are able to perceive more of the universe around them because they are able to transcend the scalar gradients imposed by a single language'.[54]

A persistent presence or a turning point?

In all the previous chapters we have considered language, and whatever preceded it, to have been a system of communication. We now find that language, and notably words, can influence how the world is perceived and what we can know and think about. Abstract concepts require abstract words, as labels and as cognitive anchors. The use of metaphor is essential for us to grasp the meaning of such concepts, even before we attempt to explain them to someone else. Not only for abstract concepts, but also for resolving any complex problem we are facing: metaphor pervades language and is the source of human creativity and scientific progress.

Has language always influenced how its users perceive and think about the world? Or is this an innovation, perhaps a turning point in the evolution of language after millions of years when its sole use had been for communication? The 'unbelievable monotony' of the Acheulean made by *Homo erectus* and the stochastic variation of Neanderthal technology through time provide little evidence for the type of cognitive and cultural diversity that arises from the multitude of languages in use today. And yet the evidence about the evolution of the vocal tract and the brain suggests that both *H. erectus* and the Neanderthals had a form of language. Might their languages have been restricted to communicating thoughts rather than doubling up as a tool for thought? Might their languages have been without metaphor? These ideas need further evidence, hopefully to be found in the final fragment of the language puzzle: the use of symbols and the creation of art.

15

SIGNS AND SYMBOLS

Writing in the 1880s the American philosopher Charles Peirce made a three-way distinction between symbols, signs and indexes. Symbols are arbitrarily related to their referents and often specific to a particular culture. Signs are iconic – they visually resemble their referent. Indexes have a factual relationship with their referent. A picture of a flame is a sign of fire while a picture of a burnt tree is an index of fire. Both pictures might also have symbolic meanings unrelated to fire. From long before the appearance of *Homo habilis*, the world has been replete with naturally occurring signs and indexes, such as the cloud of smoke from a distant fire, the footprints of an antelope and the scent of a hyena. It is only during the human past that visual symbols have been made, these now providing a pervasive mode of communication. Like visual symbols, most words are arbitrarily related to their referents; while some visual symbols are universally understood, others are similar to words by being specific to a language community. With such overlaps, we need to ask whether the first use of visual symbols by our human ancestors marked a threshold in the evolution of language.

Indexes, symbols and words

Indexes are used by all animals in their foraging activities, as well as by modern-day hunter-gatherers. Their use of footprints, scats and other types of spoors for tracking animals is universal and requires inferring meaning from the visual traces, or indexes to use Peirce's terms. To the hunter-gatherers' trained eyes and minds, they can provide an immense amount of information: the type of animal, its age and sex, how long since it passed by, how quickly it was moving and so forth. The traces may be no more than a faint, ambiguous mark in firm sand or thin snow and are often discussed at length within the group to arrive at the most plausible interpretation, which is continually modified as tracking continues.[1]

The significance of 'reading' animal tracks likely explains why their images are frequent in hunter-gatherer art, in which they can appear to be an abstract symbol to an untrained eye. By this means we can at least date their use back to the modern humans who had colonised Europe by 41,000 years ago because we find tracks depicted in their rock engravings and paintings.[2] It would be perverse, however, not to push their use much further back in time considering their value in not only finding prey but also escaping predation. My guess is that inferring meaning from animal tracks was key to the emergence of hunting and scavenging by the earliest *Homo* in the African savannah at *c*.2.5 million years ago.

Does the depiction of an animal hoofprint on a cave wall turn it from an indexical sign to a symbol? We can certainly see this transition occurring in historical times. During the Middle Ages, the skull and crossbones depicted on a grave slab represented death and can be described as either an index or an icon; when used today as a warning about the danger from hazardous substances, the skull and crossbones is a symbol. Similarly, its past use on the Jolly Roger flags of pirate ships might be thought of as an index because the pirates were out to kill you, but its derived use today, as a warning against software piracy, is a symbol.

Although it is tempting to think that a capacity to make and use visual symbols automatically implies a capacity for arbitrary words, this is problematic.[3] We readily understand the meaning of arbitrary words when placed in a sequence (*my house is on fire*) and easily reject those which are meaningless (*my fire is on house*). But we struggle with sequences of symbols: when seeing a red cross followed by a red heart and then red flames, we would not immediately recognise that sequence as either meaningful or meaningless, and it could be open to several equally valid interpretations.

A further complication is that visual symbols can be personally meaningful while words need to be shared. The shelf in front of me is littered with natural objects I've picked up when walking, including shells, pebbles and fossils from beaches, feathers, a tiny fir cone and a bird skull. These are mementos, reminding me of places I have enjoyed, although I can no longer recall where some of them are from. I'm not sure whether to call them symbols, signs or indexes, but I know their meanings are mine alone, unknown and of no interest to anyone else. Similarly, many people wear rings and necklaces or have tattoos that have symbolic meanings that are known to them alone. There is no equivalence to such individually meaningful symbols in language. To make matters even more difficult, we know that entirely unmodified natural objects, sometimes of the most undistinguished types, can be the most precious and powerful symbols. Holy relics of the Middle Ages are a prime example, scraps of human bone supposedly coming from a martyred saint, or a fragment of wood believed to come from the cross of the crucifixion. Such items can only be recognised as symbols because of their context, often placed into sealed caskets. Would we be able to recognise such scraps of bone as symbols once the casket has rotted away?

The overlaps between symbols, icons and indexes, their dependence on context and uncertain relationship with words, impose challenges when interpreting finds from the archaeological record that might be symbolic and indicative of language.

Such finds include minerals used as sources of pigment; bird claws and pierced shells that may have been worn as pendants; and engraved pieces of ochre, stone and bone. Further challenges arise from the often-uncertain date of such finds and which type of humans made them. Their significance has been vociferously debated by archaeologists during the last two decades, often without resolution. We will consider a sample of these finds to decide whether any robust inferences about language can be made.

Coloured minerals, pigments and the language of colour

Ochres, composed of iron oxides and hydroxides, varying in colour from yellow through orange to red and brown, were of great interest to early humans throughout the Middle Stone Age (MSA) of Africa.[4] Ranging in size from tiny fragments to large blocks, pieces of ochre are frequently found amid the debris of tool making, fire use and daily life. The earliest known examples come from the sites of Kathu Pan and Wonderwerk Cave in southern Africa, dating to at least 500,000 years ago. Ochre became more widespread after 300,000 years ago, and is abundant at sites such as Blombos, Pinnacle Point and Sibudu, all coastal caves in South Africa.[5]

The ochre had often been carried from a distant source, sometimes as far as 80 km from where it was discarded, and indicates a preference for strong-red, sparkly varieties. Most pieces show no sign of use, but some may have been ground to produce a fine powder or have polished surfaces that indicate rubbing against soft materials, such as animal hide, wood or human skin. Some pieces have ground facets that converge to a point, making what we would call a crayon. Occasionally, and predominantly after 100,000 years ago, pieces of ochre were deliberately scratched or engraved with a sharp point, creating sets of near-parallel lines and geometric designs. The same types of patterns were also engraved onto fragments of, and possibly whole, ostrich eggshells.

A similar use of red ochre is evident in layers from 120,000 to 90,000 years ago in Qafzeh Cave in Israel, a site associated with an early dispersal of *Homo sapiens* from Africa, a few of whom are buried in the cave.[6] Some pieces of Qafzeh's ochre had been scraped, ground, rubbed against surfaces and/or heat treated before use. Although yellow ochre was locally available, preference was given to red, acquired from a more distant source. Pieces of ochre and red staining on pierced seashells were found within the Qafzeh human burials.

The Neanderthals in Europe also showed interest in coloured minerals over the same time frame as the *Homo sapiens* in Africa. The earliest finds of red ochre come from *c*.250,000 years ago at Maastricht-Belvédère, Netherlands.[7] The Neanderthals, however, preferred minerals that could produce black rather than red pigment, notably manganese dioxide. Finds are considerably rarer than those of ochre in Africa, with the European evidence largely restricted to the period after 60,000 years ago. As with ochre in Africa, a small proportion of the manganese pieces show signs of use by either grinding or rubbing, with a few examples of 'crayons', notably from the French site of Pech-de-l'Azé.[8] Although the evidence is scant, red ochre remained of interest: marine shells possibly dating to *c*.115,000 and 50,000 years ago from sites in Spain have ochre-coloured surfaces and may have been pierced for use as pendants – although both that interpretation and the dates remain contested.[9]

The archaeologists Francesco d'Errico and João Zilhão have argued that the collecting of minerals, whether in Africa, Asia or Europe, was to produce pigment for colouring objects and bodies.[10] They assert that such use of pigment is 'evidence for symbolic thinking' and that 'body decoration can be regarded as a proxy for language abilities'.[11] How so? With no consideration of the differences between symbols, signs and indexes, they do not explain the significance of context and the problematic relationship between visual symbols and spoken words. Regarding the use of black pigments by Neanderthals in Europe, the linguist Rudolf Botha has provided a devastating critique of their chain of unwarranted claims from pigment use to body painting to symbolism and to language.[12] Many of Botha's concerns are applicable to the use of red ochre at Qafzeh and in Africa.

A key challenge is that pigment-rich minerals are known to be used by hunter-gatherers for a variety of utilitarian tasks. Experiments using manganese blocks as recovered from the Neanderthal occupation at Pech-de-l'Azé have shown that these and their powder promote the ignition and combustion of wood, which fits with the use of fire in that cave.[13] Had the Neanderthals wished to paint themselves black, soot and charcoal were readily available from those fires at far less cost than grinding up lumps of mineral. Similarly, ochre has several utilitarian roles: it is effective for softening and preserving hides; it can be used as an additive in hafting adhesives; and when applied to the skin it is an antibacterial agent, a sunscreen and an insect repellent.[14] I suspect that there are many other uses of minerals, all long forgotten in the modern world but once available to our ancestors who lived with a more intimate understanding and reliance on the natural world and its many materials.

A reasonable response to such utilitarian explanations for ochre is that they do not preclude symbolism: a face painted red to deter insects can simultaneously carry symbolic meaning depending on how the pigment is applied and who is viewing the face. The use of ochre in burials at Qafzeh, its sheer abundance at sites after 100,000 years ago in Africa, and the preference for especially deep red and glittery minerals, suggest that its use went beyond that of preparing hides and protection from the sun. Moreover, unlike black pigment, red is an especially striking colour. It provides a pervasive and potent indexical sign in nature, one that would be attractive for appropriation by those individuals and communities beginning to make signs and symbols for themselves. The power of red was explained by Nicholas Humphrey in his 1976 essay 'The colour currency of nature', returning us to the significance of context:

> My guess is that [red's] potential to disturb lies in [its] very ambiguity as a signal colour. Red toadstools, red ladybirds, red poppies are dangerous to eat, but red tomatoes, red strawberries, red apples are good. The open red mouth of an aggressive monkey is threatening, but the red bottom of a sexually receptive female is appealing. The flushed cheeks of a man or woman may indicate anger, but they may equally indicate pleasure. Thus, the colour red, of itself, can do no more than alert the viewer, preparing him to receive a potentially important message; the content of the message can be interpreted only when the context of the redness is defined. When red occurs in an unfamiliar context it becomes therefore a highly risky colour. The viewer is thrown into conflict as to what to do. All his instincts tell him to do something, but he has no means of knowing what that something ought to be.[15]

My guess is that the first non-utilitarian use of red was for indexical signs and icons, most likely as a cosmetic to heighten the redness of lips, fullness of breasts or anger of a face. Such intentional use provides a continuity with the colour currency of red in the natural world. The abundant use of red ochre by later *Homo sapiens* in Africa suggests its initial use as an indexical sign may have seamlessly merged into one of symbolic significance.

As with tools that have a deliberately imposed form, I find it difficult to resist the idea that words for red existed in whatever language was spoken in Africa at 100,000 years ago. Like those words used for the colour green by the Anangu hunter-gatherers of the Western Desert of Australia that we encountered in

Chapter 3, words for red might have labelled a package of sensations rather than just the colour. Perhaps that package encompassed the warmth, light and heat of a fire, the emotions induced by a sunset, and blood from a wound or menstruation.[16] Or maybe, following Nicholas Humphrey, the word for red and any use of its colour had simply meant 'be alert' because something is about to happen, which might be either good or bad.

Neanderthal cave painting? Most unlikely

In 2018, a research team led by Dirk Hoffmann from the Max Planck Institute for Evolutionary Anthropology claimed that Neanderthals had painted images using red pigment in three caves in Spain: amorphous red marks on a stalagmite at Ardales, a rectangular sign on a wall at La Pasiega and hand stencils at Maltravieso.[17] Previously, all cave painting had been attributed to the *H. sapiens* who, at the time of their study, were understood to have arrived in western Europe after 42,000 years ago. The cave paintings are usually believed to reflect the fully modern cognitive and linguistic capacities of those new arrivals. By implication, Hoffmann and his colleagues were now attributing those capabilities to Neanderthals.

The claim was based on uranium–thorium (U-Th) dating of calcite that had formed over the paintings. This dating method relies on the decay rate of an isotope of uranium into one of thorium. Radiocarbon dating similarly relies on radiometric decay but is time restricted to c.50,000 years ago and uses tiny amounts of charcoal removed from the paintings, the charcoal having been used as a pigment. Radiocarbon dating has produced almost 150 dates from cave paintings. All of these are more recent than 38,000 years ago, which is consistent with the widely held view that the paintings were made by modern humans after Neanderthals had become extinct.

Radiocarbon dating is not possible for the mineral-based

red pigments used in the paintings that Hoffmann and his colleagues wished to date. They used the U-Th method to date the calcite that lay over the red pigments with the oldest dates providing a minimum age for the paintings. In all three caves, these were found to be close to 65,000 years old. Because modern humans were believed not to be present until after 42,000 years ago, Neanderthals must have been the artists. This became an academic cause célèbre, heralded as a breakthrough, readily accepted by the media, public and those academics who showed insufficient caution. It is most probably wrong.

Three criticisms have arisen, challenging the dates.[18] The first is that Hoffmann and his colleagues had not verified that the red stains on the stalagmite in Ardales Cave derived from pigment deliberately placed there by a human hand: it could have been a natural deposit arising from iron-oxide-rich minerals that are carried by water flowing in the cave. Second, doubts have been expressed that the samples from all three caves were taken from calcite that covered the paintings rather than from the rockface on which the paintings had been made. Distinguishing between the two is difficult when layers of calcite are microscopically thin and one is working in the confines of a cave. Third, even if the sampled calcite did overlie the paintings, its chemical composition may have been influenced by the wet environments of the caves, causing dates to appear older than they really are.

Such methodological issues might explain why only one out of twenty-one U-Th samples from La Pasiega provided a date prior to the known presence of modern humans, giving a minimum value of 64,800 years ago; the other twenty dates ranged from 25,230 to 730 years ago. Similarly, the samples taken from the vicinity of the hand stencils in Maltravieso and the red stains in Ardales gave a wide range of values, from 66,700 to 14,700 years ago. While this was not unexpected, because calcite would have been continuing to form long after the paintings were made, such a wide spread of dates might also suggest post-formation chemical change to the calcite layers. Considering such method-ological challenges, picking out the oldest dates and especially using the single anomalous date from La Pasiega to overturn a century of study and the outcome of almost 150 radiocarbon dates requires a robust defence. Unsurprisingly, Hoffmann and his colleagues provided that in a response to their critics.[19] But the jury is out. I am confidently siding with the critics until the early dates are replicated and independently validated.

There is an urgent need to do so because it is probable that modern humans were in Western Europe significantly earlier than 42,000 years ago. In 2022 the archaeologist Ludovic Slimak and his colleagues published evidence that a group had used Mandrin Cave in the Rhône Valley in France at *c*.54,000 years ago.[20] Neanderthals had long been in the Rhône Valley and made use of the cave before modern humans left distinctive stone tools, worked bone, engraved stone and an eagle talon with cut marks suggesting its use as a pendant. The modern humans may have used the cave on one or more occasions in a single year. They had hunted with bows and arrows, discarding many triangular flint points with distinctive damage, showing they had been shot into prey. After the modern humans had moved on, the Neanderthals returned and left more of their own debris on the cave floor. They kept on using short thrusting spears, just as they had done for many thousands of years.[21]

We have no idea where the modern humans went; there is no further known trace of them at Mandrin Cave until after 42,000 years ago, their return being part of the Aurignacian dispersal into Europe. Even the 54,000-year-old modern human occupation at Mandrin Cave is still eleven millennia after the 65,000-year-old date for the cave paintings in Spain. But uncertainty about both sets of dates closes that gap, and we are left with the possibility of even earlier modern human incursions into Europe that have yet to be discovered.

Shell beads from the MSA of Africa and western Asia

While there remains dispute concerning the significance of ochre used by *H. sapiens* during the Middle Stone Age of Africa, there is no question that marine shells had been collected, transported and strung together using either natural or artificially made holes.[22] The strings of beads are reasonably assumed to have been worn as body ornaments. What they imply about society, symbolism and language remains less evident.

The most meticulously studied collection of shell beads comes from Blombos Cave.[23] Sixty-eight round shells of *Nassarius kraussianus*, commonly known as the tick shell, were recovered from within the cave associated with the Still Bay points, ochre and other debris dating to between 78,000 and 72,000 years ago. The shells are small, 10 mm at most, and their former tiny mollusc inhabitants would have been no use for food; microscopic analysis showed the shells had been deliberately perforated and have wear traces around the holes indicative of rubbing against a string. Some appear to have been heated, possibly to enhance their colour. There were five clusters of shells at different levels and areas of the cave deposits, suggesting separate strings of beads. The wear patterns on those in the lower levels indicate that the beads had been worn as bracelets, necklaces, chokers, collars and/or headbands; those from the upper levels suggest that beads had been knotted onto string to hang as pairs.

At least ten other Middle Stone Age sites in South Africa have marine shells that had likely been used for personal ornamentation. They are always fewer in number than at Blombos and come from less well-defined contexts. Most notable are two *Conus* shells from Border Cave that were deliberately perforated, associated with Howiesons Poort artefacts and the burial of an infant dating to *c.*75,000 years ago.[24]

A greater abundance of shell beads has been found 8,000 km to the north at sites on and close to the North African coast. The oldest known collection comes from Bizmoune Cave in

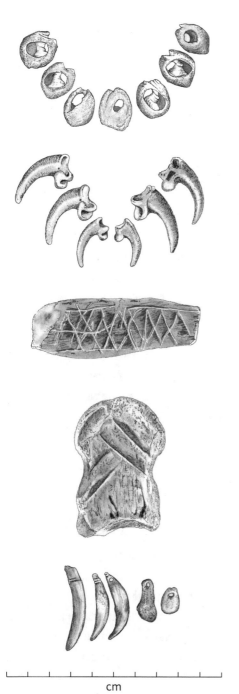

Shell beads (*Tritia gibbosula*) used by *Homo sapiens* at Bizmoune Cave, Morocco, c.142,000 years ago

White-tailed eagle talons used as pendants by Neanderthals, from Krapina, Croatia, c.130,000 years ago

Engraved nodule of ochre (MI-6) made by *Homo sapiens* at Blombos Cave, South Africa, c.73,000 years ago

Carved toe bone from a giant deer from Einhornhöhle, Germany, carved by a Neanderthal, c.51,000 years ago

Châtelperronian animal teeth pendants from Arcy-sur-Cure, France, probably made by *Homo sapiens* but found in a layer with Neanderthal skeletal remains, c.41,000 years ago

cm

Figure 15 Early signs or symbols from Europe and Africa

southwest Morocco, *c.*50 km from the sea when occupied.[25] Thirty-three shell beads were recovered from a layer dated to 142,000 years ago, also containing tanged Aterian points and blades produced from Levallois cores. With one exception, the beads were made from *Tritia gibbosula* shells. These are similar in size and shape to those of *N. kraussianus* from Blombos, both coming from the same family (Nassariidae). The perforations at Bizmoune Cave were natural, either from the shells having eroded on the beach or caused by predators that could pierce the shells. Although natural, some of the perforations had been modified by chipping and exhibit polish, suggestive of rubbing against a string. Several shells had traces of red ochre (Figure 15).

Contrebandiers Cave, close to Morocco's Atlantic coast, has the largest known collection of shell beads, numbering over 150 from layers dating to *c.*115,000–96,000 years ago.[26] They have strong similarities to the beads from Bizmoune Cave despite a time gap of over 30,000 years: the majority were made from *T. gibbosula*, the perforations were natural rather than artificial, they had polished edges indicative of having been strung and traces of ochre.

The same types of shell beads were still being made more than 20,000 years later. Thirteen were found in Grotte des Pigeons, Taforalt, Morocco, associated with Aterian stone tools and dated to 82,500 years old.[27] Ten of these have wear traces on the shell surface, suggesting they had rubbed against each other while being worn. The shells would have been white when collected; several had traces of red pigment and others had been blackened by burning to create a colourful array.

Shell beads have been found in the Mediterranean region of western Asia, at sites associated with the earliest *H. sapiens* to have dispersed from Africa, dating to between 120,000 and 100,000 years ago. Two possible *Tritia* sp. shell beads have been recovered from Skhul Cave that made use of natural holes. At Qafzeh Cave, located 40 km from the coast, marine shells were found in deposits below human burials.[28] These include seven half shells (valves) of a large, rounded bivalve known as *Glycymeris nummaria*, which have perforations at their umbos (the beak-like joint of the shell). The holes are natural but again preserve rounding suggesting the shells had been strung, also indicated by patches of polish where they had rubbed together. Traces of yellow and red staining are preserved.

How can this evidence for shell use in the Middle Stone Age of Africa and western Asia be interpreted? Daniella Bar-Yosef Mayer, an archaeologist at the Steinhardt Museum, Tel Aviv, and leading expert on shell beads, suggests a developmental sequence.[29] Her scenario begins before 150,000 years ago with the collecting of shells from beaches that are taken to sites such as Pinnacle Point Cave in South Africa and Misliya Cave, Israel, both occupied by *H. sapiens* and having unmodified bivalves. Bar-Yosef Mayer suggests that such shells may have had a subconscious and universal appeal to the human mind, and more speculatively acted as natural symbols about the origin of life – a conjecture too far for me. By at least 142,000 years ago people had begun to collect naturally perforated shells. This might reflect the date when string made from twisted plant fibres had been invented, enabling shells to be publicly displayed; still later, after 80,000 years ago, artificial perforations were made as the wearing of beads became more prominent.

Bar-Yosef Mayer's scenario can be reinterpreted as starting with the collection of curious shells from the beach, just as I do today, to the wearing of naturally pierced shells as indexical signs of personal identity, which were later attributed with shared meanings by a social group, transforming the beads from indexical signs into symbols.

The striking similarity in shape and size between the preferred *Tritia* sp. shells in the north and the *Nassarius* sp. in the south may confirm Bar-Yosef Mayer's suggestion of a subconscious and universal attraction to objects of a certain type. The persistent use of *T. gibbosula* in North Africa for tens of thousands of years is striking, especially when there had been other types of shells on the beaches that might have been collected, a few of which were occasionally used. The attraction to *T. gibbosula* might reflect the transmission from generation to generation of a particular value attached to this type of shell. This implies long-term population continuity in North Africa for more than 50,000 years, from 142,000 years ago (Bizmoune Cave) to 82,500 years ago (Grotte des Pigeons). That seems problematic considering the likely disruptions to population dynamics caused by climatic change throughout this period. An alternative is that the decisions to collect and string together shells of *T. gibbosula* might have been repeatedly reinvented, reflecting the type of universal attraction to these objects that Bar-Yosef Mayer suggests.

The long-lasting use of *Nassarius* shells through the sequence at Blombos Cave poses a similar dilemma. In this case an enduring cultural tradition appears more likely, although one that allowed a shift in the style of body ornamentation by stringing beads together in different ways through time. That shift might have reflected a transition from their use as an indexical sign of individual identity to becoming a symbol of group affiliation of social status. If so, one might further speculate that it was accompanied by a lexical shift from the use of iconic to arbitrary words to name either the shells or the body adornments.

The use of beads, feathers and talons in Europe

The evidence for Neanderthal body ornaments in Europe is much rarer, more ambiguous and hence more contested than the evidence for the use of body ornaments by *H. sapiens* in Africa, matching the relationship we found with pigment use between the two regions. It is also in dramatic contrast to that from the earliest Upper Palaeolithic culture in Europe, made by incoming *Homo sapiens*. This is denoted by the Aurignacian culture, present throughout Europe by 41,000 years ago and containing abundant beads made from shells, teeth, ivory and bone, suggestive of ornately decorated bodies and clothing.

The most controversial set of potential Neanderthal body ornaments comes from Grotte du Renne, Arcy-sur-Cure, France, excavated between 1949 and 1963.[30] Archaeologists have long argued over this collection of perforated and grooved teeth of various animals and small perforated beads of ivory (Figure 15). There is no question that these items were used as body ornaments and have strong similarities to those of the Aurignacian. But they come from a layer containing Neanderthal skeletal remains and stone tools of the Châtelperronian culture. This layer is dated to between 45,500 and 41,000 years ago, found above one with Middle Palaeolithic Neanderthal stone tools and below an Upper Palaeolithic Aurignacian horizon.[31] The Châtelperronian stone tools have similarities to both those of the Middle Palaeolithic made by Neanderthals and those of the Upper Palaeolithic made by modern humans. Which type of human made them is much debated. My own view is that they were made by Neanderthals who had observed modern humans making and using Upper Palaeolithic stone tools and adapted their own technology.[32]

Of greater controversy is who made the body ornaments in the Châtelperronian layer.[33] Francesco d'Errico and João Zilhão have made meticulous studies of the artefacts and dating evidence to argue that both the stone tools and body ornaments were made by Neanderthals before modern humans had arrived in Europe and hence could not have been copied from those being made and worn by the incomers. Their proposal came before the evidence from Mandrin Cave was discovered, but I doubt if that evidence for modern humans in Europe before 50,000 years ago would have changed their minds. Most archaeologists were already incredulous of their proposal: there is no other evidence for Neanderthals having made such body ornaments and their similarities with those from the Aurignacian cannot be mere coincidence.

Another interpretation is that the contents of the underlying Middle Palaeolithic, the Châtelperronian and the overlying Aurignacian layers at Grotte du Renne have become mixed up. This is a common occurrence on archaeological sites, arising from processes such as human trampling, root penetration, freeze–thaw events that create crevices in the sediments, and burrowing animals. Such processes and their consequences were often undetected by the excavation methods used when Grotte du Renne was excavated between 1949 and 1963. In this scenario, the body ornaments eroded from the Aurignacian layer into the underlying Châtelperronian layer.

A new twist to this long-running debate arose in 2023 with the re-analysis of the Châtelperronian skeletal remains.[34] All of these had been thought to be from Neanderthals, but one fragment was now identified as part of a hip bone from a newborn modern human. How had it got there? It might have also eroded into the Châtelperronian deposits from the overlying Aurignacian layers. Alternatively, it might indicate that successive groups of modern humans and Neanderthals used the cave before 41,000 years ago, both contributing debris (including their bones) to what became known as the Châtelperronian layer.

A third possibility is that the combination of modern human and Neanderthal skeletal remains in the same layer indicates the two human types had lived together, sharing their genes, knowledge, tool-making traditions and words. I suspect that was the case. Whatever the correct explanation, the Neanderthals had certainly not made the body ornaments by themselves and without the influence of modern humans because of their overwhelming similarity to those from the Aurignacian and their absence from any other Neanderthal occupation.

That does not mean I am unsympathetic to other evidence for Neanderthal body ornaments. A small quantity of evidence has accumulated during the last decade, none of which is definitive but it cannot be disregarded. Shells with holes and possibly coloured with red pigment have been dated to 115,000-year-old deposits at Cueva de los Aviones, Spain. They may have been worn as pendants. The dating method, however, used the same technique and might suffer the same flaws as that used for the so-called Neanderthal cave paintings.[35] An ochre-covered fossil marine shell from a *c.*46,000-year-old deposit in Grotta di Fumane, Italy, may also have been a pendant – but that is an isolated find and also after modern humans had entered Europe on at least one occasion.[36]

Far more persuasive evidence indicates that Neanderthals were making body ornaments from the talons and feathers of birds, especially large raptors. Talons from several sites, ranging from 130,000-year-old Krapina in Croatia to *c.*50,000-year-old Les Fieux in France, have cut marks and areas of polish suggesting they had been worn as pendants (Figure 15).[37] Cut marks on wing bones of eagles, vultures and falcons from 45,000–42,000-year-old deposits at Grotta di Fumane, and 40,000–38,000-year-old deposits in Gorham's Cave, Gibraltar, suggest the removal of feathers for body decoration.[38]

Had the Neanderthals used bows and arrows, the feathers might have been used for fletching their arrows, but there is no evidence they invented such hunting technology. Because most of the evidence for using feathers and talons is dated to after the time when modern human–Neanderthal interbreeding had occurred, and also because the modern humans at Mandrin Cave had been wearing eagle talon pendants, we cannot exclude the possibility that the Neanderthals had been influenced by seeing how modern humans were decorating their bodies. However, the evidence from Krapina pre-dates that of Mandrin Cave suggesting that the Neanderthals invented bird-associated regalia for themselves.

The use of talons and feathers from large birds is pervasive among modern human hunter-gatherers. It is often associated with shamanism in which the birds are believed to mediate between the human and spirit world; by wearing their feathers, people – some of whom may be designated as shamans – seek to acquire the power of such birds.[39] There is no requirement to attribute such symbolic significance to the talons and feathers that were likely worn by the Neanderthals, although they surely experienced a sense of awe and wonder when seeing eagles soaring in the sky and hunting their prey. They may have worn eagle talons and feathers for aesthetic purposes, as indexes of personal identity, or as a form of display – associating themselves with the beauty, speed and hunting prowess of an eagle, while demonstrating their own hunting skill by having acquired feathers and talons. None of these uses are symbolic in the sense of attributing arbitrary meanings, but all resonate with the behaviour we find in modern humans from the Upper Palaeolithic to the present day. Consequently, they might be taken to imply language, if only to provide words for feathers, talons and such spectacular birds.

There is a significant difference between the personal adornments of choice of the Neanderthals in Europe, feathers and talons, and those of the early *Homo sapiens* in Africa, shell beads. As hunter-gatherers, the Neanderthals would have seen birds daily, been familiar with the plumage and talons of each type and known their habits and lifestyle. They would have recognised the feathers and talons being worn and readily inferred the message being sent: the wearer wishes to be seen as a fast runner like a peregrine falcon, a powerful hunter like an eagle, or is seeking to attract a partner by wearing vibrantly coloured feathers just as a robin does in spring.

There could be no such easy inference of messages from *H. sapiens* in Africa when seeing a string of shell beads. There would have been less familiarity with shells in the wild. Even if they were seen daily by coastal foragers, there are no habits of behaviour to associate with shells of different shapes, colours and sizes. Whereas bird-derived ornaments lend themselves to becoming indexical signs, shells are suited to sending either no message at all or one that is arbitrarily defined and must be revealed rather than independently inferred.

Engravings on ochre, bone and stone

The final objects to consider are engravings made by *Homo sapiens* in Africa before their dispersal after 70,000 years ago and by the Neanderthals in Europe before the widespread arrival of modern humans after 42,000 years ago. What can these tell us about their respective linguistic capabilities?

Fifteen pieces of ochre with incised lines made by a sharp point have been recovered from Blombos Cave, South Africa, dating to between 100,000 and 72,000 years ago.[40] While some may have arisen accidently when a flat-faced piece of ochre was used as a support for cutting plants or hide, and other lines may have been made when scraping ochre for powered pigment, several pieces have lines that were deliberately engraved for their own sake. These lines are too thin or sinuous to have been made for generating powder and are sometimes arranged to make geometric designs. Moreover, because considerable force and concentration had been required to incise straight lines onto the ochre, such designs cannot be explained as casual doodling. Francesco d'Errico, Christopher Henshilwood (who led the Blombos excavation) and Ian Watts, an expert on ochre, believe the designs had symbolic meanings to the people who occupied the cave.

The most impressive piece is known as MI-6 (Figure 15). This is a nodule 7.5 cm long, 3.5 cm wide and 2.5 cm thick, which has a face that had been flattened by grinding before a cross-hatched pattern was incised. Microscopic analysis shows that a set of oblique lines was initially incised from the top right to the bottom left of the face, each made with multiple strokes of the point; these lines were then crossed by a second set made from the top left to the bottom right. Three horizontal lines were then incised, one at the top of the oblique lines, one at their bottom and one that passed through the middle, where the oblique lines intersected, to create a cross-hatched pattern. Other pieces are less elaborate, having semi-parallel lines, hints of cross-hatching, fan shapes and in one case a Y-shaped design. There is no evidence for a symbolic code because each of the patterns is unique, although that might be because the majority had been made on organic materials that have rotted away.

Did these patterns have symbolic meaning? Maybe and maybe not. How can we know? And what does 'having a symbolic meaning' mean? When I visited Blombos Cave, I was initially struck by the similarities between the patterns and the natural lines on the coastal rock formations immediately outside the cave. The rocks also had parallel and sometimes cross-hatched lines, caused by geological processes over millions of years. I suspect those patterns may have provided an unconscious inspiration for the designs on the ochre.

I was also struck by a possible analogy between the Blombos engravings and the geometric signs made in the recent art of Indigenous central Australian peoples.[41] Their signs are most frequently made as casual drawings in the sand or earth to accompany storytelling. The signs look entirely abstract but often represent events such as people sitting in a camp or making a journey. Such meanings can only be acquired when listening to the story – the signs are uninterpretable by anyone who was not present at the time. Moreover, the same geometric sign can mean different things at different stages of the story – the signs have multiple meanings which are gradually revealed. This art seems pertinent to the Blombos engravings because they also appear relatively casual, the fireplaces within the cave suggest night-time storytelling, and any meanings they may once have had are entirely lost to us today. Conversely, the effort involved to incise complex geometric designs onto a nodule of ochre contrasts with that required to scribble a mark in dry sand or even to scratch a few sub-parallel lines onto a piece of ochre.

Kristian Tylén from Aarhus University and her colleagues have made an important study of the Blombos engravings and those made on ostrich eggshell coming from Diepkloof Rock Shelter located 400 km northwest of Blombos and dating to between 109,000 and 52,000 years ago.[42] They recognised a similar chronological sequence at both sites in which sets of sub-parallel lines developed to become complex cross-hatchings, as exemplified by MI-6 from Blombos. Tylén and her colleagues undertook carefully controlled laboratory experiments with modern-day participants to test the extent to which the designs were salient, memorable and could be reproduced. All three measures increased as the designs became more structured through time, suggesting that they had evolved through a cultural transmission process that increased their learnability – not unlike that which occurs with the transmission of spoken language from one generation to the next. Designs that can be better remembered and reproduced have greater potential to be used as symbols. Consequently, the 100,000- to 50,000-year-old sequences might document the slow emergence of symbols from marks that had once lacked any shared meanings, or had been casually used as signs when telling stories. This was nicely expressed by Tylén and her colleagues as African Middle Stone Age humans acquiring an increased 'sensitivity to the potential cognitive consequences of their interventions in the material world'.

Although neither western Asia nor Europe provide equivalent collections of engravings to those from South Africa, the marked objects from these regions require our attention. We have, for instance, a 120,000-year-old fragment of bone from an aurochs (the original wild cattle) from the site of Nesher Ramla, Israel, that has six deep and nearly parallel incisions perpendicular to its length made with a flint point.[43] These can neither be explained by butchery, trampling nor inadvertent damage arising from using the bone as a chopping surface. The date suggests these incisions were made by the early dispersal of *H. sapiens* from Africa who also left burials, used red ochre and shell pendants at Qafzeh Cave. In the same region but later in time at 55,000 years ago, a piece of bone and the cortical surface of a flint flake from Quneitra in the Golan Heights were both engraved with a series of concentric arcs.[44] Whether they were made by the hands of a Neanderthal or a modern human is unknown.

A cluster of sites in the Crimea with Neanderthal remains, but dating after 40,000 years ago when modern humans may have been within that region, have produced incised stone and bone objects of potential symbolic significance. These include a set of thirteen sub-parallel lines incised onto the cortex of a flint flake from Kiik Koba dating to 37,000–33,000 years ago; parallel and fan-shaped lines incised onto animal bones and teeth from Prolom II; and seven evenly spaced notches on the wing bone of a raven from Zaskalnaya VI dated to between 43,000 and 38,000 years ago.[45] We cannot discount the possibility that some or all of these objects were made by modern humans.

The best-documented piece comes from Einhornhöhle in northern Germany, dating to 51,000 years ago (Figure 15).[46] This is the phalanx (toe bone) from a giant deer engraved with a geometric pattern formed by two interlaced sets of lines to form a chevron pattern. This small object, 5.56 cm long by 4.00 cm wide, was found at the entrance to the cave amid debris typical of that left by Neanderthals. Modern humans may have been in the south of France at that date, but there is no known trace of them in northern Europe. Dirk Leder, the lead archaeologist who excavated this object, described it as 'one of the most complex cultural expressions' to have been made by Neanderthals and believes that the engraved chevrons are of symbolic significance. I'm not so sure. Neanderthals had been imposing form onto nodules of stone and pieces of wood for more than 300,000 years. Doing the same with a piece of bone is certainly a rare find but not necessarily different in kind. As with the notched raven bone from Zaskalnaya VI, the Einhornhöhle engraved bone might have been part of a multi-component tool, its surface shaped to fit snugly against some other component.

'Of symbolic significance' is also the interpretation that the archaeologist Joaquin Rodríguez-Vidal and his colleagues attribute to a criss-cross pattern formed by thirteen deep groves cut into bedrock at Gorham's Cave, Gibraltar.[47] They argue that the grooves were deliberately engraved rather than being natural cracks as found elsewhere in the cave. Sediment overlying the design contained stone tools attributed to the Neanderthals and is dated to 38,500 years ago. Rodríguez-Vidal's team concluded that the sign had been intentionally made before that date and had a symbolic meaning for the Neanderthals who used the cave.

Doubts have been expressed.[48] The overlying deposit is evidently disturbed because it has older dates above younger dates. Might this deposit have recently eroded within the cave to cover an engraving made by modern humans long after the Neanderthals had gone extinct? Even if the deposit is in place, we know that by 38,500 years ago modern humans were in Europe and they might have made the sign before the Neanderthal use of the cave. Moreover, the attribution of the stone tools in the deposit to the Neanderthals has been questioned because they are few and lack diagnostic features. Even if all these doubts are groundless, the Neanderthals may have made the criss-cross pattern for a utilitarian purpose.

A more cautious approach has been taken by Jean-Claude Marquet and his colleagues when reporting what they describe as the 'earliest unambiguous Neanderthal engravings'.[49] These are found on a cave wall at La Roche-Cotard in central France and are dated to earlier than 57,000 years ago, after which the entrance to the cave had become blocked by natural sedimentation until it was opened again in the nineteenth century. Neanderthal stone tools were found on the floor of the cave, along with butchered bones from bison and deer. The cave had also been used by cave bears before the Neanderthal occupation and by hyenas after their time. Eighty-four groups of marks had been made on the cave wall by pressing fingers into its soft surface, sometimes dragging these to make near-parallel lines and possible geometric shapes. The term 'engravings' makes these sound rather grand, suggestive of tool use. A more accurate term, as used by Marquet throughout his description, is 'finger-flutings'. Cave bears had also marked the walls with their claws, creating parallel lines that were thinner, deeper and with V-shaped cross sections. Such marking is a common occurrence in caves where the bears make their dens and hibernate. At La Roche-Cotard, the two types of marks are often found on the same wall surface.

Marquet and his colleagues reasonably interpret the human marks as representing 'conscious design and intent', while recognising they cannot be established as having symbolic significance. The Neanderthals may have been inspired to make their marks by those left on the walls by the bears. Maybe they had arisen from activities in the cave that resulted in contact with the walls, perhaps dancing or playing, or simply exploring the cave in the darkness or with the light of a small candle. Indeed, it would be more curious if Neanderthals had been inside a cave with a soft sediment lining over its walls and not left any marks at all.

Signs or symbols?

The majority of words are arbitrary and culturally specific, sharing these features with visual symbols. We might expect, therefore, that the use of visual symbols would have evolved hand in hand with language. But despite having found evidence for a Neanderthal linguistic ability from their vocal tracts (Chapter 5), technology (Chapter 7) and brain size (Chapter 11), there is no compelling evidence that they made and used visual symbols during the 350,000 years of their existence. A careful scrutiny of the evidence must conclude that Neanderthals used manganese and ochre for practical rather than symbolic purposes, wore feathers and talons as indexical signs, that Châtelperronian beads had been made by modern humans and the dating of so-called Neanderthal cave painting is unreliable. The finds from Einhornhöhle, Gorham's Cave and La Roche-Cotard indicate that on some occasions the Neanderthals were deliberately making marks, at least after 60,000 years ago. But such interventions in the material world do not necessarily represent the making of symbols.

The case for symbol use by *Homo sapiens* in Africa is more persuasive but not definitive. Their engravings are not dissimilar to those of the Neanderthals, but more numerous and with stronger evidence for imposed geometric designs. Their repetitive use of the same type of shells for beads and the manner in which the shells were coloured suggests they were given a symbolic meaning. Similarly with the use of red ochre – this is more abundant and a more potent colour than the manganese used by Neanderthals. The association between engravings, red ochre, beads and fire in Blombos Cave seems quite different from the isolated finds relating to Neanderthals, disparate in nature and widely dispersed in time and space. And yet we cannot be sure. Any difference between the Neanderthal use of objects and marks in Europe and those by *H. sapiens* in Africa before 70,000 years ago is marginal. My evaluation is that both are primarily indexical signs but that *H. sapiens* crossed the threshold into making symbols on some occasions. But more was to come from them after they dispersed into Europe and Asia – the first figurative art was made soon after 40,000 years. Nothing more came from the Neanderthals.

16

CONCLUSION: THE EVOLUTION
OF LANGUAGE

This book is trying to solve the language puzzle: why, when and how did language evolve? I've gathered evidence from linguistics and archaeology, anthropology and genetics, from neuroscience, psychology and ethology. I've used it to assemble fourteen fragments of the puzzle that now lie piecemeal on the table.

The edge pieces remind us that we are the only surviving member of the *Homo* genus and lived as Stone Age hunter-gatherers until farming began a mere 10,000 years ago (Chapter 2). They also remind us about the nature of fully modern language, how it combines words and rules in a variety of ways to create an immense diversity of languages in the world today (Chapter 3).

A glance at the fragments waiting to be inserted provide hints as to what picture might be revealed. We can see the word-like and syntax-like qualities of chimpanzee vocalisations (Chapter 4), the modern-like vocal tract of 500,000-year-old *H. heidelbergensis* (Chapter 5), and the evolutionary significance of iconic words (Chapter 6). While we see similarities between stone tools and words, we also find the monotony of early technology (Chapter 7). We can see aspects of syntax emerging from cultural transmission (Chapter 8), and children learning language by using general-purpose learning methods (Chapter 9). On another fragment we find a step change in the use of fire after 400,000 years ago (Chapter 10). Language has been found to be distributed throughout the brain, which, in modern humans at least, has long-distance connections between areas of functional specialisation formed by neural networks that develop under the combined influence of genes and environment (Chapters 11 and 12). A particularly large fragment shows us how words are always changing what they mean, how they are pronounced and used (Chapter 13). Another illustrates the role of language in thought, especially that of metaphor to enable abstract and complex ideas to emerge and then be shared (Chapter 14). Finally, we see that both Neanderthals and *H. sapiens* used indexical signs, which may have evolved into symbol use by *H. sapiens* in Africa by 70,000 years ago (Chapter 15).

It is now time to fit these fragments together to complete the language puzzle. Here is my best shot.

The starting point

The last common ancestor (LCA) of chimpanzees and humans, living in Africa between 8 million and 6 million years ago, had mental capabilities and vocalisations not dissimilar to those of the chimpanzees today, whether *Pan paniscus* (the bonobo) or *Pan troglodytes* (chimpanzee). That remained the case with the LCA's relatives and descendants, including *Sahelanthropus tchadensis* and *Ardipithecus ramidus*.

Those forest-dwelling primates used their calls to influence the behaviour of others in their groups. Most calls expressed emotional states: loud, rapid barks and piercing screams with teeth displayed for anger and aggression; quieter, slower hoos and coos, when expressing affection and building social bonds. All the calls were holistic – any sound elements they contained had no meaning in themselves. The calls varied between species, reflecting their lifestyle and anatomy, and between communities reflecting ecological context and cultural traditions. Although these primates had some control over making a call, they had little influence over the sound itself beyond making it a little quieter or louder, longer or shorter. In comparison with the later hominins, their brains had relatively small numbers of neurons and localised neural networks. They had limited muscular control over the exhalation of breath, engagement of the vocal folds and articulations of the tongue, teeth and lips. Vocal membranes, air sacs, large teeth and protruding faces constrained the range and consistency of the sounds they made. Despite these constraints, the primate calls had some word-like and syntax-like qualities: the meanings of calls were sometimes context dependent; the same sound elements were combined in different ways in different calls; juveniles had to supplement their genetically based instincts to learn about what calls to make in which circumstances.

The emergence of words

At around 4 million years ago there was a turn in the global climate towards aridity, causing the forests to retreat and the hominins to move onto open grassland with scattered trees. The number of hominin species proliferated with a multitude of adaptations – the australopiths. Some became specialised grazers on dry plant foods while others began to pick scraps of meat, fat and marrow from the carcasses of abandoned carnivore kills. All faced the challenge of living in an open, predator-rich environment. They needed to live in larger groups for safety and interpret natural signs to work out what was happening, or about to happen, in their world: the tracks and trails of predators and prey, the flight of birds, the smell of smoke from a distant savannah fire.

As with any population, random genetic mutations created a pool of variability in cognitive skills. Some individuals were better at statistical learning, detecting recurrent associations and using these to their advantage. By watching members of their group, they learned to predict who will fight for the food and who will acquiesce when under threat, who to collaborate with and who to avoid. Such individuals became adept at negotiating the complexity of the social world. They used their vocal calls to build alliances, secure status and gain access to resources, whether from food sharing or stealing food acquired by others. By so doing, they secured reproductive advantage – their significant genes for statistical learning carried their social skills into the next generation via the extra neurons and/or neural networks that those genes developed.

Either the same individuals or others gained similar advantage by being adept at recognising regularities in the natural world and building categories of natural signs – how the fresh pawprint of a leopard is a sign of danger and the dry wisp of a withered stem indicates a fat tuber buried below. The value of making such associations further expanded the capacity for learning, and with it the size of the brain. So too did the value of knowing how to hold a stone nodule in one hand and strike it with another at the correct angle and with the correct force to detach a flake. The resultant tools enabled scraps of meat and fat to be taken more swiftly from carcasses, and bones to be cracked apart for marrow. Speed was of the essence because carnivores were also looking for nutritious food, whether morsels from a long-dead carcass or a tasty hominin to eat. Efficiency and effectiveness when out and about on the savannah were the hominin bywords for survival.

By 2.8 million years ago these selective pressures coalesced and built a larger brain for at least one, and perhaps several, hominin species, which we now call *Homo habilis*. General-purpose learning processes were used within all domains and produced mutually supportive outcomes: enhanced social alliances facilitated the transmission of tool-making skills; these provided stone flakes for scavenging and nodules for breaking bones apart to find marrow and for pounding plants; the meat, fat and marrow fuelled the brain, powering the extra neurons that enabled natural signs to be used to find the carcasses and provide success within the social world.

The larger brain of *Homo habilis*, sometimes reaching 800 cm³, was both a cause and consequence of this social–technology–diet feedback loop. As a consequence of its size, connections arose between the visual, motor, auditory and somatosensory cortices, these connections being required for statistical learning. The connections also caused cortical leakage, likely arising from insufficient myelination. Such leakage made synaesthesia the normal and ubiquitous state of mind. Sensory impressions for how predators looked, moved and smelt seeped unknowingly into their alarm calls – slithering sounds for snakes, pouncing-type calls for leopards, squawk-like cries for birds.

Such cries and calls were often effortlessly blended with iconic gestures. The calls emitted when making stone tools took on the tenor of a hammerstone strike and the sharpness of a flake; rounded sounds and cupped hands expressed bulbous tubers; tiny, piercing sounds and pinched fingers for lice and fleas. Some calls blended a throng of multimodal sensations into a single segment of sound: the vocal response to grassland fire capturing bright, flickering colours, heat, crackling sounds and the drifting smell of woodsmoke within a single synaesthetic call.

Iconicity now shaped the calls for building relationships. Those used for social bonding became quieter and calmer, expressing in sound how one would physically caress a friend. Emotional states – sensory impressions of oneself – were vocalised more frequently: sounds such as *ouch* when hurt, *oh-oh* when surprised and *tsk-tsk* when disappointed.

Such sensory impressions remained embedded within the ape-like vocalisations of *Homo habilis* for a million years or more. They were expressed as segments of sound within longer holistic calls, often combining what we call vowels and consonants in a succession of rapid exhalations of breath. Those individuals who were most sensitive to the iconic segments gained advantage: they acquired knowledge about the world from the synaesthetic calls of others – knowledge about predator danger, the whereabouts of food and the likely behaviour of other individuals. Similarly, those who, by chance, had relatively high voluntary control over embedding iconic segments into their own calls, controlled the knowledge and behaviour of others – eliciting help to protect a carcass from hyenas but ensuring no one else knew about a cluster of berries until they were ready to tell. Although acquiring their own independent meaning, the iconic segments remained too inconsistent in their sounds, lacked sufficiently well-defined beginnings and ends, and were too dependent on the stimulus being present to be considered as words.

A capability in the new art of exploiting iconicity within calls was not the only means by which reproductive advantage was gained. Multiple selective forces were driving human evolution in a myriad of ways. The value of reducing the energetic cost of locomotion, of being able to carry tools and food when walking, and reducing exposure to the sun, drove the evolution of bipedalism from the competent gait of *Homo habilis* to the efficient walking and running of *H. erectus*. This enabled longer distances to be travelled and larger areas to be searched within the course of a single day. The value of throwing rocks accurately and over long distances to scare off hyenas from a desirable carcass or ward off an aggressive lion promoted the evolution of hand–eye coordination and shoulder joints similar to those we have today.

The challenges of living within a larger group gave a social advantage to those individuals who could use their statistical learning to predict the behaviour of others in ever greater detail and in more circumstances. By knowing who to trust, collaboration increased: when some went looking for meat, others dug for tubers; their gains were pooled and shared. Selective advantage also came from carrying stone tools around, using the same tool for a variety of tasks, and as a ready source of razor-sharp flakes whenever slicing through flesh or cutting plant stems was required. Such tools were best made by shaping a nodule into a form that fitted easily into the hand by removing flakes from both of its faces.

The abilities to forage more widely and collaborate more effectively, while always having a tool at hand, enhanced the quality of the diet: more meat, riper fruits and fleshier tubers. As the need for chewing dry seeds and tough roots diminished, the robust structure of the face and size of the molar teeth did the same. Two evolutionary trends came into conflict: efficient bipedalism required a narrow pelvis, while larger brains required a wider birth canal. This was resolved in favour of the former. Babies were born in a premature state for a mammal of the size of *H. erectus* and had to maintain a foetal rate of growth outside the womb. A new stage in human life history emerged: childhood.

The non-linguistic developments had an impact on vocalisation and gesture. Anatomical changes relating to bipedalism, the carrying of a heavier brain, having smaller teeth and a flatter face, influenced the shape of the vocal tract. By 1.6 million years ago there was a different and a wider range of phonemes voiced across the African grasslands than when *H. habilis* and the australopiths had made their calls and cries. Facial muscles were now as important for making several sounds on the same exhalation of breath as they were for chewing food. The ability for muscular control that enabled hammerstones to be manipulated when making bifaces did the same for the vocal tract, enabling rapid phonetic change. Childhood enhanced the opportunity for vocal learning while creating a learning bottleneck – a narrow window of time either to supplement or to suppress genetic instincts as to what calls to make in which circumstances. The calls with iconic elements were the easiest to learn and modify, infants and children having an inbuilt perceptual bias to synaesthetic sounds.

It was after this package of anatomical, cognitive and life-history traits had emerged that iconic sound segments were gradually transformed into iconic words. The instinctive drive for efficient and effective communication was the cause. Those individuals with the most developed statistical learning identified recurrent associations between sequences of synaesthetic sounds within a call and their referents in the world, whether an approaching snake, a pool of water or an abandoned carcass. The challenge of identifying such associations was eased by the hominin making the call and the one who was listening having the same cross-modal, synaesthetic perception of the world, this narrowing down the possible referents of the call for the listener. When a recurrent association was identified, the drive for efficiency eroded the peripheral and meaningless sounds to isolate its essence into a stand-alone segment of iconic sound. Those individuals who were most adept at making, perceiving and acting on such sounds gained the most knowledge about the world and secured reproductive advantage: they were better communicators, more effective in the social world, acquired more and higher-quality food to eat and share. They had more sex, as did their offspring, enabling their genes and cultural habits to spread through the population.

As iconic sounds passed through successive learning bottlenecks from generation to generation, their acoustics were shaped into a consistent form, helped by the evolutionary loss of vocal membranes and air sacs. Because the meanings of iconic sounds were embedded in the sounds themselves, they could be used and understood by others without their referents being present. That was of considerable value – an iconic sound with a pointing gesture made by a hominin standing upon a rise to others below communicated either run away or run towards, depending on whether the iconic sound was for an approaching lion or a tree in fruit. The combination of the learning bottleneck and capabilities for statistical learning resulted in iconic sounds becoming combined into short sequences, two or three sounds strung together with each contributing to an overall meaning, their order being another aspect of iconicity by capturing the sequence of events. With their acoustic consistency, displacement and use in novel combinations, what had once been iconic sound segments embedded within ape-like calls had slowly become human-like iconic words.

The same trajectory of change was happening with stone tools. The irregularity of form and one-off use of Oldowan cores and flakes were replaced by consistently shaped, multiuse bifacial tools. The making of such tools was helped by speaking a few iconic words of instruction to oneself, and then sharing them with others to enable the transmission of technical knowledge from one generation to the next. But with the restriction imposed by verbal iconicity, a technological threshold was quickly reached that could not be overcome for more than a million years.

Like the new iconic words, bifaces could be carried around the landscape for use as needs arose. Such was the reproductive advantage gained, that by 1.5 mya what we now call Broca's area had developed within the *Homo erectus* brain, supporting the production of both words and tools. Bifaces were now labelled by using an iconic word, with the most proficient word users devising separate iconic labels for the handaxe, pick and cleaver. This turned them into objects for mental reflection, contributing to the refinements in their manufacture. As from 700,000 years ago, handaxes were made to be thinner and more elegant, often with a carefully imposed form and degree of symmetry far beyond the requirement of a butchery tool. By doing so the knappers displayed their technical skills, these combining planning, hand–eye coordination, physical strength and aesthetic sensibilities. Those watching took note, especially those of the opposite sex, whether male or female, who were looking for signs of good genes they might secure for their offspring.

With the new iconic words, language had arrived, although it sounded quite different to that which we have today. *H. erectus* had limited ability to control its breathing, and hence rarely made more than a few words on each exhalation of breath, these often remaining within utterances that contained ape-like calls and cries. But as iconic words proliferated, their combinations became longer, and their information content increased. As words acquired an independent status, the remaining calls and cries lost whatever semantic content they had once contained. They became dedicated to expressing emotional states by increasing their use of timbre, pitch and rhythm.

Such calls induced emotional states in others and were thereby effective at building social bonds of the type that cannot be achieved via words. Mothers were now rocking and singing to their babies; groups of adolescents had begun to chant together before embarking on challenging collaborative tasks. The new-found sense of rhythm not only enabled the hominins to sing and dance, but also aided with physical tasks, such as long-distance running across the savannah when following injured game. By getting into rhythm, vocally and physically, hominins evolved two distinctive forms of communication: strings of iconic words rich in information that would evolve into language; brains and bodies alive to melody and rhythm that provided the foundation for music. But there was no strict division: even at this early stage of language, the meanings of words and phrases were nuanced by changing their pitch and their stress, by adding pauses and changing their tempo. The most expressive words, those that invoked a multi-sensory experience such as the heat, aromas, crackle and flames of a savannah fire or the thunderous charge of an angry bull elephant, made use of rhythm, repetition and gesture. Linguists today would call them ideophones.

The evolution of language

Pre-iconic-word-using *H. erectus* had already dispersed from Africa as part of a mammalian diaspora, responding to climate change that had transformed northern latitudes into habitable land. *H. erectus* went further than other mammals, notably to the east, swiftly reaching what is now called Java by 1.6 mya. The later iconic-word-using *H. erectus* in Africa also dispersed when climatic conditions allowed, spreading throughout the continent and further afield. Aided by iconic words, they quickly adapted to a wider range of environments than their pre-iconic-word-using counterparts. Some left their African roots behind to enter the higher latitudes of Europe and Asia. Without the ability to make fire, there was a limit to their tolerance of cold. This caused them to abandon landscapes when the climate turned against them, only to return when it turned again. Throughout this time *H. erectus* continued to evolve, with populations becoming isolated from each other and adapting to their new environments. The tools they made were conditioned by the raw materials available and changes in need, but merely drifted around the constant themes of bifaces, flakes and cores.

As groups dispersed, their iconic words were moulded by new sensations: the hominins encountered animals in shapes and sizes, and with movements and habits never experienced before; such qualities had to be captured within a novel sequence of vowels and consonants and formed new iconic words. In North Africa, western Asia and Europe, the hominins delighted in finding flint: a material easier to knap and with far sharper edges than the basalt and quartzite they had used previously: accordingly, their iconic word for stone also took on a sharper edge. Similarly for new types of plants, weather, topographic features and whatever else was necessary and possible to communicate within the limitations of iconic words and gestures. These could only make categorical descriptions, capturing no more than generalities of sound, size and shape, speed and movement, and maybe some aspects of texture and colour.

Linguistic diversity began to emerge but was severely constrained by the reliance on iconicity. Those hominins living on the cold steppe of Europe amid mammoth, musk ox and reindeer developed a different set of iconic words to those on the grasslands of Africa who needed to invoke zebra and python. But strong similarities remained: the iconic word for mammoth was nearly identical to that for elephant because of the animals' similarities in shape and size. Likewise, the iconic words used in Europe and in Africa for lions remained much the same, these carnivores living on both the European steppe and the African savannah, having once been part of the mammalian diaspora. Within the evergreen forests of eastern Asia, *H. erectus* developed a different range of iconic words including one for bamboo, a material that could replace stone as a raw material for cutting tools.

Linguistic diversity also began to emerge via the transmission of language from one generation to the next and from routine talking and listening as iconic words were misheard, mispronounced and misunderstood. An instinctive desire for efficiency and effectiveness led to sound erosion, the merging of words and compounding whenever a new iconic word was required. The language-acquisition bottleneck effect continued with greater degrees of compositionality emerging, enabling a wider range of novel utterances to be made.

Whatever its source, linguistic diversity was tightly constrained because the *H. erectus* brain was tied to a synaesthetic perception of the world. Its size imposed another constraint: varying between 600 and 1,250 cm^3, this had a limited capacity to store new concepts and their associated words. Arbitrary elements may have been introduced into some iconic words, enabling differentiation between similarly sized and looking animals, in the manner that purely iconic words cannot achieve.

But if such hybrid words drifted too far from their iconic roots, their meaning became lost, as did the word itself.

Wherever *H. erectus* communities were located, and whatever types of environments they lived in, their languages had a high degree of similarity. A modern-day linguist travelling through their world and the million years of their existence would describe the *H. erectus* languages as unbelievably monotonous, adopting the same phrase that archaeologists use when describing their Acheulean stone tools. Just as language had reached a threshold, so too had technology: bifaces were made for more than a million years in all regions of the inhabited world.

Whereas the synaesthetic brain had been essential for the invention of words, it was now a constraint on enlarging the lexicon and evolving language. Selective advantage shifted to those individuals whose childhood synaesthesia dissipated as they became adults – just as happens in modern humans today. Advantage was also gained by those with a larger brain that had more capacity for storing words with arbitrary elements, the meaning of which had to be remembered rather than intuitively grasped. Such words naturally arose through the constant intergenerational shuffling of sounds and meanings but were unable to gain a foothold in the lexicon until the larger and less synaesthetic brain had evolved.

Both constraints, those of iconicity and insufficient storage, were partially lifted by the time of *Homo heidelbergensis* at around 750,000 years ago. Brain size reached 1,100–1,400 cm³, encompassing the lower range of later Neanderthals and modern humans. The selective pressures for brain expansion had remained the same: the need to predict the behaviour of other individuals, ideally by understanding their mental states, including the holding of false beliefs; the continuing need to understand the natural world by knowing the habits of animals, monitoring the changing seasons and reading natural signs; the need to turn raw materials into tools, not just stone but also wood, bark and plant fibres, bone, horn, hide, sinews and guts, and even the dried footpads of dead elephants that could provide ready-made trays for carrying. Although the selective pressures remained the same as in the distant past, as did the process of natural selection via genetic mutation, inheritance, competition and differential reproduction, the consequences for the structure and working of the brain were now profoundly changed.

Statistical learning and other general-purpose learning methods had reached a threshold within the brain of *H. erectus*. They were unable to improve any further because there are only so many regularities that can be identified in the world and limits on the inferences they allow. Consequently, via the combined forces of natural selection and cultural experience, the brain developed multiple areas of functional specialisation, areas of the cortex and extended neural networks that were dedicated to receiving specific types of input, storing specific types of knowledge and processing these in specific ways. Some of these may have related to specific aspects of language processing, while there were at least three bundles of specialised mental processes for engaging with the social world, the natural world and for manipulating physical materials and objects, used for making tools from wood and stone. Each delivered domain-specific ways of thinking and stores of knowledge.

The combination of reduced synaesthesia, functional specialisation and increased storage capacity liberated the lexicon. No longer was this constrained to iconic words, although such words remained essential, because they were easy for children to acquire, supported their learning of the new and more difficult arbitrary words, and were an effective means to describe the world in generic terms. Statistical learning also remained essential: not only had the number of words proliferated but also the speed at which they were said, requiring language-learning children to find the words within what had become a continuous stream of sound.

Arbitrary words emerged as the sounds and meanings of iconic words were transformed by their repeated use involving mishearing, mispronunciation and misunderstanding. While the passage of words through successive language-learning bottlenecks played a key role in this transformation, so too did the day-to-day talk within the speech community – the billions of utterances spoken and heard across thousands of years. These were made within socially dynamic groups, by the old, adolescents and the young, by those with higher and lower status, by men and women, by those with either good or poor articulation and hearing, as a means to build social bonds, to cooperate and to compete.

Driving and shaping this process was, as ever, the instinctive desire for efficient and effective communication: eroding unnecessary sounds within a word, merging words together to make them easier to say, compounding existing words when a new word was required, and the constant influence of semantic change. Although the majority of changes were unconscious and unnoticed even after they had occurred, there were occasional instances of deliberate change when individuals or groups sought to affirm their identity by how they pronounced their existing words and invented new ones – often led by female adolescents. Some iconic words became arbitrary by maintaining their sounds but changing their meanings; others did the converse, changed their sounds while maintaining what they meant. The newly evolved ability for mind reading aided the learning of arbitrary words by enabling insights into the minds who uttered them; conversely, new words enhanced the ability to understand the intentions, desires and beliefs held by other minds.

The enlarged lexicon, with its words/concepts stored throughout the brain, enriched communication and changed behaviour. Hominins could share ideas and knowledge, enabling new concepts to arise that could not have been achieved by one mind alone. This enabled *H. heidelbergensis* to further expand its range, moving to higher latitudes in Europe after 600,000 years ago. One of its innovations was the making of fire, this becoming habitual soon after 400,000 years ago. Bifacial knapping and handaxes were replaced by a diversity of tool-making methods using prepared cores to produce flakes, points and blades, providing a more efficient use of stone. The Levallois technique became widespread by 300,000 years ago. Interest was shown in coloured minerals for use as sunscreen and as an insect repellent; throwing spears were made. All these items were labelled with words. This not only enabled members of the speech community to talk about such objects but also helped individual minds to devise and manage the thought processes required for their manufacture and use.

Language evolved not only by increasing the number of words, but also by formulating words of a different kind. The previous reliance on iconicity had constrained the use of proper names, those used for specific individuals and things rather than for general categories. These now flourished, with all members of social groups acquiring their own specific name and hence identity.

As people conversed and language was acquired by each new generation, a combination of random errors, individual decisions and an underlying drive for efficiency and effectiveness transformed some object words into those for action – nouns into verbs. And the converse: action words became nouns. Further transformations followed, with nouns and verbs becoming adjectives and adverbs. These were used to enrich communication about the world – more detailed descriptions and finer categorisations of animals and plants, of weather conditions and types of materials, of colours, sounds and odours, of how people looked, moved and behaved towards others. The only constraint was that lexical words were anchored in sensations, referring only to what could be seen, heard, smelt, touched or tasted: the lexicon was constrained to concrete words.

A host of grammatical words slowly developed, the *the*-s, *him*-s, *her*-s, *they*-s and *them*-s, the *in*-s, *at*-s and *on*-s, the *this*-es and *that*-s, the *but*-s, *and*-s, *for*-s and *when*-s of language. These emerged as the cumulative outcomes of language transmission from person to person, generation to generation, over thousands and tens of thousands of years – a slow, incremental development of linguistic complexity and the required neural networks within the brain. What had been simple descriptions of past events and future plans using short strings of iconic words now became sequences of noun phrases combined with verb phrases – the first stories with actors and events.

Other than nouns and verbs, which were ubiquitous, all or none of the new types of words might have evolved within any one speech community of *Homo heidelbergensis*. Similarly, speech communities evolved different rules for how words should be ordered and modified to change their meaning. Linguistic diversity flourished wherever these large-brained, cognitively specialised, non-synaesthetic hominins found themselves at home.

The advantage gained from having a larger and more diverse lexicon had finessed the vocal tract into its modern form by the time that *H. heidelbergensis* was living and dying at the Sima de los Huesos in Spain at 450,000 years ago. The genes required to build that vocal tract were inherited by its two descendants, *H. neanderthalensis* in Europe and *H. sapiens* in Africa. Similarly for the auditory tract, although that of *H. heidelbergensis* maintained some ancestral features, and this took slightly different forms in the two descendant species creating subtle differences in the acoustic sensitivities of *H. neanderthalensis* and *H. sapiens*.

Whereas the constraint of iconicity had been lifted, two others remained, putting a brake on language evolution and leaving the languages of *H. heidelbergensis* and its immediate descendants quite different to any of those we have today. The first constraint was the ongoing cycle of climate change.

Between 500,000 and 100,000 years ago, there were four inter-glacial periods, each as warm and wet as it is today; each was followed by a gradual but choppy return to a glaciated world, with a fall in sea level, expansion of glaciers in the high latitudes and deserts in the low. The cycles of climate change caused populations to repeatedly expand, disperse and fragment, with frequent local extinctions – there was no stability into which cumulative cultural change could take hold. Just as a language was reaching a level of complexity, perhaps with its first syntactical structures, its speech community crashed, returning the language to strings of unordered words, from which the slow iterative process towards linguistic complexity had to begin again.

The second constraint was within the brain. Whereas the pruning of neuronal connections during childhood had removed the iconicity constraint in the adult brain, another had arisen: domain-specific mentality. The brain's new areas of functional specialisation took particular types of input which were processed in particular types of ways for discrete domains of knowledge, enabling complex computations of a type that statistical learning could never achieve. These came at a cost, however, because the neural networks for each domain were isolated from each other. Knowledge and the ways of thinking required for activity in the social world were detached from those used for engaging with the natural world and for making tools. The only means to make connections was via the neural networks still used for statistical learning, but these struggled to access the expert knowledge generated by each functional specialisation.

Domain-specific mentality constrained the nature of hominin thought and behaviour. Although *H. heidelbergensis*, followed by *H. neanderthalensis* in Europe and *H. sapiens* in Africa, were able to excel in working and shaping raw materials, building diverse and complex social relationships, and in understanding the natural world, they were unable to blend together the knowledge and ways of thinking used within each domain. This severely constrained their capacity for creative thought.

The Neanderthals exhibited outstanding skills in knapping stone, shaping wood and combining materials to make spears with hafted points and no doubt a multitude of other multi-component tools that have left no trace. But whether in Europe or western Asia, whether living in ice age steppe or in woodland, whether at 250,000 or 50,000 years ago, they made the same range of stone flakes, blades and tools. While this was not the monotony of the Acheulean it amounted to millennia of minor variations around a single theme of prepared core technology with no directional change.

The Neanderthals' domain-specific mentality prevented them bringing together knowledge of their prey with their ability to shape materials to invent the bow and arrow and the spear thrower – wooden tools with a hook to provide extra leverage when throwing a spear. They were unable to devise hunting weapons that would not only target specific types of prey in specific circumstances but would also do so from a safe distance. Without the spear thrower, their long spears were thrown with insufficient power to make a clean kill and were more frequently used for close-quarter stabbing. Many Neanderthals died young from hunting injuries acquired by having to attack their prey with general-purpose short thrusting spears, often being tossed or trampled in the process.

Similarly, Neanderthals struggled to use material culture for social interaction, such as to denote personal and group identity, or to send messages about status, relationships and personality. The most they achieved was the use of ready-made products from the natural world, the feathers and talons of birds, and, most likely, visually striking coloured furs. General-purpose learning sufficed for deciding that wearing eagle feathers identified oneself with the power of that bird but could not blend social and technical knowledge to devise objects such as strings of beads that carried their own custom-made messages. Marks were occasionally made on cave walls, on bits of bone and stone, but these were merely from their habit of intervening in the material world, usually applied to functional tasks such as making tools or building shelters. The marks were enjoyable to make and pleasing to see, but they had no meaning and were as unintelligible to other Neanderthals as they are to us today.

Neanderthal languages, along with those of the Denisovans in central Asia and the earliest *H. sapiens* in Africa, evolved under the constraint of this domain-specific mentality and the disruptive impacts of environmental change. Their multitude of languages had significant similarities and important differences to ours today. They all had iconic, hybrid and arbitrary words; they all differed in their range of phonemes, word classes, rules of morphology and syntax, just as we find among languages in the modern-day world. They used prosody, gestures and body language to nuance and sometimes change the meaning of their words entirely. The content of their lexicons reflected what was of importance to each speech community: Neanderthals living in northern environments had more words for snow than those in southern climes.

The extent to which languages shared words and grammatical structures reflected how long their speech communities had been apart. As groups fragmented and contact diminished, dialects developed which were eventually replaced by languages that were unintelligible to each other. Four major language families evolved in the Neanderthal world associated with demographic clusters in western Europe, southern Europe, western Asia, and one in the east that stretched into central Asia. Within these clusters the Neanderthals lived in small groups that frequently required inbreeding to survive, the languages themselves also becoming inbred. Because so much knowledge was already shared within the close-knit speech communities, little had to be said to communicate a thought, pragmatics doing much of the work. Neanderthal words evolved to have few phonemes; they became long, morphologically complex and placed within inefficient grammatical structures. Bottlenecks during language learning hardly existed because children would hear the entire lexicon and utterances of their language before they came of age. Neanderthals from one community struggled to understand and learn the language of another.

If we could have listened to Neanderthal chat, we would have been struck by how nasal it sounded, how their plosives – their /t/ees, /p/ees and /b/ees – were relatively loud, and how their utterances were of such long duration. These features arose from having larger nasal cavities and lung capacities than found in modern humans. The frequency of ideophones, those easily recognised iconic words that most vividly express multi-sensory experience, would have also been striking. Moreover, if a translator was available, we would have noticed something different from all the languages found in the world today: an absence of metaphor and abstract words, those whose meanings cannot be defined by sensory experience alone. Abstract words were absent because the domain-specific mentality was unable to support abstract concepts, these requiring the use of analogies and metaphors that draw on multiple domains of knowledge. Because of their domain-specific mentality and just like their *Homo heidelbergensis* forebears and early *Homo sapiens* in Africa, the Neanderthals remained constrained to the use of concrete words – those anchored in the external world.

The flowering of language

Both the Neanderthals in Europe and *H. sapiens* in Africa experienced a constant trickle of genetic mutations, chance switches of one of the four (A, C, G, T) types of nucleotides for another in the 3 billion pairs of nucleotides that made up each and every human genome. The majority of mutations had no impact; some were marginally beneficial or detrimental; others were profound. A few that occurred within and spread through the populations of *H. sapiens* in Africa influenced the still evolving brain: the cerebellum became larger and changed shape; the occipital lobes reduced in size, constraining visual processing but releasing neural capacity for other functions; neural networks extended their reach. Other mutations slowed down the rate of child development enabling a longer period for the new long-distance neural networks to be sparked into action.

Major changes occurred after 300,000 years ago: the *H. sapiens* brain became more globular in shape, a process that required at least another 200,000 years to complete. While brain size may have increased marginally, long-distance neural networks and new patterns of wave-like neuronal activation through the globular brain now connected functionally specialised areas and turned the brain into a global workspace. Stores of knowledge and ways of thinking that had evolved to deliver specialised functions, those which had remained isolated in the *H. heidelbergensis* and Neanderthal brains, could now come together. With such cognitive fluidity came an enhanced level of verbal fluency, delivered by the enlarged cerebellum that influenced a wide range of linguistic processes in the brain.

Cognitive fluidity was initially sparse and transient but provided sufficient connections to trigger a linguistic and cognitive revolution, one that needed language itself to complete. Two new features were added to the linguistic repertoire. The first was metaphor, whether spoken or signed: the use of knowledge about one conceptual domain, the source, to inform that of another, the target. By definition, the domain-specific mentality of Neanderthals had prevented the use of metaphor as a tool for communicating and thinking about the world. With cognitive fluidity, a *Homo sapiens* mother could describe her daughter as being as brave as a lion, while believing that lions had human-like thoughts and desires; time could be described as space; and space by words derived from the human body. The translation of metaphorical concepts into spoken utterances, and the reverse by those who were listening, had been enabled by changes in the human brain: the release of neural resources from visual processing into that of language, the enhanced linguistic functions from the enlarged cerebellum, and the extended connectivity that turned the brain into a global workspace.

With metaphor came symbolic thought – the use of one thing, whether a natural object or a made artefact, to represent something else. Metaphor also enhanced communicative power, including the ability to describe and explain complex technological skills and ideas to others. These could be more easily shared, their development becoming a collaborative process between minds. Spear throwers and bows were quickly invented, soon followed by a host of new types of tools that could only be devised and made by the meeting of metaphorical minds. As with language itself, there was a new dynamic to technological change – one that continues apace today. After beginning to live by metaphors, we have been doing so ever since.

As metaphors were used, the required neural connections became stronger by the repeated firing of their synapses: language was now building the brain it required. The use of metaphor released a degree of creativity and complex thought that was impossible to achieve within a domain-specific mind and provided a new dynamic to language change. Once used, each metaphor began to lose its power and eventually had to be replaced by another to create the desired social effect: the lexicon expanded exponentially; new metaphors were born as soon as others died.

Abstract concepts could now flourish within the human mind, these requiring metaphors to be understood and explained to others. Abstract concepts needed to be anchored in the mind with their own labels: abstract words. The Neanderthals had none of these, being reliant on concrete words, whose meanings were defined by experience – the sight of a mammoth, sound of the wind, smell of a lion, touch of a stone, or the taste of a ripe berry. As such, their lexicons had little impact on how they perceived the world. Modern humans used linguistic metaphors to explain and understand abstract concepts – ideas about other worlds and ancestral beings, notions of justice, identity and ownership.

Such concepts could be assembled into new categories with no grounding in the world, which then eased the invention, learning, recall and explanation of further abstract concepts that now had a category to sit within. Once the category of the supernatural was conceived, this provided a home for new concepts and their associated words, such as ghosts, spirits and spells. The lexicon continued to expand and stories took on new dimensions with complex plots, heroes and villains, magical events and imaginary worlds. Storytelling became an art form for impressing, persuading, educating and entertaining by the skilful use of words.

As the globular, cognitively fluid brain evolved, such developments were happening throughout Africa, from the far north to the south, and from east to west. Communities were connected by social networks enabling a continent-wide flow of new ideas, words, objects and genes. Cultural and linguistic diversity flourished at a level unimaginable for both the Neanderthals and pre-modern *H. sapiens*, evident from new ways and new traditions for making stone tools. Nevertheless, with the impact of climatic instabilities on low and often fragile populations, it took time for metaphors and abstract words to become sustained within language – for a while they came and went, as did communities and whatever languages they spoke and artefacts they made.

Their impact only becomes materially evident after 200,000 years ago as red ochre gradually shifted its role from providing indexical signs to harbouring symbolic meaning. By 150,000 years ago, shells with natural holes were being worn on strings; by 80,000 years ago, shells were being deliberately perforated and coloured to provide more extravagant displays and send symbolic messages about identity and belief. By 70,000 years ago, slabs of ochre provided platforms for incising geometric images, replete with symbolic significance.

Such objects played an active role in building the required neural networks for metaphorical phrases and abstract concepts to further develop. Shells, with their seemingly crafted shapes and shiny colours, coming from the mysterious sea and once housing strange creatures, lent themselves to stories and use as symbols. As shells beads were made, worn and seen by others, they repeatedly sparked the synapses of the incipient neural networks that delivered the cognitively fluid mind. So too did the use of ochre, engraved geometric images and fire. While fire had been managed and habitually used for over 300,000 years, by 100,000 years ago its night-time flames were supporting a linguistic workout for the use of metaphor and abstract words as talk of the mundane was replaced by that about the world of spirits and daemons. Not just of the weird and wonderful, but also the funny and absurd. Puns, double entendres and innuendos, all reliant on metaphor and the verbal fluency of the modern mind, now pervaded language. These gave modern humans a joy of words that remained absent among the domain-specific Neanderthals. *Homo sapiens* laughed their way into modernity.

This was the flowering of language. Its roots had been ape-like holistic utterances made by australopiths and *Homo habilis*; its stem, the iconic words first spoken by *Homo erectus* at 1.6 million years ago. Arbitrary and grammatical words had grown as shoots from that stem within the minds and voices of *Homo heidelbergensis* and its descendants, the Neanderthals in Europe and early *Homo sapiens* in Africa. But only at 100,000 years ago did the blossoms appear: a multitude of metaphors and abstract words coming in a wealth of colours, textures and

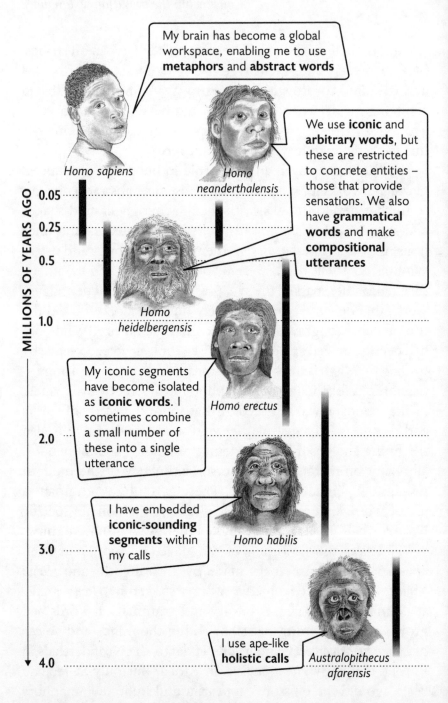

Figure 16 The evolution of language

shapes, spreading their seed and cross-fertilising throughout Africa – and soon the world.

With that knowledge of language evolution, we can now put some flesh onto what had been the dry and silent bones of our human ancestors (Figure 16).

Global diaspora

The next part of the human story is well known. By 60,000 years ago, at least one and probably several communities of *Homo sapiens* had dispersed from Africa, travelling north into the Levant and east into Arabia. Their descendants kept on moving while their bodies and brains kept evolving, creating genetic diversity within modern humans as populations fragmented and went their separate ways on the greatest ever diaspora of humankind.

Some interbred with Neanderthals, gaining useful genes for living in new northern climes and adopting a few new words and phrases along the way, especially the expressive ideophones that the Neanderthals used to such good effect. Small exploratory incursions were made into Europe, before groups wearing elaborate beads, geared up with an assortment of expertly designed hunting weapons, and prone to carving animal figurines from bone and stone, swept across the region at around 41,000 years ago. The resident Neanderthals responded with a final flourish of their culture influenced by the visual symbols and the new technology they saw but struggled to comprehend. They interbred with the new arrivals in an effort to boost their numbers. But Neanderthal populations dwindled into extinction, unable to survive the combined pressures of climatic instabilities and their new neighbours who hunted more effectively and protected themselves from the cold with stitched garments and substantial dwellings. The modern humans decorated cave walls with paintings of animals and imaginary beings, epitomising the cognitively fluid, meta-phorical and abstract-concept-rich modern human mind. A mind underpinned and sustained by its flow of words.

Some of the modern humans who had travelled east from Africa encountered and interbred with the Denisovans and possibly other human types that science has yet to discover, which also went extinct. Moving on, boats were built to reach Australia by 60,000 years ago and snowshoes were made for crossing the Bering Strait into the Americas at 30,000 years ago. Human groups swiftly dispersed throughout both continents, adapting their lifestyle to steppe, mountains and woodlands, deserts, rain-forests and coastlines. More species went extinct, not humans but the megafauna of Australia and the Americas – mammoths, mastodons, giant sloths and giant flightless birds. As such losses occurred, human populations flourished, as did cultural, linguistic and genetic diversity.

Throughout the world, technology and all aspects of material culture were on the move, constantly changing under a myriad of environmental and cultural factors. Language change was both a cause and a consequence: a stream of new words and new ways to put them together, influencing how people perceived and thought about the world, enabling them to assemble new concepts about life, the universe and everything.

People spoke, thought and acted within the constraints and opportunities of their environments. Regions of low biomass could support only sparse populations, and hence there were fewer minds to share their words, build new concepts and drive cumulative technological change. Elsewhere, the seasonal abundance of resources, such as migrating shoals of salmon along the Pacific Rim and migrating reindeer herds on northern tundra, required new technology to exploit and provided opportunities for social gatherings. These were occasions for the exchange of information, ideas, artefacts, foodstuffs, genes and people who moved from one social group to another. The gatherings provided opportunities for new words, new meanings and new pronunciations to spread, especially when people from distant communities came together speaking dialects or who had already transitioned into languages of their own.

All hunter-gatherers, whether digging up tubers in the Australian deserts, trapping seal in the Arctic or hunting peccary within the Amazon, continued to live in a world of dramatic climate change. The repeated glacial–interglacial cycle that had persisted for over 2 million years was in a downturn at 40,000 years ago, pushing the planet into the depths of a glaciation. The nadir arrived at 20,000 years ago causing environmental disruption throughout the world. Once again, human populations became fragmented, declined in number, and were entirely lost from the most severely affected regions. But with metaphorically rich language and cognitively fluid minds, human communities had a new resilience to environmental change: dwellings were built with mammoth bones and covered with hides and turfs; bone needles sewed reindeer furs together; fires warmed the insides of caves.

Climatic amelioration arrived soon after 20,000 years ago. First, there was a gradual and then a rapid rise in temperature, achieving the levels that we enjoy today by 14,500 years ago. Human communities responded, once again expanding their range and inventing new technology, now to exploit the warm-loving animals and plants that were also dispersing from glacial refugia as deserts became grassland and woodland replaced tundra. But the climate soon crashed again. The planet was plunged into another 1,000 years of aridity and cold, causing populations to retract and demographic decline.

Relief came at 11,500 years ago with a second rapid bout of global warming. That ushered in the warm, wet and stable post-glacial climate that we continue to enjoy today and call the Holocene. Once again, hunter-gatherers expanded their range and developed new technologies for exploiting the warm-adapted animals and plants that were soon beginning to thrive.

This was the first time that humans were living within such clement and stable conditions since the fully modern linguistic capacity had evolved: a combination of iconic, arbitrary and grammatical words, coming in both concrete and abstract forms, a pervasive use of metaphor, a range of word classes, morphological and syntactical structures, all manifest in a multitude of constantly changing languages throughout the world. The last time humans had lived in such environmental conditions was between 130,000 and 115,000 years ago, before cognitive fluidity had fully emerged within *Homo sapiens* and when Neanderthals could only reluctantly respond rather than actively adapt to a warmer world. This time it was different.

The modern humans now viewed their world through the metaphorical eye of cognitively fluid and abstract ideas, all sustained by the words they used. As we know from those hunter-gatherers who survived until the modern day, rivers, mountains, waterfalls and caves were the work of ancestral beings who created the world. The animals they hunted were not only sources of meat and fat but could also be brothers and sisters, or even the ancestral beings themselves in another guise. The animals and plants were kindly given to humans for consumption, a gift that had to be repaid by ritual and respect for the natural world.

In most regions of the world, such attitudes had little material impact on the environment and human lifestyles. It was a cultural veneer: hunting and gathering continued, within the constraints and exploiting the opportunities that environmental conditions allowed. In some regions, however, the nature of the plants and animals was susceptible to change. Revolutionary consequences for human history were soon to follow.

How we talked our way out of the Stone Age

One of these regions was an arc of grassland, woodland, river valleys and uplands in western Asia that is known today as the Fertile Crescent. It spans from what is now the Gulf of Aqaba of the Red Sea, northwards along the Mediterranean coast to what is now southern Turkey, turns to the east before swinging south through the Zagros mountains towards the Persian Gulf. Hunter-gatherers had exploited this region ever since the first dispersals of *H. erectus* from Africa 1.8 million years ago. It had been the shared home of Neanderthals and early *Homo sapiens* before fully modern language had evolved. Throughout those times, the Fertile Crescent had harboured wild wheat and barley, wild peas, lentils and chickpeas, along with wild goats and sheep: the progenitors of domesticated varieties in the world today.

By 20,000 years ago, modern humans were living in the Fertile Crescent by hunting gazelle on the plains and wild goats in the mountains; they collected tiny seeds from wild grasses, nuts, berries, fruits and tubers; they went fishing and fowling in the surviving wetlands. They lived in small communities that were always on the move as the animals and plants in the vicinities of their camps became depleted.

When climatic conditions improved, the grassland flourished and human populations grew. By 14,500 years ago, the communities could remain within the same settlement for longer periods of time. They invested in architecture, made massive stone mortars to grind the wild grain they gathered, and placed their dead into cemeteries to claim the land. Then, at 12,500 years ago came the millennium-long climate crash.

The failure of winter rains, a drop in temperature and shortened growing seasons depleted the plant foods that could be gathered. Gazelle and other animals became scarce; fishing and fowling were unproductive. Earlier humans would have adapted by returning to a fully mobile lifestyle, living in smaller groups, retreating to any environmental refuge they could find. The modern humans – the fully modern language-using, cognitively fluid, metaphorically rich humans – were different.

They began to treat the plants around them as if they were their children: removing pests, providing water, transplanting them to better soils. By so doing, the plants, notably wild wheat and barley, responded and their productivity increased. After many cycles of cultivation and harvesting, their grain was of a larger size and remained attached to what had become a bulbous ear. That enabled the grain to be collected with ease rather than each tiny piece having to be picked from the ground as their forebears of 20,000 years ago had to do. The hunter-gatherers were soon harvesting the cereals, using sickles made with flint blades inserted in handles of bone.

By the time the Holocene arrived at 11,500 years ago, the hunter-gatherers were not only committed to cultivating cereals but were also on their way to full dependency. That was soon realised because the new rainfall and longer growing seasons further enhanced the yields. The hunter-gatherers could not stop tending their plants, because the plants had become dependent on human intervention: they were no longer able to seed themselves if the harvesting were to halt. Communities elsewhere, notably in the Zagros, were bringing the same caring attitudes to the wild goats they hunted, resulting in the same mutual dependency. Orphaned baby goats were looked after within the human community. As they grew and bred with other captive goats, the animals became more placid, losing their massive horns and becoming amenable to human control.

While such domestication of plants and animals had been reliant on metaphorical thought and language – using the caring words and concepts previously restricted to the social world – their management was made possible by the ability to invent new concepts, new words and new tools, these bootstrapping each other into existence. A new and ever-shifting lexicon about goats emerged, words for different types of wild, herded and fully domesticated goats, words describing their differences by sex, age, size, health, colour, behaviour and so forth. Some words may have been entirely new, such as that for a fully domesticated goat; others changed in their frequency from the time when goats were entirely hunted in the hills. A new set of words were invented for the new lifestyle of herding: those for pens, tethers, diseases, cuts of meat, types of fat, slaughter and so forth.

The same for plants. Just as modern-day rice cultivators have many words for rice, so too did the emergent Neolithic farmers invent many words for wheat and barley, distinguishing between the wild and domestic forms, stages of growth, healthy and disease-ridden plants. Not just for the plants themselves, but also for the processes of sowing, watering, weeding, grinding and storing. New words were required for new tools such as for hoes, grinding stones and granaries, these words not just describing such artefacts but anchoring new concepts in the minds.

As new words were emerging, others were being forgotten and then entirely lost. There was less interest in tracking wild animals and gathering wild plants, causing the words for the subtle differences in footprints, scats and ripening fruits to be spoken less frequently and gradually forgotten. Similarly, for words about odours that, when living as hunter-gatherers, had been as important as the words for colour and taste. Conversely, the words for colour expanded as craft activities increased and trade began between communities with the need for attractive items, notably words for green and blue to describe the stone beads that were now being made from copper-rich stone. The new diet of finely ground cereals reduced tooth wear and subtly altered the sounds of language, increasing the frequencies of /f/s and /v/s as words continued to be invented and changed.

By 10,000 years ago, the hunter-gatherers of the Fertile Crescent had become farmers. They had talked their way into this new lifestyle by using words to build both the concepts and the technology it required. The key had not only been the use of metaphor and abstract words, but also the compulsion for humans to talk and their joy in using words that had been gradually instilled within the human mind ever since language began to evolve more than 2.5 million years ago. By 10,000 years ago, people simply could not help themselves from talking about the world around them, about their relationships and new ideas. Nor could they help themselves from having fun with words. With such constant talk and chatter, dialogues and gossip, speeches, conversations and tête-à-têtes, it was inevitable that new concepts would arise, inventions be made, and lifestyles would change. It remains the same today.

Neolithic farming settlements flourished, populations grew and soon became towns of several thousand people, providing homes for craftsmen and traders, priests and politicians. Languages shifted from being esoteric, as characteristic of small, self-sufficient communities, to being outward facing, able to facilitate contact with people from speech communities elsewhere. Words became shorter and combined with a more consistent set of rules, enabling swift language learning by adult outsiders – those who came to trade, to find partners, to join what were now centres of innovation, entertainment and social change. With such shifts in language and population numbers, bilingualism and multilingualism became more frequent, if not the norm. This further enhanced the creativity and innovation found within the new farming communities.

The Fertile Crescent was not the only location in the world for this to happen. Soon after the Holocene had begun, farming also arose in China, where millet and rice, chickens and pigs were domesticated. A little later, so too in Mesoamerica with maize, squash and peppers, and then in New Guinea with taro and bananas. From these and other centres of innovation, farming spread throughout the world. This was largely by the dispersal of families who left their homelands to find new land to cultivate, sometimes interbreeding with the hunter-gatherers they encountered on their way. Hunter-gatherers who resisted the change in lifestyle were forced into the marginal zones that were too dry or cold for farming to take a hold.

Talking continued in the farming villages, towns and urban centres. Metaphors abounded as new concepts were shaped, new words invented and languages evolved through the ever-present learning bottlenecks from generation to generation, via the multitude of utterances spoken and heard every day, and under the ongoing desire for effective and efficient communication. Metallurgy was invented, bringing the Stone Age to its formal end. That was followed by writing, which took language on a new phase of its journey, one that had started with iconic words on the African savannah 1.6 million years ago.

It was during the long durée of language evolution that we became entirely dependent on words for every aspect of our lives. To maintain such dependency, evolution not only gave us the joy of words but made language the life force of being human.

Figure sources

All figures are drawn from multiple sources in the public domain via the internet, with addition of the following:

Figure 2: modified and extended from Bae et al. (2017).
Figure 8: artefacts redrawn (top down) from Kiura (2019), Roberts et al. (1997) and Shimelmitz and Kuhn (2013).
Figure 9: modified from Clark, J.D. (1993) and McBrearty and Brooks (2000), with artefacts redrawn from (top left, clockwise) Usik et al. (2013), Tryon and Faith (2013), Bushozi et al. (2020), Larsson (2000), Villa and Soriano (2010), Villa et al. (2009b), Bader et al. (2015), Mesfin et al. (2020) and Barton and d'Errico (2012).
Figure 14: redrawn from Max Planck Gesellschaft: www.mpg. de/11883269/homo-sapiens-brain-evolution
Figure 15: redrawn (top down) from Sehasseh et al. (2021), Radovčić et al. (2015), Henshilwood et al. (2009), Leder et al. (2021) and Soressi and Roussel (2014).

Acknowledgements

The Language Puzzle has had a long gestation. I feel I should thank everyone I've spoken to ever since acquiring my own first words. For the sake of brevity, I will be selective.

I would like to thank Giovanni Bennardo for inviting me to talk at the workshop on 'Cultural Model Theory: Shaping a New Anthropology' that took place in May 2022 at Northern Illinois University. Preparing for and participating in that workshop helped me formulate my thoughts about how language may have evolved. I met the linguist Stephen Levinson at the workshop, whose work I greatly admire. He encouraged me with an offhand remark after my presentation saying, 'I think you may have something there.' My conversations with Stephen made me realise how little I knew about language, as he was evidently aware, and sent me on a journey into as much linguistics as I could manage. I hope it has been sufficient.

The University of Reading and its Department of Archaeology have always provided a stimulating and enjoyable place to work, with tremendously supportive colleagues. Rob Hosfield, my Head of Department, kindly read my chapters about archaeology and human evolution, correcting some matters of fact, and providing me with more to think about. Annemieke Milks did likewise while answering my queries about prehistoric hunting and kindly appeared in my office brandishing a two-metre replica of a Neanderthal spear to literally prove a point.

I am grateful to Andrew Franklin of Profile Books for asking me to write a book about farming and then acquiescing to my request to write about language instead. He provided great encouragement and helpful editorial advice, as did Janice Audet from Basic Books. The team at Profile Books have been exceptionally helpful and supportive. I am grateful to Philippa Logan for excellent copy-editing and to Rachel Becker for preparing the figures, both patiently coping with my persistent tweaking.

I am also indebted to my twin brother, Richard, who was the first person I ever talked to and whose wise words of advice, about this book and other things, are greatly valued. Nick, my son, drew my interest to the Enlightenment and the enduring value of its debates about language. I am also grateful to Heather, my daughter, for our discussions about children's minds. Words cannot express my gratitude to Sue, my wife. She patiently listened to my ideas, read and commented on my text, discussed and debated with me. Most of all, Sue shared her own joy of words, inspiring me to find my own.

Bibliography

Ackermann, L., Hepach, R. and Mani, N. (2020). Children learn words easier when they are interested in the category to which the word belongs. *Developmental Science* 23: e12915.

Adler, D.S., Wilkinson, K.N. et al. and Gasparian, B. (2014). Early Levallois technology and the Lower to Middle Paleolithic transition in the Southern Caucasus. *Science* 345, 1609–13.

Aiello, L. (1996). Terrestriality, bipedalism and the origin of language. *Proceedings of the British Academy* 88, 269–89.

Aiello, L.C. and Dunbar, R.I.M. (1993). Neocortex size, group size, and the evolution of language. *Current Anthropology* 34, 184–93.

Akhilesh, K., Pappu, S. et al. and Singhvi, A.K. (2018). Early Middle Palaeolithic culture in India around 385–172 ka reframes Out of Africa models. *Nature* 554, 97–101.

Akinnaso, F.N. (1982). On the differences between spoken and written language. *Language and Speech* 25, 97–125.

Akita, K. (2013). The lexical iconicity hierarchy and its grammatical correlates. In *Iconic Investigations* (eds L. Elleström, O. Fischer and C. Ljungberg), pp. 331–50. Amsterdam: John Benjamins.

Albert, R.M., Berna, F. and Goldberg, P. (2012). Insights on Neanderthal fire use at Kebara Cave (Israel) through high resolution study of prehistoric combustion features: evidence from phytoliths and thin sections. *Quaternary International* 247, 278–93.

Alemseged, Z., Spoor, F. et al. and Wynn, J.G. (2006). A juvenile early hominin skeleton from Dikika, Ethiopia. *Nature* 443, 296–301.

Alperson-Afil, N. (2008). Continual fire-making by hominins at Gesher Benot Ya'aqov, Israel. *Quaternary Science Reviews* 27, 1733–9.

Alpher, B. (1994). Yir-Yoront ideophones. In *Sound Symbolism* (eds L. Hinton, J. Nichols and J.J. Ohala). pp. 161–77. Cambridge: Cambridge University Press.

Aranguren, B., Revedin, A. et al. and Santaniello, F. (2018). Wooden tools and fire technology in the early Neanderthal site of Poggetti Vecchi (Italy). *Proceedings of the National Academy of Sciences* 115, 2054–9.

Arbib, M.A. (2005). From monkey-like action recognition to human language: an evolutionary framework for neurolinguistics. *Behavioral and Brain Sciences* 28, 105–67.

Arbib, M. and Bickerton, D. (eds) (2010). *The Emergence of Protolanguage*. Amsterdam: John Benjamins.

Arbib, M.A., Liebal, K. and Pika, S. (2008). Primate vocalization, gesture, and the evolution of human language. *Current Anthropology* 49, 1053–76.

Ardesch, D.J., Scholtens, L.H. et al. and van den Heuvel, M.P. (2019). Evolutionary expansion of connectivity between multimodal association areas in the human brain compared with chimpanzees. *Proceedings of the National Academy of Sciences* 116, 7101–6.

Arensburg, B., Harrell, M. and Nathan, H. (1981). The human middle ear ossicles: morphometry, and taxonomic implications. *Journal of Human Evolution* 10, 199–205.

Arensburg, B., Schepartz, L.A. et al. and Rak, Y. (1990). A reappraisal of the anatomical basis for speech in Middle Palaeolithic hominids. *American Journal of Physical Anthropology* 83, 137–46.

Aronoff, M., Meir, I., Padden, C.S. and Sandler, W. (2008). The roots of linguistic organization in a new language. *Interaction Studies* 9, 133–53.

Ashton, N. and Davis, R. (2021). Cultural mosaics, social structure, and identity: the Acheulean threshold in Europe. *Journal of Human Evolution* 156, 103011.

Ashton, N. and Stringer, C. (2023). Did our ancestors nearly die out? *Science* 381, 947–8.

Atkinson, E.G., Audesse, A.J. et al. and Henn, B.M. (2018). No evidence for recent selection at *FOXP2* among diverse human populations. *Cell* 174, 1424–35.e15.

Atran, S. (1990). *Cognitive Foundations of Natural History: Towards an Anthropology of Science*. Cambridge: Cambridge University Press.

Aubert, M., Brumm, A. et al. and Dosseto, A. (2014). Pleistocene cave art from Sulawesi, Indonesia. *Nature* 514, 223–7.

Aubert, M., Brumm, A. and Huntley, J. (2018). Early dates for 'Neanderthal cave art' may be wrong. *Journal of Human Evolution* 125, 215–17.

Bader, G.D., Will, M. and Conard, N.J. (2015). The lithic technology of Holley Shelter, KwaZulu-Natal and its place within the MSA of Southern Africa. *South African Archaeological Bulletin* 70, 149–65.

Bae, C.J., Douka, K. and Petraglia, M.D. (2017). On the origin of modern humans: Asian perspectives. *Science* 358, eaai9067.

Bankieris, K. and Simner, J. (2015). What is the link between synaesthesia and sound symbolism? *Cognition* 136, 186–95.

Barham, L., Duller, G.A.T. et al. and Nkombwe, P. (2023). Evidence for the earliest structural use of wood at least 476,000 years ago. *Nature* 622, 107–11.

Barrett, L.F. (2017). *How Emotions Are Made: The Secret Life of the Brain*. Boston: Houghton Mifflin Harcourt.

Barton, N. and d'Errico, F. (2012). North African origins of symbolically mediated behaviour and the Aterian. *Developments in Quaternary Sciences* 16, 23–34.

Barton, R. and Venditti, C. (2014). Rapid evolution of the cerebellum in humans and other great apes. *Current Biology* 24, 2440–4.

Bar-Yosef, O. and Bordes, J.-G. (2010). Who were the makers of the Châtelperronian culture? *Journal of Human Evolution* 59, 586–93.

Bar-Yosef Mayer, D. and Porat, N. (2008). Green stone beads at the dawn of agriculture. *Proceedings of the National Academy of Sciences* 105, 8548–51.

Bar-Yosef Mayer, D.E., Vandermeersch, B. and Bar-Yosef, O. (2009). Shells and ochre in Middle Paleolithic Qafzeh Cave, Israel: indications for modern behavior. *Journal of Human Evolution* 56, 307–14.

Bar-Yosef Mayer, D., Groman-Yaroslavski, I. et al. and Weinstein-Evron, M. (2020). On holes and strings: earliest displays of human adornment in the Middle Palaeolithic. *PLoS One* 15, e0234924.

Beals, M.E., Frayer, D.W., Radovčić, J. and Hill, C.A. (2016). Cochlear labyrinth volume in Krapina Neandertals. *Journal of Human Evolution* 90, 176–82.

Begus, K. and Southgate, V. (2012). Infant pointing serves an interrogative function. *Developmental Science* 15, 611–17.

Begus, K., Gilga, T. and Southgate, V. (2014). Infants learn what they want to learn: responding to infant pointing leads to superior learning. *PLoS One* 9, e108817.

Beier, J., Anthes, N., Wahl, J. and Harvati, K. (2018). Similar cranial trauma prevalence among Neanderthals and Upper Palaeolithic modern humans. *Nature* 563, 686–90.

Beller, S. and Bender, A. (2008). The limits of counting: numerical cognition between evolution and culture. *Science* 319, 213–15.

Benítez-Burraco, A., Torres-Ruiz, R. et al. and García-Bellido, P. (2022). Human-specific changes in two functional enhancers of *FOXP2*. *Cellular and Molecular Biology* 68, 16–19.

Bergelson, E. and Swingley, D. (2012). At 6–9 months, human infants know the meanings of many common nouns. *Proceedings of the National Academy of Sciences* 109, 3253–8.

Berger, T.D. and Trinkaus, E. (1995). Patterns of trauma among the Neandertals. *Journal of Archaeological Sciences* 22, 841–52.

Berk, L.E. (1994). Why children talk to themselves. *Scientific American* 271(5), 78–83.

Berlin, B. (1992). *Ethnobiological Classification: Principles of Categorization of Plants and Animals in Traditional Societies*. Princeton: Princeton University Press.

Berlin, B. (2006). The First Congress of Ethnozoological Nomenclature. *Journal of the Royal Anthropological Institute* 12, 523–44.

Berlin, B. and Kay, P. (1969). *Basic Color Terms: Their Universality and Evolution*. Berkeley and Los Angeles: University of California Press.

Berna, F., Goldberg, P. et al. and Chazan, M. (2012). Microstratigraphic evidence of in situ fire in the Acheulean strata of Wonderwerk Cave, Northern Cape province, South Africa. *Proceedings of the National Academy of Sciences* 109, E1215–20.

Berwick, R.C. and Chomsky, N. (2017). *Why Only Us: Language and Evolution*. Cambridge, MA: MIT Press.

Bickerton, D. (2009). *Adam's Tongue: How Humans Made Language, How Language Made Humans*. New York: Hill and Wang.

Bickerton, D. (2014). *More than Nature Needs: Language, Mind, and Evolution*. Cambridge, MA: Harvard University Press.

Blank, A. (1999). Why do new meanings occur? A cognitive typology of the motivations for lexical semantic change. In *Historical Semantics and Cognition* (eds A. Blank and P. Koch), pp. 61–89. Berlin/New York: De Gruyter Mouton.

Blasi, D., Wichmann, S. et al. and Christiansen, M. (2016). Sound–meaning association biases evidenced across thousands of languages. *Proceedings of the National Academy of Sciences* 113, 10818–23.

Blasi, D.E., Moran, S.E. et al. and Bickel, B. (2019). Human sound systems are shaped by post-Neolithic changes in bite configuration. *Science* 363, 1192.

Bloom, P. (2000). *How Children Learn the Meanings of Words*. Cambridge, MA: MIT Press.

Bloom, P. (2001). Précis of *How Children Learn the Meaning of Words*. *Behavioral and Brain Sciences* 24, 1095–1134.

Blumenthal, S.A., Levin, N.E. et al. and Cerling, T. (2017). Aridity and hominin environments. *Proceedings of the National Academy of Sciences* 114, 7331–6.

Boaretto, E., Hernandez, M.H., Goder-Goldberger, M. and Barzilai, O. (2021). The absolute chronology of Boker Tachtit (Israel) and implications for the Middle to Upper Paleolithic transition in the Levant. *Proceedings of the National Academy of Sciences* 118, e2014657118.

Boas, F. (1911). Introduction in *Handbook of American Indian Languages*, Vol. 1. Government Print Office (Smithsonian Institution, Bureau of American Ethnology, Bulletin 40), pp. 1–18.

Boas, F. (1920). The methods of ethnology. In F. Boas (1940/1966) *Race, Language and Culture*, pp. 281–9. New York: Free Press/Macmillan.

Boë, L.-J., Heim, J.-L., Honda, K. and Maeda, S. (2002). The potential Neandertal vowel space was as large as that of modern humans. *Journal of Phonetics* 30, 465–84.

Boë, L-J., Berthommier, F. et al. and Fagot, J. (2017). Evidence of a vocalic proto-system in the baboon (*Papio papio*) suggests pre-hominin speech precursors. *PLoS One* 12, e0169321.

Boesch, C. and Boesch, H. (1990). Tool use and tool making in wild chimpanzees. *Folia Primatologica* 54, 86–99.

Bohn, M., Call, J. and Tomasello, M. (2016). Comprehension of iconic gestures by chimpanzees and human children. *Journal of Experimental Child Psychology* 142, 1–17.

Bohsali, A. and Crosson, B. (2016). The basal ganglia and language: a tale of two loops. In *The Basal Ganglia: Novel Perspectives on Motor and Cognitive Functions* (ed. J.-J. Soghomonian), pp. 217–42. Cham, Switzerland: Springer International Publishing.

Borghi, A.M., Barca, L., Binkofski, F. and Tummolini, L. (2018). Abstract concepts, language and sociality: from acquisition to inner speech. *Philosophical Transactions of the Royal Society of London B* 373, 20170134.

Boroditsky, L. (2000). Metaphoric structuring: understanding time through spatial metaphors. *Cognition* 75, 1–28.

Boroditsky, L. (2001). Does language shape thought? Mandarin and English speakers' conceptions of time. *Cognitive Psychology* 43, 1–22.

Boroditsky, L. (2011). How language shapes thought. *Scientific American* 304, 62–5.

Boroditsky, L., Schmidt, L. and Phillips, W. (2003). Sex, syntax, and semantics. In *Language in Mind: Advances in the Study of Language and Thought* (eds D. Gentner and S. Goldin-Meadow), pp. 61–78. Cambridge, MA: MIT Press.

Botha, R. (2020). *Neanderthal Language: Demystifying the Linguistic Powers of Our Extinct Cousins*. Cambridge: Cambridge University Press.

Botha, R. and Knight, C. (eds) (2009). *The Prehistory of Language*. Oxford: Oxford University Press.

Boutonnet, B. and Lupyan, G. (2015). Words jump-start vision: a label advantage in object recognition. *Journal of Neuroscience* 35, 9329–35.

Bouzouggar, A., Barton, N. et al. and Stambouli, A. (2007). 82,000-year-old shell beads from North Africa and implications for the origins of modern human behavior. *Proceedings of the National Academy of Sciences* 104, 9964–9.

Brakke, K.E. and Savage-Rumbaugh, E.S. (1995). The development of language skills in bonobo and chimpanzee – I. Comprehension. *Language and Communication* 15, 121–48.

Brighton, H. (2002). Compositional syntax from cultural transmission. *Artificial Life* 1, 25–54.

Brighton, H., Smith, K. and Kirby, S. (2005). Language as an evolutionary system. *Physics of Life Reviews* 2, 177–226.

Brittingham, A., Hren, M.T. et al. and Adler, D. (2019). Geochemical evidence for the control of fire by Middle Palaeolithic hominins. *Scientific Reports* 9, 15368.

Brodbeck, C., Hong, L.E. and Simon, J.Z. (2018). Rapid transformation from auditory to linguistic representations of continuous speech. *Current Biology* 28, 3976–83.

Brooks-Gunn, J. and Lewis, M. (1979). 'Why mama and papa?': the development of social labels. *Child Development* 50, 1203–6.

Brosseau-Liard, P., Penney, D. and Poulin-Dubois, D. (2015). Theory of mind selectively predicts preschoolers' knowledge-based selective word learning. *British Journal of Developmental Psychology* 33, 464–75.

Brysbaert, M., Warriner, A.B. and Kuperman, V. (2013). Concreteness ratings for 40 thousand generally known English word lemmas. *Behavior Research Methods* 46, 904–11.

Brysbaert, M., Stevens, M., Mandera, P. and Keuleers, E. (2016). How many words do we know? Practical estimates of vocabulary size dependent on word definition, the degree of language input and the participant's age. *Frontiers in Psychology* 7, 1116.

Budil, I. (1994). Functional reconstruction of the supralaryngeal vocal tract of fossil human. *Human Evolution* 9, 35–52.

Buisan, R., Moriano, J., Andirkó, A. and Boeckx, C. (2022). A brain region-specific expression profile for genes within large introgression deserts and under positive selection in *Homo sapiens*. *Frontiers in Cell and Developmental Biology* 10, 824740.

Bullinger, A.F., Zimmerman, F., Kaminski, J. and Tomasello, M. (2011). Different social motives in the gestural communication of chimpanzees and human children. *Developmental Science* 14, 58–68.

Burling, R. (2012). Words came first: adaptations for word-learning. In *The Oxford Handbook of Language Evolution* (eds M. Tallerman and K.R. Gibson), pp. 401–16. Oxford: Oxford University Press.

Bushozi, P.M., Skinner, A. and de Luque, L. (2020). The Middle Stone Age (MSA) technological patterns, innovations, and behavioral changes at Bed VIA of Mumba Rockshelter, Northern Tanzania. *African Archaeological Review* 37, 293–310.

Bybee, J. (2012). Domain-general processes as the basis for grammar. In *The Oxford Handbook of Language Evolution* (eds M. Tallerman and K.R. Gibson), pp. 528–36. Oxford: Oxford University Press.

Call, J. and Tomasello, M. (1999). A nonverbal false belief task: the performance of children and great apes. *Child Development* 70, 381–95.

Call, J. and Tomasello, M. (2008). Does the chimpanzee have a theory of mind? 30 years later. *Trends in Cognitive Sciences* 12, 187–92.

Callaway, E. (2023). Oldest genetic data from a human relative found in 2-million-year-old teeth. *Nature* 619, 446.

Calude, A. (2021). The history of number words in the world's languages – what have we learnt so far? *Philosophical Transactions of the Royal Society of London B* 376, 20200206.

Cangelosi, A. (2012). Robotics and embodied agents modelling of the evolution of language. In *The Oxford Handbook of Language Evolution* (eds M. Tallerman and K.R. Gibson), pp. 605–20. Oxford: Oxford University Press.

Cangelosi, A. and Parisi, D. (1998). The emergence of a 'language' in an evolving population of neural networks. *Connection Science* 10, 83–97.

Cangelosi, A. and Parisi, D. (eds) (2002). *Simulating the Evolution of Language*. Berlin: Springer-Verlag.

Capitani, E., Laiacona, K., Mahon, B. and Caramazza, A. (2003). What are the facts of semantic category-specific deficits? A critical review of the clinical evidence. *Cognitive Neuropsychology* 20, 213–61.

Caramazza, A. and Shelton, J.R. (1998). Domain-specific knowledge systems in the brain: the animate–inanimate distinction. *Journal of Cognitive Neuroscience* 10, 1–34.

Caramazza, A., Chialant, D., Capasso, R. and Miceli, G. (2000). Separable processing of consonants and vowels. *Nature* 403, 428–30.

Caron, F., d'Errico, F., Del Moral, P., Santos, F. and Zilhão, J. (2011). The reality of Neandertal symbolic behavior at the Grotte du Renne, Arcy-sur-Cure, France. *PLoS One* 6, e21545.

Carrion-Castillo, A., Franke, B. and Fisher, S.E. (2013). Molecular genetics of dyslexia: an overview. *Dyslexia* 19, 214–40.

Carruthers, P. (1996). *Language, Thought and Consciousness: An Essay in Philosophical Psychology*. Cambridge: Cambridge University Press.

Carruthers, P. (2002). The roots of scientific reasoning: infancy, modularity and the art of tracking. In *The Cognitive Basis of Science* (eds P. Carruthers, S. Stich and M. Siegal), pp. 73–95. Cambridge: Cambridge University Press.

Carruthers, P. (2017). The illusion of conscious thought. *Journal of Consciousness Studies* 24(9–10), 228–52.

Castelvecchi, D. (2023). The human brain's characteristic wrinkles help to drive how it works. *Nature* 618, 223–4.

Cataldo, D.M., Migliano, A.B. and Vinicius, L. (2018). Speech, stone tool-making and the evolution of language. *PLoS One* 13, e0191071.

Chater, N. and Christiansen, M.H. (2012). A solution to the logical problem of language evolution: language as an adaptation to the human brain. In *The Oxford Handbook of Language Evolution* (eds M. Tallerman and K.R. Gibson), pp. 627–39. Oxford: Oxford University Press.

Cheney, D.L. and Seyfarth, R.M. (1990). *How Monkeys See the World: Inside the Mind of Another Species*. Chicago: Chicago University Press.

Cheng, Q., Roth, A., Halgren, E. and Mayberry, R.I. (2019). Effects of early language deprivation on brain connectivity: language pathways in deaf native and late first-language learners of American Sign Language. *Frontiers in Human Neuroscience* 13, 320.

Childs, G.T. (1994). African ideophones. In *Sound Symbolism* (eds L. Hinton, J. Nichols and J.J. Ohala), pp. 178–204. Cambridge: Cambridge University Press.

Ching, M.S.L., Shen, Y. et al. and Wu, B.L. (2010). Deletions of *NRXN1* (Neurexin-1) predispose to a wide spectrum of developmental disorders. *American Journal of Medical Genetics* 153B, 937–47.

Chomsky, N. (1957a). Review of Skinner's 'Verbal Behavior'. *Language* 35, 26–58.

Chomsky, N. (1957b). *Syntactic Structures*. Berlin: De Gruyter Mouton.

Chomsky, N. (2011). Language and other cognitive systems: what is special about language? *Language Learning and Development* 7, 263–78.

Christiansen, M.H. and Chater, N. (2008). Language as shaped by the brain. *Behavioral and Brain Sciences* 31, 489–558.

Christiansen, M.H. and Kirby, S. (eds) (2003). *Language Evolution*. Oxford: Oxford University Press.

Clark, A. (1998). Magic words: how language augments human computation. In *Language and Thought: Interdisciplinary Themes* (eds P. Carruthers and J. Boucher), pp. 162–83. Cambridge: Cambridge University Press.

Clark, B. (2013). Syntactic theory and the evolution of syntax. *Biolinguistics* 7, 169–97.

Clark, E.V. and Clark, H.H. (1979). When nouns surface as verbs. *Language* 55, 767–811.

Clark, J.D. (1993). African and Asian perspectives on the origins of modern humans. In *The Origin of Modern Humans and the Impact of Chronometric Dating* (eds M.J. Aitken, C.B. Stringer and P.A. Mellars), pp. 148–78. Princeton: Princeton University Press.

Clarkson, C., Jacobs, Z. et al. and Pardoe, C. (2017). Human occupation of northern Australia by 65,000 years ago. *Nature* 547, 306–10.

Clay, Z. and Zuberbühler, K. (2011). Bonobos extract meaning from call sequences. *PLoS One* 6, e18786.

Clay, Z., Archbold, J. and Zuberbühler, K. (2015). Functional flexibility in wild bonobo vocal behaviour. *PeerJ* 3, e1124.

Conde-Valverde, M., Martínez, I. et al. and Arsuaga, J.L. (2019). The cochlea of the Sima de los Huesos hominins (Sierra de Atapuerca, Spain): new insights into cochlear evolution in the genus *Homo*. *Journal of Human Evolution* 136, 102641.

Conde-Valverde, M., Martínez, I. et al. and Arsuaga, J.L. (2021). Neanderthals and *Homo sapiens* had similar auditory and speech capacities. *Nature Ecology and Evolution* 5, 609–15.

Condillac, E.B. (1746). *Essai sur l'origine des connaissances humaines*. Amsterdam.

Conklin, H. (1957). *Hanunóo Agriculture*. FAO Forestry Development Paper No. 12. Rome FAO: United Nations.

Coopmans, C.W., de Hoop, H. et al. and Martin, A.E. (2022). Hierarchy in language interpretation: evidence from behavioural experiments and computational modelling. *Language, Cognition and Neuroscience* 37, 420–39.

Corballis, M.C. (2002). *From Hand to Mouth: The Origins of Language*. Princeton: Princeton University Press.

Coski, R.C. (2003). Condillac: language, thought and morality in the man and animal debate. *French Forum* 28, 57–75.

Crockford, C., Herbinger, I., Vigilant, L. and Boesch, C. (2004). Wild chimpanzees produce group-specific calls: a case for vocal learning? *Ethology* 110, 221–43.

Crockford, C., Wittig, R.M., Mundry, R. and Zuberbühler, K. (2012). Wild chimpanzees inform ignorant group members of danger. *Current Biology* 22, 142–6.

Crockford, C., Wittig, R.M. and Zuberbühler, K. (2017). Vocalizing in chimpanzees is influenced by social-cognitive processes. *Science Advances* 3, e1701742.

Crosson, B., Benjamin, M. and Levy, I. (2007). Role of the basal ganglia in language and semantics: supporting cast. In *Neural Basis of Semantic Memory* (eds J. Hart Jr. and M.A. Kraut), pp. 219–43. Cambridge: Cambridge University Press.

Crow, T.J. (1997). Is schizophrenia the price that *Homo sapiens* pays for language? *Schizophrenia Research* 28, 127–41.

Crozier, W.R. (2014). Differentiating shame from embarrassment. *Emotion Review* 6, 269–76.

Cruz, M. and Frota, S. (eds) (2022). *Prosodic Variation with(in) Languages: Intonation, Phrasing and Segments*. Sheffield: Equinox Publishing.

Crystal, D. (ed.) (2003). *The Cambridge Encyclopedia of the English Language*. Cambridge: Cambridge University Press.

Cummins, F., Gers, F. and Schmidhuber, J. (1999). Comparing prosody across many languages. I.D.S.I.A Technical Report IDSIA-07-99.

Cuskley, C. and Kirby, S. (2013). Synesthesia, cross-modality, and language evolution. In *The Oxford Handbook of Synesthesia* (eds J. Simner and E. Hubbard). Oxford: Oxford University Press.

Dąbrowska, E. (2015). What exactly is Universal Grammar, and has anyone seen it? *Frontiers in Psychology* 6, 852.

Dahan, D. (2015). Prosody and language comprehension. *Wiley Interdisciplinary Reviews – Cognitive Science* 6, 441–52.

de Boer, B. (2009). Acoustic analysis of primate air sacs and their effect on vocalization. *Journal of the Acoustical Society of America* 126, 3329–43.

de Boer, B. (2012). Self-organization and language evolution. In *The Oxford Handbook of Language Evolution* (eds M. Tallerman and K.R. Gibson), pp. 613–20. Oxford: Oxford University Press.

de Boer, B. and Fitch, W.T. (2010). Computer models of vocal tract evolution: an overview and critique. *Adaptive Behavior* 18, 36–47.

Dediu, D. and Ladd, D.R. (2007). Linguistic tone is related to the population frequency of the adaptive haplogroups of two brain size genes, ASPM and Microcephalin. *Proceedings of the National Academy of Sciences* 104, 10944–9.

Dediu, D. and Levinson, S.C. (2018). Neanderthal language revisited: not only us. *Current Opinion in Behavioral Sciences* 21, 49–55.

Dediu, D. and Moisik, S.R. (2019). Pushes and pulls from below: anatomical variation, articulation and sound change. *Glossa: A Journal of General Linguistics* 4, 7.

Defleur, A.R. and Desclaux, E. (2019). Impact of the last interglacial climate change on ecosystems and Neanderthals behavior at Baume Moula-Guercy, Ardèche, France. *Journal of Archaeological Science* 104, 114–24.

Dehaene, S. (1997). *The Number Sense: How the Mind Creates Mathematics*. Oxford: Oxford University Press.

Dehaene, S. (2009). *Reading in the Brain: The Science and Evolution of a Human Invention*. New York: Penguin.

Deino, A.L., Behrensmeyer, A.K. et al. and Potts, R. (2018). Chronology of the Acheulean to Middle Stone Age transition in eastern Africa. *Science* 360, 95–8.

de la Torre, I., Martínez-Moreno, J. and Mora, R. (2013). Change and stasis in the Iberian Middle Paleolithic. *Current Anthropology* 54, S320–36.

Dellert, J., Johansson, N.E., Frid, J. and Carling, G. (2021). Preferred sound groups of vocal iconicity reflect evolutionary mechanisms of sound stability and first language acquisition: evidence from Eurasia. *Philosophical Transactions of the Royal Society of London B* 376, 20200190.

Demoule, J.-P. (2023), translated from the 2019 French edition. *The Indo-Europeans: Archaeology, Language, Race, and the Search for the Origins of the West*. Oxford: Oxford University Press.

Demuro, M., Arnold, L.J. et al. and Arsuaga, J.L. (2019). New bracketing luminescence ages constrain the Sima de los Huesos hominin fossils (Atapuerca, Spain) to MIS 12. *Journal of Human Evolution* 131, 76–95.

Dennell, R. (2018). The Acheulean assemblages of Asia: a review. In *The Emergence of the Acheulean in East Africa and Beyond: Contributions in Honor of Jean Chavaillon* (eds R. Gallotti and M. Mussi), pp. 195–212. Berlin: Springer.

Dennett, D. (1994). Learning and Labeling. *Mind and Language* 8, 540–48.

d'Errico, F. (2008). Le rouge et le noir: implications of early pigment use in Africa, the Near East and Europe for the origin of cultural modernity. *Current Themes in Middle Stone Age Research* 10, 168–74.

d'Errico, F. and Blackwell, L. (2016). Earliest evidence of personal ornaments associated with burial: the Conus shells from Border Cave. *Journal of Human Evolution* 93, 91–108.

d'Errico, F. and Soressi, M. (2002). Systematic use of manganese pigment by the Pech-de-l'Azé Neandertals: implications for the origin of behavioral modernity. *Journal of Human Evolution* 42, A13.

d'Errico, F., Zilhão, J. et al. and Pelegrin, J. (1998). Neanderthal acculturation in Western Europe? A critical review of the evidence and its interpretation. *Current Anthropology* 39, S1–44.

d'Errico, F., Henshilwood, C., Lawson, G. et al. and Julien, M. (2003). Archaeological evidence for the emergence of language, symbolism, and music – an alternative multidisciplinary perspective. *Journal of World Prehistory* 17, 1–70.

d'Errico, F., Henshilwood, C.S., Vanhaeren, M. and van Niekerk, K. (2005). *Nassarius kraussianus* shell beads from Blombos Cave: evidence for symbolic behaviour in the Middle Stone Age. *Journal of Human Evolution* 48, 3–24.

d'Errico, F., Vanhaeren, M. Henshilwood, C. et al. and van Niekerk, K. (2009). From the origin of language to the diversification of languages: what can archaeology and palaeoanthropology say? In *Becoming Eloquent: Advances in the Emergence of Language, Human Cognition, and Modern Cultures* (eds F. d'Errico and J.-M. Hobert), pp. 13–68. Amsterdam / Philadelphia: John Benjamins.

de Saussure, F. (1916). *Cours de linguistique générale*. Paris: Payot.

DeSilva, J.M., Traniello, J.F.A., Claston, A.G. and Fannin, L.D. (2021). When and why did human brains decrease in size? A new change-point analysis and insights from brain evolution in ants. *Frontiers in Ecology and Evolution* 9, 742639.

Deutscher, G. (2005). *The Unfolding of Language*. London: Penguin.

Deutscher, G. (2010). *Through the Language Glass: Why the World Looks Different in Other Languages*. London: Penguin.

Deutscher, G. (2012). The grammaticalization of quotatives. In *The Oxford Handbook of Grammaticalization* (eds B. Heine and H. Narrog), pp. 646–55. Oxford: Oxford University Press.

de Villiers, J. (2000). Language and theory of mind: what are the developmental relationships? In *Understanding Other Minds: Perspectives from Developmental Cognitive Neuroscience* (eds S. Baron-Cohen, H. Tager-Flusberg and D.J. Cohen), pp. 83–123. Oxford: Oxford University Press.

Diaz, R.M. and Berk, L.E. (eds) (1992). *Private Speech: From Social Interaction to Self-Regulation*. Hillsdale, NJ: Erlbaum.

Dibble, H.L., Aldeias, V. et al. and El-Hajraoui, M. (2012). New excavations at the site of Contrebandiers Cave, Morocco. *PaleoAnthropology* 2012, 145–201.

Dibble, H.L., Aldeias, V. et al. and El-Hajraoui, M. (2013). On the industrial attributions of the Aterian and Mousterian of the Maghreb. *Journal of Human Evolution* 64, 194–201.

Dingemanse, M., Blasi, D. et al. and Monaghan, P. (2015). Arbitrariness, iconicity, and systematicity in language. *Trends in Cognitive Sciences* 19, 603–15.

Dingemanse, M. 2018. Redrawing the margins of languages: lessons from research on ideophones. *Glossa: A Journal of General Linguistics* 3, 1–30.

Donald, M. (1991). *Origins of the Modern Mind: Three Stages in the Evolution of Culture and Cognition*. Cambridge, MA: Harvard University Press.

Driscoll, K. (2015). Animals, mimesis, and the origin of language. In *Des animaux et des hommes: Savoirs, représentations, interactions* (eds A. Choné and C.

Repussard), pp. 173–94. Special issue of *Recherches germaniques* hors-série 10. Strasbourg: Presses Universitaires de Strasbourg.

Du, A. and Wood, B.A. (2020). Brain size evolution in the hominin clade. In *Landscapes of Human Evolution: Contributions in Honour of John Gowlett* (eds J. Cole, J. McNabb, M. Grove and R. Hosfield), pp. 9–17. Oxford: Archaeopress.

Du, A., Zipkin, A.M. et al. and Wood, B A. (2018). Pattern and process in hominin brain size evolution are scale-dependent. *Proceedings of the Royal Society B* 285, 20172738.

Dunbar, R.I.M. (1992). Neocortex size as a constraint on group size in primates. *Journal of Human Evolution* 22, 469–93.

Dunbar, R.I.M. (1995). Neocortex size and group size in primates: a test of the hypothesis. *Journal of Human Evolution* 28, 287–96.

Dunbar, R.I.M. (1997). *Grooming, Gossip, and the Evolution of Language*. Cambridge, MA: Harvard University Press.

Eckert, P. (1988). Adolescent Social Structure and the Spread of Linguistic Change. *Language in Society* 17, 183–207.

Edmiston, P. and Lupyan, G. (2015). What makes words special? Words as unmotivated cues. *Cognition* 143, 93–100.

Ekman, P. and Davidson, R. (eds) (1994). *The Nature of Emotion: Fundamental Questions*. Oxford: Oxford University Press.

Ember, C.R. and Ember, M. (2007). Climate, econiche, and sexuality: influences on sonority in language. *American Anthropologist* 109, 180–5.

Enard, W., Przeworski, M. et al. and Pääbo, S. (2002). Molecular evolution of *FOXP2*, a gene involved in speech and language. *Nature* 418, 869–72.

Enard, W., Gehre, S. et al. and Pääbo, S. (2009). A humanized version of *FOXP2* affects cortico-basal ganglia circuits in mice. *Cell* 137, 961–71.

Endress, A. and Hauser, M.D. (2010). Word segmentation with universal prosodic cues. *Cognitive Psychology* 61, 177–99.

Evans, N. and Levinson, S. (2009). The myth of language universals: language diversity and its importance for cognitive science. *Behavioral and Brain Sciences* 32, 429–92.

Everett, C. (2017). Languages in drier climates use fewer vowels. *Frontiers in Psychology* 8, 1285.

Fabre, V., Condemi, S. and Degioanni A. (2009). Genetic evidence of geographical groups among Neanderthals. *PLoS One* 4, e5151.

Faivre, J.-P., Discamps, E. et al. and Lenoir, M. (2014). The contribution of lithic production systems to the interpretation of Mousterian industrial variability in south-western France: the example of Combe-Grenal (Dordogne, France). *Quaternary International* 350, 227–40.

Falk, D., Hildebolt, C. et al. and Prior, F. (2005). The brain of LB1, *Homo floresiensis*. *Science* 308, 242–5.

Fedorenko, E., Duncan, J. and Kanwisher, N. (2012). Language-selective and domain-general regions lie side by side within Broca's area. *Current Biology* 22, 2059–62.

Fedurek, P., Slocombe, K.E., Hartel, J.A. and Zuberbühler, K. (2015). Chimpanzee lip-smacking facilitates cooperative behaviour. *Scientific Reports* 5, 13460.

Fedurek, P., Zuberbühler, K. and Dahl, C.D. (2016). Sequential information in a great ape utterance. *Scientific Reports* 6, 38226.

Fernández-Domínguez, J. (2019). The onomasiological approach. In *The Oxford Research Encyclopedia of Linguistics*. doi.org/10.1093/acrefore/9780199384655.013.579

Fernández Peris, J., Barciela González, V. et al. and Verdasco, C. (2012). The earliest evidence of hearths in Southern Europe: the case of Bolomor Cave (Valencia, Spain). *Quaternary International* 247, 267–77.

Ferretti, F. (2022). *Narrative Persuasion: A Cognitive Perspective on Language Evolution*. Cham, Switzerland: Springer International.

Finlayson, C., Brown, K. et al. and Negro, J.J. (2012). Birds of a feather: Neanderthal exploitation of raptors and corvids. *PLoS One* 7, e45927.

Fischer, J., Wheeler, B.C. and Higham, J.P. (2015). Is there any evidence for vocal learning in chimpanzee food calls? *Current Biology* 25, R1028–9.

Fisher, S.E. (2017). Evolution of language: lessons from the genome. *Psychonomic Bulletin and Review* 24, 34–40.

Fisher, S.E. and Vernes, S.C. (2015). Genetics and the language sciences. *Annual Review of Linguistics* 1, 289–310.

Fitch, W.T. (2009). Fossil cues to the evolution of speech. In *The Cradle of Language* (eds R. Botha and C. Knight), pp. 112–34. Oxford: Oxford University Press.

Fitch, W.T. (2010). *The Evolution of Language*. Cambridge: Cambridge University Press.

Fitch, W.T. (2017). Empirical approaches to the study of language evolution. *Psychonomic Bulletin and Review* 24, 3–33.

Fitch, W.T. (2019). Animal cognition and the evolution of human language: why we cannot focus solely on communication. *Philosophical Transactions of the Royal Society of London B* 375, 20190046.

Fontaine, L. (2017). The early semantics of the neologism BREXIT: a lexicogrammatical approach. *Functional Linguistics* 4, 6.

Fort, M., Lammertink, I. et al. and Tsuji, S. (2018). Symbouki: a meta-analysis on the emergence of sound symbolism in early language acquisition. *Developmental Science* 21, e12659.

Foster, F. and Collard, M. (2013). A reassessment of Bergmann's rule in modern humans. *PLoS One* 8, e72269.

Fouts, R. (1998). *Next of Kin: My Conversations with Chimpanzees*. New York: Harper Collins.

Frayer, D. and Nicolay, C. (2000). Fossil evidence for the origin of speech sounds.

In *The Origins of Music* (eds N. Wallin, B. Merker and S. Brown), pp. 217–34. Cambridge, MA: MIT Press.

Friederici, A.D. (2011). The brain basis of language processing: from structure to function. *Physiological Review* 91, 1357–92.

Friederici, A.D. (2017). Evolution of the neural language network. *Psychonomic Bulletin and Review* 24, 41–7.

Froehle, A. and Churchill, S.E. (2009). Energetic competition between Neandertals and anatomically modern humans. *PaleoAnthropology*, 2009, 96–116.

Galaburda, A.M., Sherman, G.F. et al. and Geschwind, N. (1985). Developmental dyslexia: four consecutive patients with cortical anomalies. *Annals of Neurology* 18, 222–33.

Gao, X., Zhang, S., Zhang, Y. and Chen, F. (2017). Evidence of hominin use and maintenance of fire at Zhoukoudian. *Current Anthropology* 58, S267–77.

Garber, M. (2013). "Friend", as a verb, is 800 years old. *The Atlantic*, 25 July 2013

García-Martínez, D., Torres-Tamayo, N. et al. and Bastir, M. (2018). Ribcage measurements indicate greater lung capacity in Neanderthals and Lower Pleistocene hominins compared to modern humans. *Communications Biology* 1, 117.

García-Medrano, P., Ollé, A., Ashton, N. and Roberts, M.B. (2019). The mental template in handaxe manufacture: new insights into Acheulean lithic technological behavior at Boxgrove, Sussex, UK. *Journal of Archaeological Method and Theory* 26, 396–422.

Gardner, R.A. and Gardner, B.T. (1969). Teaching sign language to a chimpanzee. *Science* 165, 664–72.

Gardner, R.A., Gardner, B.T. and Van Cantfort, T.E. (eds) (1989). *Teaching Sign Language to Chimpanzees*. Albany: State University of New York Press.

Garrett, A. (2014). The sound change. In *The Routledge Handbook of Historical Linguistics* (eds C. Bowern and B. Evans), pp. 227–48. London: Routledge.

Gasser, M. (2004). The origins of arbitrariness in language. *Proceedings of the 26th Annual Conference of the Cognitive Science Society*, 434–9.

Gaudzinski-Windheuser, S. and Roebroeks, W. (2011). On Neanderthal subsistence in last interglacial forested environments in Northern Europe. In *Neanderthal Lifeways, Subsistence and Technology: One Hundred Fifty Years of Neanderthal Study* (eds N.J. Conrad and J. Richter), pp. 61–71. New York: Springer.

Gelman, S.A. and Meyer. M. (2011). Child categorization. *Wiley Interdisciplinary Reviews – Cognitive Science* 2, 95–105.

Ghazanfar, A.A. and Takahashi, D.Y. (2014). The evolution of speech: vision, rhythm, cooperation. *Trends in Cognitive Sciences* 18, 543–53.

Ghazanfar, A.A., Turesson, H.K. et al. and Logothetis, N.K. (2007). Vocal-tract resonances as indexical cues in rhesus monkeys. *Current Biology* 17, 425–430.

Giancarlo, M. (2001). The rise and fall of the great vowel shift? The changing ideological intersections of philology, historical linguistics, and literary history. *Representations* 76, 27–60.

Gibson, E., Futrell, R. et al. and Conway, B.R. (2017). Color naming across languages reflects color use. *Proceedings of the National Academy of Sciences* 114, 10785–90.

Gicqueau, A., Schuh, A. et al. and Maureille, B. (2023). Anatomically modern human in the Châtelperronian hominin collection from the Grotte du Renne (Arcy-sur-Cure, Northeast France). *Scientific Reports* 13, 12682.

Gillespie-Lynch, K., Greenfield, P., Lyn, H. and Savage-Rumbaugh, S. (2014). Gestural and symbolic development among apes and humans: support for a multimodal theory of language evolution. *Frontiers in Psychology* 5, 1228.

Girard-Buttoz, C., Zaccarella, E. et al. and Crockford, C. (2022). Chimpanzees produce diverse vocal sequences with ordered and recombinatorial properties. *Communications Biology* 5, 410.

Givón, T. (1971). Historical syntax and synchronic morphology: an archaeologist's field trip. *Papers from the 7th Meeting of the Chicago Linguistic Society*, pp. 394–415. Chicago: Chicago Linguistic Society.

Golestani, N. (2014). Brain structural correlates of individual differences at low to high levels of the language processing hierarchy: a review of new approaches to imaging research. *International Journal of Bilingualism* 18, 6–34.

Gómez, R. (2002). Variability and detection of invariant structure. *Psychological Science* 13, 431–6.

Gong, T., Shuai, L. and Zhang, M. (2014). Modelling language evolution: examples and predictions. *Physics of Life Reviews* 11, 280–302.

Goodall, J. (1964). Tool-using and aimed throwing in a community of free-living chimpanzees. *Nature* 201, 1264–6.

Goren-Inbar, N., Alperson, N. et al. and Werker, E. (2004). Evidence of hominin control of fire at Gesher Benot Ya'aqov, Israel. *Science* 304, 725–7.

Gowlett, J.A.J. (2016). The discovery of fire by humans: a long and convoluted process. *Philosophical Transactions of the Royal Society B* 371, 20150164.

Gowlett, J.A.J., Harris, J.W.K., Walton, D. and Wood, B.A. (1981). Early archaeological sites, hominid remains and traces of fire from Chesowanja, Kenya. *Nature* 294, 125–9.

Graf Estes, K. Evans, J.L., Alibali, M.W. and Saffran, J.R. (2007). Can infants map meaning to newly segmented words? Statistical segmentation and word learning. *Psychological Science* 18, 254–60.

Gravina, B., Bachellerie, F. et al. and Bordes J.-G. (2018). No reliable evidence for a Neanderthal–Châtelperronian association at La Roche-à-Pierrot, Saint-Césaire. *Scientific Reports* 8, 15134.

Green, R.E., Krause, J., Briggs, A.W. et al. and Pääbo, S. (2010). A draft sequence of the Neandertal genome. *Science* 328, 710–22.

Gruber, T. and Zuberbühler, K. (2013). Vocal recruitment for joint travel in wild chimpanzees. *PLoS One* 8, e76073.

Gruber, T., Frick, A. et al. and Biro, D. (2019). Spontaneous categorization of tools based on observation in children and chimpanzees. *Scientific Reports* 9, 18256.

Grünberg, J.M. (2002). Middle Palaeolithic birch-bark pitch. *Antiquity* 76, 15–16.

Grund, C., Neumann, C., Zuberbühler, K. and Gruber, T. (2019). Necessity creates opportunities for chimpanzee tool use. *Behavioral Ecology* 30, 1136–44.

Grzega, J. (2015). Word-formation in onomasiology. In *Word Formation: An International Handbook of the Languages of Europe* (eds P. Müller, I. Ohnheiser, S. Olsen and F. Rainer), pp. 79–93. Berlin: De Gruyter Mouton.

Gunz, P., Neubauer, S. et al. and Hublin, J.-J. (2012). A uniquely modern human pattern of endocranial development. Insights from a new cranial reconstruction of the Neandertal newborn from Mezmaiskaya. *Journal of Human Evolution* 62, 300–13.

Gunz, P., Tilot, A.K. et al. and Fisher, S.E. (2019). Neandertal introgression sheds light on modern human endocranial globularity. *Current Biology* 29, 120–7.

Gunz, P., Neubauer, S. et al. and Alemseged, Z. (2020). *Australopithecus afarensis* endocasts suggest ape-like brain organization and prolonged brain growth. *Science Advances* 6, eaaz4729.

Haak, W., Lazaridis, I. et al. and Reich, D. (2015). Massive migration from the steppe was a source for Indo-European languages in Europe. *Nature* 522, 207–11.

Haaland, M., Miller. C. et al. and Henshilwood, C. (2021). Geoarchaeological investigation of occupation deposits in Blombos Cave in South Africa indicate changes in site use and settlement dynamics in the southern Cape during MIS 5b–4. *Quaternary Research* 100, 170–223.

Haidle, M.N. (2009). How to think a simple spear. In *Cognitive Archaeology and Human Evolution* (eds S.A. de Beaune, F.L. Coolidge and T. Wynn), pp. 57–73. Cambridge: Cambridge University Press.

Haiman, J. (1998). *Talk Is Cheap: Sarcasm, Alienation, and the Evolution of Language*. Oxford: Oxford University Press.

Hale, C.M. and Tager-Flusberg, H. (2003). The influence of language on theory of mind: a training study. *Developmental Science* 6, 346–59.

Hardy, B.L., Moncel, M.-H. et al. and Gallotti, R. (2013). Impossible Neanderthals? Making string, throwing projectiles and catching small game during Marine Isotope Stage 4 (Abri du Maras, France). *Quaternary Science Reviews* 82, 23–40.

Hardy, B.L., Moncel, M.-H., Kerfant, C. et al. and Mélard, N. (2020). Direct evidence of Neanderthal fibre technology and its cognitive and behavioral implications. *Scientific Reports* 10, 4889.

Hare, B., Call, J. and Tomasello, M. (2001). Do chimpanzees know what conspecifics know? *Animal Behaviour* 61, 139–51.

Hare, B., Call, J. and Tomasello, M. (2006). Chimpanzees deceive a human competitor by hiding. *Cognition* 101, 495–514.

Harmand, S., Lewis, J.E. et al. and Roche, H. (2015). 3.3-million-year-old stone tools from Lomekwi 3, West Turkana, Kenya. *Nature* 521, 310–15.

Harris, K. and Nielsen, R. (2016). The genetic cost of Neanderthal introgression. *Genetics* 203, 881–91.

Harvati, K. and Reyes-Centeno, H. (2022). Evolution of *Homo* in the Middle and Late Pleistocene. *Journal of Human Evolution* 173, 103279.

Harvati, K., Röding, C. et al. and Kouloukoussa, M. (2019). Apidima Cave fossils provide earliest evidence of *Homo sapiens* in Eurasia. *Nature* 571, 500–4.

Haun, D.B.M., Rapold, C.J., Janzen, G. and Levinson, S. (2011). Plasticity of human spatial cognition: spatial language and cognition covary across cultures. *Cognition* 119, 70–80.

Hauser, M.D., Newport, E.L. and Aslin, R. (2001). Segmentation of the speech stream in a non-human primate: statistical learning in cotton-top tamarins. *Cognition* 78, B53–64.

Hay, J. and Bauer, L. (2007). Phoneme inventory size and population size. *Language* 83, 388–400.

Hayes, K.J. and Hayes, C. (1951). The intellectual development of a home-raised chimpanzee. *Proceedings of the American Philosophical Society* 95, 105–9.

Heffner, R.S. (2004). Primate hearing from a mammalian perspective. *Anatomical Record Part A* 281A, 1111–22.

Heggarty, P. (2013). Ultraconserved words and Eurasiatic? The 'faces in the fire' of language prehistory. *Proceedings of the National Academy of Sciences* 110, E3254.

Heine, B. and Kuteva, T. (2002). *World Lexicon of Grammaticalization*. Cambridge: Cambridge University Press.

Heine, B. and Kuteva, T. (2007). *The Genesis of Grammar: A Reconstruction*. Oxford: Oxford University Press.

Heine, B. and Narrog, H. (2015). Grammaticalization and linguistic analysis. In *The Oxford Handbook of Linguistic Analysis* (eds B. Heine and H. Narrog), pp. 401–24. Oxford: Oxford University Press.

Henshilwood, C.S., Sealy, J.C. et al. and Watts, I. (2001). Blombos Cave, Southern Cape, South Africa: preliminary report on the 1992–1999 excavations of the Middle Stone Age levels. *Journal of Archaeological Science* 28, 421–48.

Henshilwood, C.S, d'Errico, F. and Watts, I. (2009). Engraved ochres from the Middle Stone Age levels at Blombos Cave, South Africa. *Journal of Human Evolution* 57, 27–47.

Herbert, M.R., Ziegler, D.A. et al. and Caviness, V.S. (2004). Brain asymmetries in autism and developmental language disorder: a nested whole-brain analysis. *Brain* 128, 213–26.

Herculano-Houzel, S. (2012). The remarkable, yet not extraordinary, human brain as a scaled-up primate brain and its associated cost. *Proceedings of the National Academy of Sciences* 109, 10661–8.

Herder, J.G. (1986 [1771]). *On the Origin of Language*. Translated, with afterwords by J.H. Moran and A. Gode. Chicago: University of Chicago Press.

Hershkovitz, I., Weber, G.W. et al. and Weinstein-Evron, M. (2018). The earliest modern humans outside Africa. *Science* 359, 456–9.

Hewes, G.W. (1973). Primate communication and the gestural origin of language. *Current Anthropology* 14, 5–254.

Hewitt, G., MacLarnon, A. and Jones, K.E. (2002). The functions of laryngeal air sacs in primates: a new hypothesis. *Folia Primatologica* 73, 70–94.

Heyes, P.J., Anastasakis, K. et al. and Soressi, M. (2016). Selection and use of manganese dioxide by Neanderthals. *Scientific Reports* 6, 22159.

Higham, T., Jacobi, R. et al. and Ramsey, C.B. (2010). Chronology of the Grotte du Renne (France) and implications for the context of ornaments and human remains within the Châtelperronian. *Proceedings of the National Academy of Sciences* 107, 20234–9.

Higham, T., Brock, F. et al. and Basell, L. (2012). Chronology of the site of Grotte du Renne, Arcy-sur-Cure, France: implications for Neanderthal symbolic behaviour. *Before Farming*, 2011, 1–9.

Hinton, L., Nichols, J. and Ohala, J.J. (eds) (1994). *Sound Symbolism*. Cambridge: Cambridge University Press.

Hirschfeld, L.A. and S.A. Gelman (eds) (1994). *Mapping the Mind: Domain Specificity in Cognition and Culture*. Cambridge: Cambridge University Press.

Hlubik, S., Cutts, R. et al. and Harris, J.W.K. (2019). Hominin fire use in the Okote member at Koobi Fora, Kenya: new evidence for the old debate. *Journal of Human Evolution* 133, 214–29.

Hobaiter, C. and Byrne, R.W. (2014). The meanings of chimpanzee gestures. *Current Biology* 24, 1596–1600.

Hockett, C.F. (1958). *A Course in Modern Linguistics*. New York: Macmillan.

Hodge, C.T. (1970). The linguistic cycle. *Language Sciences* 13, 1–7.

Hodgskiss, T. (2012). An investigation into the properties of the ochre from Sibudu, KwaZulu-Natal, South Africa. *Southern African Humanities* 24, 99–120.

Hodgskiss, T. (2020). Ochre use in the Middle Stone Age. In *The Oxford Research Encyclopedia of Anthropology*. https://oxfordre.com/anthropology/display/10.1093/acrefore/9780190854584.001.0001/acrefore-9780190854584-e-51

Hoffman, R.R. (2018). Metaphor in science. In *Cognition and Figurative Language* (eds R.P. Honeck and R.R. Hoffman), pp. 393–415. London: Routledge.

Hoffmann, D.L., Angelucci, D.E. et al. and Zilhão, J. (2018a). Symbolic use of marine shells and mineral pigments by Iberian Neandertals 115,000 years ago. *Science Advances* 4, eaar5255.

Hoffmann, D.L., Standish, C.D. et al. and Pike, A.W.G. (2018b). U-Th dating of carbonate crusts reveals Neandertal origin of Iberian cave art. *Science* 359, 912–15.

Hoffmann, D.L. Standish, C.D. et al. and Pike, A.W.G. (2020). Response to White et al.'s reply: 'Still no archaeological evidence that Neanderthals created Iberian cave art'. *Journal of Human Evolution* 144, 102810.

Holloway, R.L. (1981). Volumetric and asymmetry determinations on recent hominid endocasts: Spy I and II, Djebel Ihroud I, and the Sale Homo erectus specimens, with some notes on Neanderthal brain size. *American Journal of Physical Anthropology* 55, 385–93.

Holloway, R.L., Hurst, S.D. et al. and Hawks, J. (2018). Endocast morphology of *Homo naledi* from the Dinaledi Chamber, South Africa. *Proceedings of the National Academy of Sciences* 115, 5738–43.

Hopkins, W.D., Taglialatela, J. and Leavens, D.A. (2007). Chimpanzees differently produce novel vocalizations to capture the attention of a human. *Animal Behaviour* 73, 281–6.

Hopper, P. J. and Traugott, E.C. (2003). *Grammaticalization*. Cambridge: Cambridge University Press.

Hosfield, R. (2020). *The Earliest Europeans – A Year in the Life: Seasonal Survival Strategies in the Lower Palaeolithic*. Oxford: Oxbow Books.

Hosfield, R. (2022). Variations by degrees: Western European paleoenvironmental fluctuations across MIS 13–11. *Journal of Human Evolution* 169, 103213.

Hovers, E., Ilani, S., Bar-Yosef, O. and Vandermeersch, B. (2003). An early case of color symbolism: ochre use by modern humans in Qafzeh Cave. *Current Anthropology* 44, 491–522.

Hu, W., Hao, Z. et al. and Li, H. (2023). Genomic inference of a severe human bottleneck during the Early to Middle Pleistocene transition. *Science* 381, 979–84.

Hublin, J.-J., Talamo, S. et al. and Richards, M.P. (2012). Radiocarbon dates from the Grotte du Renne and Saint-Césaire support a Neandertal origin for the Châtelperronian. *Proceedings of the National Academy of Sciences* 109, 18743–8.

Hublin, J.-J., Neubauer, S. and Gunz, P. (2015). Brain ontogeny and life history in Pleistocene hominins. *Philosophical Transactions of the Royal Society B* 370, 20140062.

Hublin, J.-J., Ben-Ncer, A. et al. and Gunz, P. (2017). New fossils from Jebel Irhoud, Morocco and the pan-African origin of *Homo sapiens*. *Nature* 546, 289–92.

Humphrey, N. (1976). The colour currency of nature. In *Colour for Architecture* (eds T. Porter and B. Mikellides), pp. 95–8. London: Studio Vista.

Humphrey, N. (2012). This chimp will kick your ass at memory games – but how the hell does he do it? *Trends in Cognitive Sciences* 16, 353–5.

Hurcombe, L. (2014). *Perishable Material Culture in Prehistory: Investigating the Missing Majority*. London: Routledge.

Hurford, J. (2012). *The Origins of Grammar: Language in the Light of Evolution.* Oxford: Oxford University Press.

Huth, A., de Heer, W.A. et al. and Gallant, J.L. (2016). Natural speech reveals the semantic maps that tile human cerebral cortex. *Nature* 532, 453–8.

Imai, M. and Kita, S. (2014). The sound symbolism bootstrapping hypothesis for language acquisition and language evolution. *Philosophical Transactions of the Royal Society of London B* 369, 20130298.

Imai, M., Kita, S., Nagumo, M. and Okado, H. (2008). Sound symbolism facilitates early verb learning. *Cognition* 109, 54–65.

Ingrey, L., Duffy, S.M. et al. and Pope, M. (2023). On the discovery of a late Acheulean 'giant' handaxe from the Maritime Academy, Frindsbury, Kent. *Internet Archaeology* 61.

Iovita, R. (2011). Shape variation in Aterian tanged tools and the origins of projectile technology: a morphometric perspective on stone tool function. *PLoS One* 6, e29029.

Isern, N. and Fort, J. (2014). Language extinction and linguistic fronts. *Journal of the Royal Society Interface* 11, 20140028.

Jackendoff, R. (1972). *Semantic Interpretation in Generative Grammar.* Cambridge, MA: MIT Press.

Jackendoff, R. (1999). Possible stages in the evolution of the language capacity. *Trends in Cognitive Sciences* 3, 272–9.

Jackendoff, R. (2002). *Foundations of Language: Brain, Meaning, Grammar, Evolution.* Oxford: Oxford University Press.

Jackendoff, R. (2011). What is the human language faculty? Two views. *Language* 87, 586–624.

Jäkel, S. and Dimou, L. (2017). Glial cells and their function in the adult brain: a journey through the history of their ablation. *Frontiers in Cellular Neuroscience* 11, 24.

Jakobson, R. (1960). Why 'mama' and 'papa'? In *Perspectives in Psychological Theory: Essays in Honor of Heinz Werner* (eds B. Kaplan and S. Wapner), pp. 124–34. New York: International Universities Press.

James, W. (1890). *The Principles of Psychology.* New York: Henry Holt.

Jaubert, J., Verheyden, S. et al. and Santos, F. (2016). Early Neanderthal constructions deep in Bruniquel Cave in southwestern France. *Nature* 534, 111–14.

Jespersen, O. (1919). The symbolic value of the vowel i. *Philosophical Journal of Comparative Philology* 1, 15–23.

Jespersen, O. (1922). *Language: Its Nature, Development and Origin.* London: George Allen & Unwin.

Kadosh, R.C., Henik, A. and Walsh, V. (2009). Synaesthesia: learned or lost? *Developmental Science* 12, 484–91.

Kano, F., Krupenye, C. et al. and Call, J. (2019). Great apes use self-experience to anticipate an agent's action in a false-belief test. *Proceedings of the National Academy of Sciences* 116, 20904–9.

Karkanas, P., Shahack-Gross, R. et al. and Stiner, M.C. (2007). Evidence for habitual use of fire at the end of the Lower Paleolithic: site-formation processes at Qesem Cave, Israel. *Journal of Human Evolution* 53, 197–212.

Karmiloff-Smith, A. (1992). *Beyond Modularity: A Developmental Perspective on Cognitive Science.* Cambridge, MA: MIT Press.

Karmiloff-Smith, A. (1998). Development itself is the key to understanding developmental disorders. *Trends in Cognitive Sciences* 2, 389–98.

Karmiloff-Smith, A. (2015). An alternative to domain-general or domain-specific frameworks for theorizing about human evolution and ontogenesis. *AIMS Neuroscience* 2, 91–104.

Kay, P. and Kempton, W. (1984). What is the Sapir–Whorf hypothesis? *American Anthropologist* 86, 65–78.

Kay, P. and Regier, T. (2003). Resolving the question of color naming universals. *Proceedings of the National Academy of Sciences* 100, 9085–9.

Keil, F.C. (1994). The birth and nurturance of concepts by domains: the origin of concepts of living things. In *Mapping the Mind: Domain Specificity in Cognition and Culture* (eds L.A. Hirschfeld and S.A. Gelman), pp. 234–54. Cambridge: Cambridge University Press.

Keller, S. (2000). An interpretation of Plato's *Cratylus*. *Phronesis* 45, 284–305.

Kendon, A. (2017). Reflections on the 'gesture-first' hypothesis of language origins. *Psychonomic Bulletin and Review* 24, 163–70.

Kennedy, M.B. (2016). Synaptic signaling in learning and memory. *Cold Spring Harbor Perspectives in Biology* 8, a016824.

Key, A., Proffitt, T. and de la Torre, I. (2020). Raw material optimization and stone tool engineering in the Early Stone Age of Olduvai Gorge (Tanzania). *Journal of the Royal Society Interface* 17, 20190377.

Kirby, S. (2001). Spontaneous evolution of linguistic structure – an iterated learning model of the emergence of regularity and irregularity. *IEEE Transactions on Evolutionary Computation* 5, 102–10.

Kirby, S. (2002). Learning, bottlenecks and the evolution of recursive syntax. In *Linguistic Evolution Through Language Acquisition* (ed. T. Briscoe), pp. 173–204. Cambridge: Cambridge University Press.

Kirby, S. (2012). Language as an adaptive system: the role of cultural evolution in the origins of structure. In *The Oxford Handbook of Language Evolution* (eds M. Tallerman and K.R. Gibson), pp. 589–604. Oxford: Oxford University Press.

Kirby, S. (2017). Culture and biology in the origins of linguistic structure. *Psychonomic Bulletin and Review* 24, 118–37.

Kirby, S., Cornish. H. and Smith, K. (2008). Cumulative cultural evolution in the laboratory: an experimental approach to the origins of structure in human language. *Proceedings of the National Academy of Sciences* 105, 10681–6.

Kita, S. (2009). Cross-cultural variation of speech-accompanying gesture: a review. *Language and Cognitive Processes* 24, 145–67.

Kiura, W.P. (2019). Stone Age cultures of East Africa. *South African Archaeological Bulletin* 74, 70–5.

Klein, R.G. and Edgar, B. (2002). *The Dawn of Human Culture*. New York: Wiley.

Knecht, S., Deppe, M. et al. and Henningsen, H. (2000). Language lateralization in healthy right-handers. *Brain* 123, 74–81.

Knight, C., Power, C. and Watts, I. (1995). The human symbolic revolution: a Darwinian account. *Cambridge Archaeological Journal* 5, 75–114.

Kochiyama, T., Ogihara, N. et al. and Akazawa, T. (2018). Reconstructing the Neanderthal brain using computational anatomy. *Scientific Reports* 8, 6296.

Köhler, W. (1929). *Gestalt Psychology*. New York: Liveright.

Kohn, M. and Mithen, S.J. (1999). Handaxes: products of sexual selection? *Antiquity* 73, 518–26.

Kojima, S., Izumi, A. and Ceugniet, K. (2003). Identification of vocalizers by pant hoots, pant grunts and screams in a chimpanzee. *Primates* 44, 225–30.

Kokocińska-Kusiak A., Woszczyło, M. et al. and Dzięcioł M. (2021). Canine olfaction: physiology, behavior, and possibilities for practical applications. *Animals (Basel)* 11, 2463.

Körtvélyessy, L., Štekauer, P. and Kačmár. P. (2021). On the role of creativity in the formation of new complex words. *Linguistics* 59, 1017–55.

Kovelman, I., Baker, S.A. and Petitto, L.-A. (2008). Bilingual and monolingual brains compared: a functional magnetic resonance imaging investigation of syntactic processing and a possible 'neutral signature' of bilingualism. *Journal of Cognitive Neuroscience* 20, 153–69.

Koziol, L.F, Budding, D. et al. and Yamazaki, T. (2014). Consensus paper: the cerebellum's role in movement and cognition. *Cerebellum* 13, 151–77.

Kozowyk, P.R.B., Soressi, M., Pomstra, D. and Langejans, G.H.J. (2017). Experimental methods for the Palaeolithic dry distillation of birch bark: implications for the origin and development of Neandertal adhesive technology. *Scientific Reports* 7, 8033.

Krause, J., Lalueza-Fox, C. et al. and Pääbo, S. (2007a). The derived *FOXP2* variant of modern humans was shared with Neandertals. *Current Biology* 17, 1908–12.

Krause, J., Orlando, L. et al. and Pääbo, S. (2007b). Neanderthals in central Asia and Siberia. *Nature* 449, 902–4.

Krupnik, I. and Müller-Wille, L. (2010). Franz Boas and Inuktitut terminology for ice and snow: from the emergence of the field to the 'Great Eskimo Vocabulary Hoax'. In *SIKU: Knowing Our Ice: Documenting Inuit Sea Ice Knowledge and Use* (eds I. Krupnik, C. Aporta et al.), pp. 377–400. Berlin: Springer Verlag.

Kuhl, P. (2000). A new view of language acquisition. *Proceedings of the National Academy of Sciences* 97, 11850–7.

Kuhlwilm, M. and Boeckx, C. (2019). A catalog of single nucleotide changes distinguishing modern humans from archaic hominins. *Scientific Reports* 9, 8463.

Kuhn, S. (1995). *Mousterian Lithic Technology: An Ecological Perspective*. Princeton: Princeton University Press.

Kuhn, S. (2006). Trajectories of change in the Middle Paleolithic of Italy. In *Transitions before the Transition: Evolution and Stability in the Middle Paleolithic and Middle Stone Age* (eds S. Kuhn and E. Hovers), pp. 109–20. New York: Springer.

Labov, W. (1994). *Principles of Linguistic Change, Vol. 1: Internal Factors*. Oxford: Blackwell.

Labov, W. (2001). *Principles of Linguistic Change, Vol. 2: Social Factors*. Oxford: Blackwell.

Labov, W. (2007). Transmission and diffusion. *Language* 83, 344–87.

Labov, W. (2010). *Principles of Linguistic Change, Vol. 3: Cognitive and Cultural Factors*. Oxford: Blackwell.

Labov, W., Ash, S. and Boberg, C. (2006). *The Atlas of North American English: Phonetics, Phonology and Sound Change*. Berlin: De Gruyter Mouton.

Lacruz, R.S., Stringer, C.B. et al. and Arsuaga, J.L. (2019). The evolutionary history of the human face. *Nature Ecology and Evolution* 3, 726–36.

Laitman, J. (1984). The anatomy of human speech. *Natural History* 93, 20–7.

Laitman, J. and Heimbuch, R.C. (1982). The basicranium of Plio-Pleistocene hominids as an indicator of their upper respiratory systems. *American Journal of Physical Anthropology* 59, 323–43.

Laitman, J., Heimbuch, R. and Crelin, E. (1979). The basicranium of fossil hominids as an indicator of their upper respiratory systems. *American Journal of Physical Anthropology* 51, 15–34.

Lakoff, G. (1987). *Women, Fire, and Dangerous Things: What Categories Reveal About the Mind*. Chicago: University of Chicago Press.

Lakoff, G. and Johnson, M. (1980). *Metaphors We Live By*. Chicago: University of Chicago Press.

Lamendella, J.T. (1976). Relations between the ontogeny and phylogeny of language: a neo-recapitulationist view. *Annals of the New York Academy of Sciences* 280, 396–412.

Lameira, A. (2017). Bidding evidence for primate vocal learning and the cultural substrates for speech evolution. *Neuroscience and Biobehavioral Reviews* 83, 429–39.

Lameira, A. and Call, J. (2018). Time–space-displaced responses in the orangutan vocal system. *Science Advances* 4, eaau3401.

Lameira, A.R., Hardus, M. et al. and Wich, S.A. (2013). Orangutan (*Pongo* spp.) whistling and implications for the emergence of an open-ended call repertoire: a replication and extension. *Journal of the Acoustical Society of America* 134, 2326–35.

Lameira, A.R., Hardus, M.E. et al. and Menken, S.B.J. (2015). Speech-like rhythm in a voiced and voiceless orangutan call. *PLoS One* 10, e116136.

Lameira, A., Vicente, R. et al. and Hardus, M.E. (2017). Proto-consonants were information-dense via identical bioacoustic tags to proto-vowels. *Nature Human Behaviour* 1, 0044.

Lameira, A., Santamaría-Bonfil, G. et al. and Wich, S.A. (2022). Sociality predicts orangutan vocal phenotype. *Nature Ecology and Evolution* 6, 644–52.

Lancy, D.F. (1983). *Cross-cultural Studies in Cognition and Mathematics*. New York: Academic Press.

Langus, A., Mehler, J. and Nespor, M. (2017). Rhythm in language acquisition. *Neuroscience and Biobehavioral Reviews* 81, 158–66.

Laporte, M.N.C. and Zuberbühler, K. (2010). Vocal greeting behaviour in wild chimpanzee females. *Animal Behaviour* 80, 467–73.

Larsson, L. (2000). The Middle Stone Age of Northern Zimbabwe in a Southern African perspective. *Lund Archaeological Review* 6, 61–84.

Leakey, M. (1984). *Disclosing the Past: An Autobiography*. New York: McGraw Hill.

Leavitt, J. (2011). *Linguistic Relativities: Language Diversity and Modern Thought*. Cambridge: Cambridge University Press.

Leder, D., Hermann, R. et al. and Terberger, T. (2021). A 51,000-year-old engraved bone reveals Neanderthals' capacity for symbolic behaviour. *Nature Ecology and Evolution* 5, 1273–82.

Lee, R.B. (1979). *The !Kung San: Men, Women, and Work in a Foraging Society*. Cambridge: Cambridge University Press.

Leivada, E., Mitrofanova, N. and Westergaard, M. (2021). Bilinguals are better than monolinguals in detecting manipulative discourse. *PLoS One* 16, e0256173.

LeMaster, B. and Monaghan, L. (2004). Variation in sign languages. In *A Companion to Linguistic Anthropology* (ed. A. Duranti), pp. 141–65. Oxford: Blackwell.

Levinson, S.C. (2003a). Language and mind: let's get the issues straight! In *Language in Mind: Advances in the Study of Language and Thought* (eds D. Gentner and S. Goldin-Meadow), pp. 25–46. Cambridge, MA: MIT Press.

Levinson, S.C. (2003b). *Space in Language and Cognition: Explorations in Cognitive Diversity*. Cambridge: Cambridge University Press.

Lewis, S.G., Ashton, N. et al. and Sier, M.J. (2019). Human occupation of northern Europe in MIS 13: Happisburgh Site 1 (Norfolk, UK) and its European context. *Quaternary Science Reviews* 211, 34–58.

Li, J., Wang, L. and Fischer, K. (2004). The organisation of Chinese shame concepts. *Cognition and Emotion* 18, 767–97.

Liebenberg, L. (1990). *The Art of Tracking: The Origin of Science*. Cape Town: David Philip.

Lieberman, P. (2007). The evolution of human speech: its anatomical and neural bases. *Current Anthropology* 48, 39–66.

Lieberman, P. and Crelin, E.S. (1971). On the speech of Neanderthal man. *Linguistic Inquiry* 2, 203–22.

Lieberman, P., Crelin, E.S. and Klatt, D.H. (1972). Phonetic ability and related anatomy of the newborn and adult human, Neanderthal man, and the chimpanzee. *American Anthropologist* 74, 287–307.

Lifschitz, A. (2012). *Language and Enlightenment: The Berlin Debates of the Eighteenth Century*. Oxford: Oxford University Press.

Lindquist, K.A., MacCormack, J.K. and Shablack, H. (2015). The role of language in emotion: predictions from psychological constructionism. *Frontiers in Psychology* 6, 444.

Liu, W., Martinón-Torres, M. et al. and Wu, X.-J. (2015). The earliest unequivocally modern humans in southern China. *Nature* 526, 696–9.

Ljubicic, G., Okpakok, S., Robertson, S. and Mearns, R. (2018). Inuit approaches to naming and distinguishing caribou: considering language, place, and homeland toward improved co-management. *Arctic* 71, 249–363.

Lloyd, E.A. (2004). Kanzi, evolution, and language. *Biology and Philosophy* 19, 577–88.

Lombao, D., Guardiola, M. and Mosquera, M. (2017). Teaching to make stone tools: new experimental evidence supporting a technological hypothesis for the origins of language. *Scientific Reports* 7, 14394.

Lombard, M. (2022). Re-considering the origins of Old World spearthrower-and-dart hunting. *Quaternary Science Reviews* 293, 107677.

Ludwig, V.U., Adachi, I. and Matsuzawa, T. (2011). Visuoauditory mappings between high luminance and high pitch are shared by chimpanzees (*Pan troglodytes*) and humans. *Proceedings of the National Academy of Sciences* 108, 20661–5.

Lupyan, G. (2012). What do words do? Toward a theory of language-augmented thought. *Psychology of Learning and Motivation* 57, 255–97.

Lupyan, G. (2015). The centrality of language in human cognition. *Language Learning* 66, 516–53.

Lupyan, G. and Bergen, B. (2015). How language programs the mind. *Topics in Cognitive Science* 8, 408–24.

Lupyan, G. and Clark, A. (2015). Words and the world: predictive coding and the language–perception–cognition interface. *Current Directions in Psychological Science* 24, 279–84.

Lupyan, G. and Dale, R. (2010). Language structure is partly determined by social structure. *PLoS One* 5, e8559.

Lupyan, G. and Dale, R. (2016). Why are there different languages? The role of adaptation in linguistic diversity. *Trends in Cognitive Science* 20, 649–60.

Lupyan, G. and Thompson-Schill, S.L. (2012). The evocative power of words: activation of concepts by verbal and non-verbal means. *Journal of Experimental Psychology: General* 141, 170–86.

Lupyan, G. and Winter, B. (2018). Language is more abstract than you think, or, why aren't languages more iconic? *Philosophical Transactions of the Royal Society B* 373, 20170137.

Lynn, C. (2014). Hearth and campfire influences on arterial blood pressure: defraying the costs of the social brain through fireside relaxation. *Evolutionary Psychology* 12, 983–1003.

Machin, A., Hosfield, R. and Mithen, S.J. (2007). Why are some handaxes symmetrical? Testing the influence of handaxe morphology on butchery effectiveness. *Journal of Archaeological Science* 34, 883–93.

Mackey, J.L. (2015). New evidence for the Epicurean theory of the origin of language: Philodemus, on Poems 5. *Cronache Ercolanesi* 45, 67–84.

MacLarnon, A.M. and Hewitt, G.P. (1999). The evolution of human speech: the role of enhanced breathing control. *American Journal of Physical Anthropology* 109, 341–63.

MacNeilage, P. (2010). *The Origin of Speech*. Oxford: Oxford University Press.

Maddieson, I. (1984). *Patterns of Sounds*. Cambridge: Cambridge University Press.

Madupe, P.P., Koenig, C. et al. and Cappellini, E. (2023). Enamel proteins reveal biological sex and genetic variability within southern African *Paranthropus*. Preprint at *BioRxiv*. doi.org/10.1101/2023.07.03.547326

Mahon, B.Z. and Caramazza, A. (2003). There are facts . . . and then there are facts. *Trends in Cognitive Sciences* 7, 481–2.

Mahon, B.Z. and Caramazza, A. (2011). What drives the organization of object knowledge in the brain? *Trends in Cognitive Sciences* 15, 97–103.

Majid, A. and Burenhult, N. (2014). Odors are expressible in language, as long as you speak the right language. *Cognition* 130, 266–70.

Majid, A. and Kruspe, N. (2018). Hunter-gatherer olfaction is special. *Current Biology* 28, 409–13.

Majid, A., Roberts, S.G. et al. and Levinson, S. (2018). Differential coding of perception in the world's languages. *Proceedings of the National Academy of Sciences* 115, 11369–76.

Majkić, A., Evans, S. et al. and d'Errico, F. (2017). A decorated raven bone from the Zaskalnaya VI (Kolosovskaya) Neanderthal site, Crimea. *PLoS One* 12, e0173435.

Majkić, A., d'Errico, F. and Stepanchuk, V. (2018). Assessing the significance of Palaeolithic engraved cortexes. A case study from the Mousterian site of Kiik-Koba, Crimea. *PLoS One* 13, e0195049.

Mallory, J.P. and Adams, D.Q. (2006). *The Oxford Introduction to Proto-Indo-European and the Proto-Indo-European World*. Oxford: Oxford University Press.

Malt, B.C., Ameel, E. et al. and Majid, A. (2011). Do words reveal concepts? *Proceedings of the 33rd Annual Meeting of the Cognitive Science Society*, 519–24.

Mamiya, P.C., Richards, T.L. et al. and Kuhl, P.K. (2016). Brain white matter structure and COMT gene are linked to second-language learning in adults. *Proceedings of the National Academy of Sciences* 113, 7249–54.

Marcazzan, D., Miller, C.E. et al. and Peresani, M. (2023). Middle and Upper Paleolithic occupations of Fumane Cave (Italy): a geoarchaeological investigation of the anthropogenic features. *Journal of Anthropological Sciences* 101, 37–62.

Margari, V., Hodell, D.A. et al. and Tzedakis, P.C. (2023). Extreme glacial cooling likely led to hominin depopulation of Europe in the Early Pleistocene. *Science* 381, 693–9.

Margiotoudi, K. and Pulvermüller, F. (2020). Action sound-shape congruencies explain sound symbolism. *Scientific Reports* 10, 12706.

Margiotoudi, K., Allritz, M., Bohn, M. and Pulvermüller, F. (2019). Sound symbolic congruency detection in humans but not in great apes. *Scientific Reports* 9, 12705.

Margiotoudi, K., Bohn, M. et al. and Allritz, M. (2022). Bo-NO-bouba-kiki: picture-word mapping but no spontaneous sound symbolic speech-shape mapping in a language trained bonobo. *Proceedings of the Royal Society of London B* 289, 20211717.

Marian, V. (2023). *The Power of Language: Multilingualism, Self and Society*. London: Pelican.

Mariën, P., Ackermann, H. et al. and Ziegler, W. (2014). Consensus paper: language and the cerebellum: an ongoing enigma. *Cerebellum* 13, 386–410.

Markman, E.M. (1990). Constraints children place on word meanings. *Cognitive Science* 14, 55–77.

Marquet, J.-C., Freiesleben, T.H. et al. and Jaubert, J. (2023). The earliest unambiguous Neanderthal engravings on cave walls: La Roche-Cotard, Loire Valley, France. *PLoS One* 18, e0286568.

Martin, L. (1986). Eskimo words for snow: a case study in the genesis and decay of an anthropological example. *American Anthropologist* 88, 418–23.

Martínez, I., Arsuaga, J.L. et al. and Rodríguez, L. (2008). Human hyoid bones from the Middle Pleistocene site of the Sima de los Huesos (Sierra de Atapuerca, Spain). *Journal of Human Evolution* 54, 118–24.

Matzinger, T., Ritt, N. and Fitch, W.T. (2021). The influence of different prosodic cues on word segmentation. *Frontiers in Psychology* 12, 622042.

Maurer, D. and Mondloch, C.J. (2006). The infant as synesthete? In *Attention and Performance XXI: Processes of Change in Brain and Cognitive Development* (eds Y. Munakata and M.H. Johnson), pp. 449–71. Oxford: Oxford University Press.

Mazza, P., Martini, F. et al. and Ribechini, E. (2006). A new Palaeolithic discovery: tar-hafted stone tools in a European Mid-Pleistocene bone-bearing bed. *Journal of Archaeological Science* 33, 1310–18.

McBrearty, S. (2003). Patterns of technological change at the origin of *Homo sapiens*. *Before Farming* 3, 1–5.

McBrearty, S. and Brooks, A.S. (2000). The revolution that wasn't: a new interpretation of the origin of modern human behavior. *Journal of Human Evolution* 39, 453–563.

McCoy, R.C., Wakefield, J. and Akey, J.M. (2017). Impacts of Neanderthal-introgressed sequences on the landscape of human gene expression. *Cell* 168, 916–27.

McGrew, W.C. (1992). *Chimpanzee Material Culture: Implications for Human Evolution*. Cambridge: Cambridge University Press.

McMullen, E. and Saffran, J.R. (2004). Music and language: a developmental comparison. *Music Perception* 21, 289–311.

McWhorter, J. (2001). *The Power of Babel: A Natural History of Language*. New York: Times Books.

McWhorter, J. (2016). *Words on the Move: Why English Won't – and Can't – Sit Still (Like, Literally)*. New York: Picador.

Melchionna, M., Profico, A. et al. and Raia, P. (2020). From smart apes to human brain boxes. A uniquely derived brain shape in late hominins clade. *Frontiers in Earth Science* 8, 273.

Mellars, P. (2010). Neanderthal symbolism and ornament manufacture: the bursting of a bubble? *Proceedings of the National Academy of Sciences* 107, 20147–8.

Mervis, C.B. and Rosch, E. (1981). Categorization of natural objects. *Annual Review of Psychology* 32, 89–115.

Mesfin, I., Leplongeon, A., Pleurdeau, D. and Borel, A. (2020). Using morphometrics to reappraise old collections: the study case of the Congo Basin Middle Stone Age bifacial industry. *Journal of Lithic Studies* 7, 1–38.

Metz, L., Lewis, J.E. and Slimak, L. (2023). Bow-and-arrow technology of the first modern humans in Europe 54,000 years ago at Mandrin, France. *Science Advances* 9, eadd4675.

Meyer, M., Arsuaga, J.L. et al. and Pääbo, S. (2016). Nuclear DNA sequences from the Middle Pleistocene Sima de los Huesos hominins. *Nature* 531, 504–7.

Michel, V., Valladas, H. et al. and Bae, C.J. (2016). The earliest modern *Homo sapiens* in China? *Journal of Human Evolution* 101, 101–4.

Mickan, A., Schiefke, M. and Stefanowitsch, A. (2014). Key is a llave is a Schlüssel: a failure to replicate an experiment from Boroditsky et al. 2003. *Yearbook of the German Cognitive Linguistics Association* 2, 39–50.

Milks, A.G., Lehmann, J. et al. and Terberger, T. (2023). A double-pointed wooden throwing stick from Schöningen, Germany: results and new insights from a multianalytical study. *PLoS One* 18, e0287719.

Miller, C.A. (2006). Developmental relationships between language and theory of mind. *American Journal of Speech-Language Pathology* 15, 142–54.

Millward, C.M. and Hayes, M. (2012). *A Biography of the English Language*. London: Wadsworth.

Mithen, S.J. (1988). Looking and learning: Upper Paleolithic art and information gathering. *World Archaeology* 19, 297–327.

Mithen, S.J. (1996). *The Prehistory of the Mind: A Search for the Origins of Art, Religion and Science*. London: Thames and Hudson.

Mithen, S.J. (2003). *After the Ice: A Global Human History, 20,000–5000 BC*. London: Weidenfeld and Nicolson.

Mithen, S.J. (2005). *The Singing Neanderthals: The Origins of Music, Language, Mind and Body*. London: Weidenfeld and Nicolson.

Mithen, S.J. (2007). Did farming arise from a misapplication of social intelligence? *Philosophical Transactions of the Royal Society of London B* 362, 705–18.

Mithen, S.J. (2019). Becoming Neolithic in words, thoughts and deeds. *Journal of Social Archaeology* 19, 67–91.

Miyagawa, S. and Clarke, E. (2019). Systems underlying human and Old World monkey communication: one, two, or infinite. *Frontiers in Psychology* 10, 1911.

Moggi-Cecchi, J. and Collard, M. (2002). A fossil stapes from Sterkfontein, South Africa, and the hearing capabilities of early hominids. *Journal of Human Evolution* 42, 259–65.

Moisik, S.R. and Dediu, D. (2017). Anatomical biasing and clicks: evidence from biomechanical modeling. *Journal of Language Evolution* 2, 37–51.

Monaghan, P., Mattock, K. and Walker, P. (2012). The role of sound symbolism in language learning. *Journal of Experimental Psychology* 38, 1152–64.

Monaghan, P., Shillcock, R.C., Christiansen, M.H. and Kirby, S. (2014). How arbitrary is language? *Philosophical Transactions of the Royal Society of London B* 369, 20130299.

Mora-Bermúdez, F., Badsha, F. et al. and Huttner, W.B. (2016). Differences and similarities between human and chimpanzee neural progenitors during cerebral cortex development. *eLife* 5, e18683.

Morales, M., Mundy, P. et al. and Schwartz, H.K. (2000). Responding to joint attention across the 6- through 24-month age period and early language acquisition. *Journal of Applied Developmental Psychology* 21, 283–98.

Moran, S., McCloy, D. and Wright, R. (2012). Revisiting population size vs phoneme inventory size. *Language* 88, 877–93.

Morgan, T.H., Uomini, N.T. et al. and Laland, K.N. (2015). Experimental evidence for the co-evolution of hominin tool-making teaching and language. *Nature Communications* 6, 6029.

Moriano, J. and Boeckx, C. (2020). Modern human changes in regulatory regions implicated in cortical development. *BMC Genomics* 21, 304.

Morin, E. and Laroulandie, V. (2012). Presumed symbolic use of diurnal raptors by Neanderthals. *PLoS One* 7, e32856.

Morley, I. (2013). *The Prehistory of Music: Human Evolution, Archaeology, and the Origins of Musicality*. Oxford: Oxford University Press.

Morphy, H. (1999). Encoding the Dreaming – a theoretical framework for the analysis of representational processes in Australian Aboriginal art. *Australian Archaeology* 49, 13–22.

Moss, H.E. and Tyler, L.K. (2003). Weighing up the facts of category-specific semantic deficits. *Trends in Cognitive Sciences* 7, 480–1.

Muehlenbein, M. (ed.) (2015). *Basics in Human Evolution*. New York: Academic Press.

Muller, A. and Clarkson, C. (2016). Identifying major transitions in the evolution of lithic cutting-edge production rates. *PLoS One* 11, e0167244.

Müller, P., Ohnheiser, I., Olsen, S. and Rainer, F. (2015). *Word Formation: An International Handbook of the Languages of Europe*. Berlin: De Gruyter Mouton.

Munroe, R. L., Fought, J.G. and Macaulay, R.K.S. (2009). Warm climates and sonority classes: not simply more vowels and fewer consonants. *Cross-Cultural Research* 43, 123–33.

Murdock, G.P. (1959). Cross-language parallels in parental kin terms. *Anthropological Linguistics* 1, 1–5.

Murphree, W.C. and Aldeias, V. (2022). The evolution of pyrotechnology in the Upper Palaeolithic of Europe. *Archaeological and Anthropological Sciences* 14, 202.

Musgrave, S., Lonsdorf, E. et al. and Sanz, C. (2020). Teaching varies with task complexity in wild chimpanzees. *Proceedings of the National Academy of Sciences* 117, 969–76.

Namboodiripad, S., Lenzen, D., Lepic, R. and Verhoef, T. (2016). Measuring conventionalization in the manual modality. *Journal of Language Evolution* 1, 109–18.

Narrog, H. and Heine, B. (2012). Introduction. In *The Oxford Handbook of Grammaticalization* (eds B. Heine and H. Narrog), pp. 1–16. Oxford: Oxford University Press.

Nettle, D. (1998). Explaining global patterns of language diversity. *Journal of Anthropological Archaeology* 17, 354–74.

Nettle, D. (1999). Is the rate of linguistic change constant? *Lingua* 108, 119–36.

Nettle, D. (2012). Social scale and structural complexity in human languages. *Philosophical Transactions of the Royal Society of London B* 367, 1829–36.

Neubauer, S., Gunz, P. and Hublin, J.-J. (2010). Endocranial shape changes during growth in chimpanzees and humans: a morphometric analysis of unique and shared aspects. *Journal of Human Evolution* 59, 555–66.

Neubauer, S., Hublin, J.-J. and Gunz, P. (2018). The evolution of modern human brain shape. *Science Advances* 4, eaao5961.

Nevalainen, T. and Palander-Collin, M. (2012). Grammaticalization and sociolinguistics. In *The Oxford Handbook of Grammaticalization* (eds B. Heine and H. Narrog), pp. 118–29. Oxford: Oxford University Press.

Nevalainen, T. and Traugott, E.C. (eds) (2012). *The Oxford Handbook of the History of English*. Oxford: Oxford University Press.

Newport, E.L. and Aslin, R.N. (2004). Learning at a distance I. Statistical learning of non-adjacent dependencies. *Cognitive Psychology* 48, 127–62.

Newport, E.L., Hauser, M.D., Spaepen, G. and Aslin, R.N. (2004). Learning at a distance II. Statistical learning of non-adjacent dependencies in a non-human primate. *Cognitive Psychology* 49, 85–117.

Nishimura, T., Mikami, A., Suzuki, J. and Matsuzawa, T. (2006). Descent of the hyoid in chimpanzees: evolution of face flattening and speech. *Journal of Human Evolution* 51, 244–54.

Nishimura, T., Tokuda, I. et al. and Fitch, W.T. (2022). Evolutionary loss of complexity in human vocal anatomy as an adaptation for speech. *Science* 377, 760–3.

Nygaard, L.C., Cook, A.E. and Namy, L.L. (2009). Sound to meaning correspondences facilitate word learning. *Cognition* 112, 181–6.

Oakley, K. (1969). Man the skilled tool-maker. *Antiquity* 43, 222–4.

Oakley, K.P. (1972). *Man the Tool-maker*. London: British Museum.

Oakley, K.P., Andrews, P., Keeley, L.H. and Clark, J.D. (1977). A reappraisal of the Clacton spearpoint. *Proceedings of the Prehistoric Society* 43, 13–30.

Ohala, J.J. (1981). The listener as a source of sound change. In *Papers from the Para-session on Language and Behavior* (eds C.S. Masek, R.A. Hendrick and M.F. Miller), pp. 178–203. Chicago: Chicago Linguistic Society.

Ohala, J.J. (2003). Phonetics and historical phonology. In *The Handbook of Historical Linguistics* (eds B.D. Joseph and R.D. Janda), pp. 669–86. Oxford: Blackwell.

Ohnuma, K., Aoki, K. and Akazawa, T. (1997). Transmission of toolmaking through verbal and non-verbal communication – preliminary experiments in Levallois flake production. *Anthropological Science* 105, 159–68.

Ortega, G. (2017). Iconicity and sign lexical acquisition: a review. *Frontiers in Psychology* 8, 1280.

Ouattara, K., Lemasson, A. and Zuberbühler, K. (2009). Campbell's monkeys use affixation to alter call meaning. *PLoS One* 4, e7808.

Pagel, M., Atkinson, Q.D., Calude, A.S. and Meade, A. (2013). Ultraconserved words point to deep language ancestry across Eurasia. *Proceedings of the National Academy of Sciences* 110, 8471–6.

Pan, D. (2004). J.G. Herder, the *Origin of Language*, and the possibility of transcultural narratives. *Language and Intercultural Communication* 4, 10–20.

Pang, J. and Fornito, A. (2023). Have we got the brain all wrong? A new study shows its shape is more important than its wiring. *The Conversation*, theconversation.com, 31 May 2023.

Pang, J., Aquino, K.M. et al. and Fornito, A. (2023). Geometric constraints on human brain function. *Nature* 618, 566–74.

Paradiso, S., Andreasen, N.C. et al. and Robinson, R.G. (1997). Cerebellar size and cognition: correlations with IQ, verbal memory and motor dexterity. *Neuropsychiatry, Neuropsychology and Behavioral Neurology* 10, 1–8.

Parker, S.T. and Gibson, K.R. (1979). A developmental model for the evolution of language and intelligence in early hominids. *Behavioral and Brain Sciences* 2, 367–81.

Partanen, E., Kujala, T. et al. and Huotilainen, M. (2013). Learning-induced neural plasticity of speech processing before birth. *Proceedings of the National Academy of Sciences* 110, 15145–50.

Paulesu, E., McCrory, E. et al. and Frith, U. (2000). A cultural effect on brain function. *Nature Neuroscience* 3, 91–6.

Pavlenko, A. (2014). *The Bilingual Mind: And What It Tells Us About Language and Thought*. Cambridge: Cambridge University Press.

Pearce, D.G. and Bonneau, A. (2018). Trouble on the dating scene. *Nature Ecology and Evolution* 2, 925–6.

Pearce, E., Stringer, C. and Dunbar, R.I.M. (2013). New insights into differences in brain organization between Neanderthals and anatomically modern humans. *Proceedings of the Royal Society of London B* 280, 20130168.

Pelaez, M., Borroto, A.R. and Carrow, J. (2018). Infant vocalizations and imitation as a result of adult contingent imitation. *Behavioral Development* 23, 81–8.

Pelucchi, B., Hay, J.F. and Saffran, J.R. (2009). Learning in reverse: 8-month-old infants track backward transitional probabilities. *Cognition* 113, 244–7.

Pereira, A.S., Kavanagh, E. et al. and Lameira, A.R. (2020). Chimpanzee lip-smacks confirm primate continuity for speech-rhythm evolution. *Biology Letters* 16, 20200232.

Peresani, M., Fiore, I. et al. and Tagliacozzo, A. (2011). Late Neandertals and the intentional removal of feathers as evidenced from bird bone taphonomy at Fumane Cave 44 ky BP, Italy. *Proceedings of the National Academy of Sciences* 108, 3888–93.

Peresani, M., Vanhaeren, M. et al. and d'Errico, F. (2013). An ochered fossil marine shell from the Mousterian of Fumane Cave, Italy. *PLoS One* 8, e68572.

Perniss, P. and Vigliocco, G. (2014). The bridge of iconicity: from a world of experience to the experience of language. *Philosophical Transactions of the Royal Society of London B* 369, 20130300.

Perry, L.K., Perlman, M. and Lupyan, G. (2015). Iconicity in English and Spanish and its relation to lexical category and age of acquisition. *PLoS One* 10, e0137147.

Pike, A.W.G., Hoffmann, D.L. et al. and Zilhão, J. (2012). U-series dating of Paleolithic art in 11 caves in Spain. *Science* 336, 1409–13.

Pinker, S. (1994). *The Language Instinct: How the Mind Creates Language*. New York: William Morrow.

Pinker, S. and Bloom, P. (1990). Natural language and natural selection. *Behavioral and Brain Sciences* 13, 707–84.

Pinker, S. and Jackendoff, R. (2009). The reality of a universal language faculty. *Behavioral and Brain Sciences* 32, 465–6.

Pinson, A., Xing, L. et al. and Huttner, W.B. (2022). Human *TKTL1* implies greater neurogenesis in frontal neocortex of modern humans than Neanderthals. *Science* 377, eabl6422.

Plummer, T.W., Oliver, J. et al. and Potts, R. (2023). Expanded geographic distribution and dietary strategies of the earliest Oldowan hominins and *Paranthropus*. *Science* 379, 561–6.

Poeppel, D., Emmorey, K., Hickok, G. and Pylkkänen, L. (2012). Towards a new neurobiology of language. *Journal of Neuroscience* 32, 14125–31.

Pollick, A.S. and de Waal, F.B.M. (2007). Ape gestures and language evolution. *Proceedings of the National Academy of Sciences* 104, 8184–9.

Ponce de León, M.S., Bienvenu, T. et al. and Zollikofer, C.P.E. (2021). The primitive brain of early *Homo*. *Science* 372, 165–71.

Posth, C., Yu, H. et al. and Krause, J. (2023). Palaeogenomics of Upper Palaeolithic to Neolithic European hunter-gatherers. *Nature* 615, 117–26.

Povinelli, D.J. and Vonk, J. (2003). Chimpanzee minds: suspiciously human? *Trends in Cognitive Sciences* 7, 157–60.

Povinelli, D.J., Eddy, T.J., Hobson, R.P. and Tomasello, M. (1996). What young chimpanzees know about seeing. *Monographs of the Society for Research in Child Development* 61, 1–189.

Preece, R.C., Gowlett, J.A.J. et al. and Lewis, S.G. (2006). Humans in the Hoxnian: habitat, context and fire use at Beeches Pit, West Stow, Suffolk, UK. *Journal of Quaternary Science* 21, 485–96.

Premack, D. (1976). *Intelligence in Ape and Man*. Hillsdale, NJ: Erlbaum.

Premack, D. (1983). The codes of man and beasts. *Behavioral and Brain Sciences* 6, 125–67.

Premack, D. (1990). Words: what are they, and do animals have them? *Cognition* 37, 197–212.

Premack, D. and Premack, A.J. (1983). *The Mind of an Ape*. New York: Norton.

Premack, D. and Woodruff, G. (1978). Does the chimpanzee have a theory of mind? *Behavioral and Brain Sciences* 4, 515–26.

Prévost, M., Groman-Yaroslavski, I. et al. and Zaidner, Y. (2021). Early evidence for symbolic behavior in the Levantine Middle Paleolithic: a 120 ka old engraved aurochs bone shaft from the open-air site of Nesher Ramla, Israel. *Quaternary International* 624, 80–93.

Price, T., Wadewitz, P. et al. and Fischer, J. (2015). Vervets revisited: a quantitative analysis of alarm call structure and context specificity. *Scientific Reports* 5, 13220.

Pritchard, E.T. (2013). *Bird Medicine: The Sacred Power of Bird Shamanism*. Rochester, VT: Bear & Company.

Proffitt, T., Reeves, J.S., Pacome, S.S. and Luncz, L.V. (2022). Identifying functional and regional differences in chimpanzee stone tool technology. *Royal Society Open Science* 9, 220826.

Prüfer, K., Racimo, F. et al. and Pääbo, S. (2014). The complete genome sequence of a Neanderthal from the Altai Mountains. *Nature* 505, 43–9.

Pullum, G. (1991). *The Great Eskimo Vocabulary Hoax and Other Irreverent Essays on the Study of Language*. Chicago: University of Chicago Press.

Quam, R. and Rak, Y. (2008). Auditory ossicles from southwest Asian Mousterian sites. *Journal of Human Evolution* 54, 414–33.

Quam, R.M., de Ruiter, D.J. et al. and Moggi-Cecchi, J. (2013a). Early hominin auditory ossicles from South Africa. *Proceedings of the National Academy of Sciences* 110, 8847–51.

Quam, R., Martínez, I. and Arsuaga, J.L. (2013b). Reassessment of the La Ferrassie 3 Neandertal ossicular chain. *Journal of Human Evolution* 64, 250–62.

Quam, R.M., Coleman, M.N. and Martínez, I. (2015a). Evolution of the auditory ossicles in extant hominids: metric variation in African apes and humans. *Journal of Anatomy* 225, 167–96.

Quam, R., Martínez, I. et al. and Arsuaga, J.L. (2015b). Early hominin auditory capacities. *Science Advances* 1, e1500355.

Quine, W. (1960). *Word and Object*. Cambridge, MA: MIT Press.

Radovčić, D., Sršen, A.O., Radovčić, J. and Frayer, D.W. (2015). Evidence for Neandertal jewelry: modified white-tailed eagle claws at Krapina. *PLoS One* 10, e0119802.

Rak, Y. and Clarke, R.J. (1979). Ear ossicle of *Australopithecus robustus*. *Nature* 279, 62–3.

Ralls, K., Fiorelli, P. and Gish, S. (1985). Vocalizations and vocal mimicry in captive harbor seals, *Phoca vitulina*. *Canadian Journal of Zoology* 63, 1050–6.

Ramachandran, V.S. and Hubbard, E.M. (2001). Synaesthesia – a window into perception, thought and language. *Journal of Consciousness Studies* 8, 3–34.

Ravignani, A. and Sonnweber, R. (2017). Chimpanzees process structural isomorphisms across sensory modalities. *Cognition* 161, 74–9.

Raviv, L., Meyer, A. and Lev-Ari, S. (2019). Larger communities create more systematic languages. *Proceedings of the Royal Society of London B* 286, 20191262.

Rayner, K. and Clifton Jr., C. (2009). Language processing in reading and speech perception is fast and incremental: implications for event-related potential research. *Biological Psychology* 80, 4–9.

Rees, D.A. (2000). The refitting of lithics from unit 4C, Area Q2/D excavations at Boxgrove, West Sussex, England. *Lithic Technology* 25, 120–34.

Regalado, A. (2011) Who coined "cloud computing." *MIT Technology Review*, 31 October 2011

Regier, T. and Kay, P. (2009). Language, thought, and color: Whorf was half right. *Trends in Cognitive Sciences* 13, 439–46.

Regier, T., Kay, P. and Cook, R.S. (2005). Focal colors are universal after all. *Proceedings of the National Academy of Sciences* 102, 8386–91.

Regier, T., Carstensen, A. and Kemp, C. (2016). Languages support efficient communication about the environment: words for snow revisited. *PloS One* 11, e0151138.

Reich, D., Green, R. et al. and Pääbo, S. (2010). Genetic history of an archaic hominin group from Denisova Cave in Siberia. *Nature* 468, 1053–60.

Revedin, A., Grimaldi, S. et al. and Aranguren, B. (2020). Experimenting the use of fire in the operational chain of prehistoric wooden tools: the digging sticks of Poggetti Vecchi (Italy). *Journal of Paleolithic Archaeology* 3, 525–36.

Revill, K.P., Namy, L.L., DeFife, L.C. and Nygaard, L.C. (2014). Cross-linguistic sound symbolism and crossmodal correspondence: evidence from fMRI and DTI. *Brain and Language* 128, 18–24.

Richter, D. and Krbetschek, M. (2015). The age of the Lower Paleolithic occupation at Schöningen. *Journal of Human Evolution* 89, 46–56.

Rijkhoff, J. (2007). Word Classes. *Language and Linguistics Compass* 1, 709–26.

Rinaldi, L.J., Smees, R., Carmichael, D.A. and Simner, J. (2020). Personality profile of child synaesthetes. *Frontiers in Bioscience* 12, 162–82.

Roach, N.T. and Richmond, B.G. (2015). Clavicle length, throwing performance and the reconstruction of the *Homo erectus* shoulder. *Journal of Human Evolution* 80, 107–13.

Roberson, D., Davies, I.R.L. and Davidoff, J. (2000). Color categories are not universal: replications and new evidence from a Stone-Age culture. *Journal of Experimental Psychology: General* 129, 369–98.

Roberson, D., Davidoff, J., Davies, I.R.L. and Shapiro, L.R. (2005). Color categories: evidence for the cultural relativity hypothesis. *Cognitive Psychology* 50, 378–411.

Roberts, M.B., Parfitt, S.A. et al. and Stewart, J.R. (1997). Boxgrove, West Sussex: rescue excavations of a Lower Palaeolithic landsurface (Boxgrove Project B, 1989–91). *Proceedings of the Prehistoric Society* 63, 303–58.

Robinson, J. (2012). A sociolinguistic approach to semantic change. In *Current Methods in Historical Semantics* (eds K. Allan and J.A. Robinson), pp. 199–232. Berlin: De Gruyter Mouton.

Rodríguez, L., Cabo, L. and Egocheaga, J. (2003). Breve nota sobre el hioides Neandertalense de Sidrón (Piloña, Asturias). In *Antropología y Biodiversida* (eds M. Aluja, A. Malgosa and R. Nogués), pp. 484–93. Barcelona: Edicions Bellaterra.

Rodríguez-Vidal, J., d'Errico, F. et al. and Finlayson, C. (2014). A rock engraving made by Neanderthals in Gibraltar. *Proceedings of the National Academy of Sciences* III, 13301–6.

Roebroeks, W. and Villa, P. (2011). On the earliest evidence for habitual use of fire in Europe. *Proceedings of the National Academy of Sciences* 108, 5209–14.

Roebroeks, W., Sier, M.J. et al. and Mücher, H.J. (2012). Use of red ochre by early Neandertals. *Proceedings of the National Academy of Sciences* 109, 1889–94.

Roffman, I., Savage-Rumbaugh, S. et al. and Nevo, E. (2012). Stone tool production and utilization by bonobo-chimpanzees (*Pan paniscus*). *Proceedings of the National Academy of Sciences* 109, 14500–3.

Roksandic, M., Radović, P., Wu, X.-J. and Bae, C.J. (2022). Resolving the 'muddle in the middle': the case for *Homo bodoensis* sp. nov. *Evolutionary Anthropology* 31, 20–9.

Romandini, M., Peresani, M., Laroulandie, V. et al. and Slimak, L. (2014). Convergent evidence of eagle talons used by late Neanderthals in Europe: a further assessment on symbolism. *PLoS One* 9, e101278.

Romberg, A.R. and Saffran, J.R. (2010). Statistical learning and language acquisition. *Wiley Interdisciplinary Reviews – Cognitive Science* 1, 906–14.

Rosch Heider, E. (1972). Probabilities, sampling and ethnographic method: the case of Dani colour names. *Man* 7, 448–66.

Ross, D. (1955). The date of Plato's *Cratylus*. *Revue Internationale de Philosophie* 32, 187–96.

Rots, V. (2013). Insights into early Middle Palaeolithic tool use and hafting in Western Europe. The functional analysis of level IIa of the early Middle Palaeolithic site of Biache-Saint-Vaast (France). *Journal of Archaeological Science* 40, 497–506.

Rowley-Conwy, P. (2007). *From Genesis to Prehistory: The Archaeological Three Age System and Its Contested Reception in Denmark, Britain, and Ireland.* Oxford: Oxford University Press.

Ruan, J., Timmermann, A. et al. and Melchionna, M. (2023). Climate shifts orchestrated hominin interbreeding events across Eurasia. *Science* 381, 699–704.

Ruebens, K. (2013). Regional behaviour among late Neanderthal groups in Western Europe: a comparative assessment of late Middle Palaeolithic bifacial tool variability. *Journal of Human Evolution* 65, 341–62.

Ruhlen, M. (1994). *On the Origin of Languages: Studies in Linguistic Taxonomy.* Stanford: Stanford University Press.

Saffran, J.R. and Kirkham, N.Z. (2018). Infant statistical learning. *Annual Review of Psychology* 69, 181–203.

Saffran, J.R. and Wilson, D.P. (2003). From syllables to syntax: multilevel statistical learning by 12-month-old infants. *Infancy* 4, 273–84.

Saffran, J.R., Aslin, R.N. and Newport, E.L. (1996). Statistical learning by 8-month-old infants. *Science* 274, 1926–8.

Salmons, J. (2021). *Sound Change*. Edinburgh: Edinburgh University Press.

Sandgathe, D.M., Dibble. H.L. et al. and Hodgkins, J. (2011). On the role of fire in Neandertal adaptations in Western Europe: evidence from Pech de l'Azé IV and Roc de Marsal, France. *PaleoAnthropology* 2011, 216–42.

Sankararaman, S., Mallick, S. et al. and Reich, D. (2014). The genomic landscape of Neanderthal ancestry in present-day humans. *Nature* 507, 354–7.

Sansalone, G., Profico, A. et al. and Raia, P. (2023). *Homo sapiens* and Neanderthals share high cerebral cortex integration into adulthood. *Nature Ecology and Evolution* 7, 42–50.

Sanz, C.M. and Morgan, D.B. (2013). Ecological and social correlates of chimpanzee tool use. *Philosophical Transactions of the Royal Society B* 368, 20120416.

Sanz, M., Daura, J. et al. and Zilhão, J. (2020). Early evidence of fire in south-western Europe: the Acheulean site of Gruta da Aroeira (Torres Novas, Portugal). *Scientific Reports* 10, 12053.

Sapir, E. (1929a). A study in phonetic symbolism. *Journal of Experimental Psychology* 12, 225–39.

Sapir, E. (1929b). The status of linguistics as a science. *Language* 5, 207–14. Reprinted in (ed.) D. Mandelbaum (1949). *Selected Writings of Edward Sapir in Language, Culture and Personality*. Berkeley: University of California Press.

Savage, P.E., Loui, P. et al. and Fitch, W.T. (2021). Music as a coevolved system for social bonding. *Behavioral and Brain Sciences* 44, e59.

Savage-Rumbaugh, S. and Lewin, R. (1994). *Kanzi: The Ape at the Brink of the Human Mind*. New York: John Wiley and Sons.

Savage-Rumbaugh, S., Rumbaugh, D.M., Smith, S.T. and Lawson, J. (1980). Reference: the linguistic essential. *Science* 201, 922–5.

Savage-Rumbaugh, S., McDonald, K. et al. and Rupert, E. (1986). Spontaneous symbol acquisition and communicative use by pygmy chimpanzees (*Pan paniscus*). *Journal of Experimental Psychology* 115, 211–35.

Savage-Rumbaugh, E.S., Shanker, S.G. and Taylor, T.J. (1998). *Apes, Language, and the Human Mind*. New York: Oxford University Press.

Scerri, E.M.L. (2017). The North African Middle Stone Age and its place in recent human evolution. *Evolutionary Anthropology* 26, 119–35.

Schaefer, N.K., Shapiro, B. and Green, R.E. (2021). An ancestral recombination graph of human, Neanderthal, and Denisovan genomes. *Science Advances* 7, eabc0776.

Schel, A.M., Townsend, S.W., Machanda, Z. and Zuberbühler, K. (2013).

Chimpanzee alarm call production meets key criteria for intentionality. *PLoS One* 8, e76674.

Schenker, N.M., Hopkins, W.D. et al. and Sherwood, C.C. (2010). Broca's area homologue in chimpanzees (*Pan troglodytes*): probabilistic mapping, asymmetry, and comparison to humans. *Cerebral Cortex* 20, 730–42.

Schlenker, P., Chemla, E. and Zuberbühler, K. (2016). What do monkey calls mean? *Trends in Cognitive Sciences* 20, 894–904.

Schmid, H.-J. (2015). The scope of word-formation research. In *Word Formation: An International Handbook of the Languages of Europe* (eds P. Müller, I. Ohnheiser, S. Olsen and F. Rainer), pp. 1–20. Berlin: De Gruyter Mouton.

Schmidt, P., Blessing, M. et al. and Tennie, C. (2019). Birch tar production does not prove Neanderthal behavioral complexity. *Proceedings of the National Academy of Sciences* 116, 17707–11.

Schoch, W.H., Bigga, G. et al. and Terberger, T. (2015). New insights on the wooden weapons from the Paleolithic site of Schöningen. *Journal of Human Evolution* 89, 214–25.

Schoenemann, P.T. (2006). Evolution of the size and functional areas of the human brain. *Annual Review of Anthropology* 35, 379–406.

Schoenemann, P.T. (2022). Evidence of grammatical knowledge in apes: an analysis of Kanzi's performance on reversible sentences. *Frontiers in Psychology* 13, 885605.

Schoenemann, P.T., Sheehan, M.J. and Glotzer, L.D. (2005). Prefrontal white matter volume is disproportionately larger in humans than in other primates. *Nature Neuroscience* 8, 242–52.

Sehasseh, E.M., Fernandez, P. et al. and Bouzouggar, A. (2021). Early Middle Stone Age personal ornaments from Bizmoune Cave, Essaouira, Morocco. *Science Advances* 7, eabi8620.

Semenza, C. (2009). The neuropsychology of proper names. *Mind and Language* 24, 347–69.

Seuren, P.A.M. (2016). Saussure and his intellectual environment. *History of European Ideas* 42, 819–47.

Sevim-Erol, A., Begun, D.R. et al. and Cihat Alçiçek, M. (2023). A new ape from Türkiye and the radiation of late Miocene hominines. *Communications Biology* 6, 842.

Seyfarth, R.M., Cheney, D.L. and Marler. P. (1980). Vervet monkey alarm calls: semantic communication in a free-ranging primate. *Animal Behaviour* 28, 1070–1094.

Shahack-Gross, R., Berna, F. et al. and Barkai, R. (2014). Evidence for the repeated use of a central hearth at Middle Pleistocene (300 ky ago). Qesem Cave, Israel. *Journal of Archaeological Science* 44, 12–21.

Shaham, D., Belfer-Cohen, A., Rabinovich, R. and Goren-Inbar, N. (2019). A

Mousterian engraved bone: principles of perception in Middle Paleolithic art. *Current Anthropology* 60, 708–16.

Shi, L., Hu, E. et al. and Su, B. (2017). Regional selection of the brain size regulating gene *CASC5* provides new insight into human brain evolution. *Human Genetics* 136, 193–204.

Shimelmitz, R. and Kuhn, S.L. (2013). Early Mousterian Levallois technology in Unit IX of Tabun Cave. *PaleoAnthropology* 2013, 1–27.

Shimelmitz, R., Kuhn, S.L. et al. and Weinstein-Evron, M. (2014). 'Fire at will': the emergence of habitual fire use 350,000 years ago. *Journal of Human Evolution* 77, 196–203.

Shoaib, A., Wang, T., Hay, J.F. and Lany, J. (2018). Do infants learn words from statistics? Evidence from English-learning infants hearing Italian. *Cognitive Science* 42, 3083–99.

Shukla, M., Nespor, M. and Mehler, J. (2007). An interaction between prosody and statistics in the segmentation of fluent speech. *Cognitive Psychology* 54, 1–32.

Sidhu, D.M. and Pexman, P.M. (2018). Five mechanisms of sound symbolic association. *Psychonomic Bulletin and Review* 25, 1619–43.

Sidhu, D.M., Westbury, C., Hollis, G. and Pexman, P.M. (2021). Sound symbolism shapes the English language: the maluma/takete effect in English nouns. *Psychonomic Bulletin and Review* 28, 1390–8.

Simon, N. (1978). Kaspar Hauser's recovery and autopsy: a perspective on neurological and sociological requirements for language development. *Journal of Autism and Child Schizophrenia* 8, 209–17.

Simonti, C.N., Vernot, B. et al. and Capra, J.A. (2016). The phenotypic legacy of admixture between modern humans and Neandertals. *Science* 351, 737–41.

Skinner, B.F. (1957). *Verbal Behavior*. New York: Appleton-Century-Crofts.

Skinner, D. (2014). How did cool become such a big deal? *Humanities* 35(4).

Slimak, L. (2023). The three waves: rethinking the structure of the first Upper Paleolithic in Western Eurasia. *PLoS One* 18, e0277444.

Slimak, L., Zanolli, C. et al. and Metz, L. (2022). Modern human incursion into Neanderthal territories 54,000 years ago at Mandrin, France. *Science Advances* 8, eabj9496.

Slocombe, K.E. and Zuberbühler, K. (2005). Functionally referential communication in a chimpanzee. *Current Biology* 15, 1779–84.

Slocombe, K.E., Kaller, T. et al. and Zuberbühler, K. (2010). Production of food-associated calls in wild male chimpanzees is dependent on the composition of the audience. *Behavioral Ecology and Sociobiology* 64, 1959–66.

Smith, A. (2012). Grammaticalization and language evolution. In *The Oxford Handbook of Grammaticalization* (eds B. Heine and H. Narrog), pp. 142–52. Oxford: Oxford University Press.

Smith, K. (2012). Why formal models are useful for evolutionary linguists. In *The*

Oxford Handbook of Language Evolution (eds M. Tallerman and K.R. Gibson), pp. 581–8. Oxford: Oxford University Press.

Smith, L.B. and Yu, C. (2008). Infants rapidly learn word-referent mappings via cross-situational statistics. *Cognition* 106, 1558–68.

Smith, T.M., Tafforeau, P. et al. and Hublin, J.-J. (2010). Dental evidence for ontogenetic differences between modern humans and Neanderthals. *Proceedings of the National Academy of Sciences* 107, 20923–8.

Sonnweber, R., Ravignani, A. and Fitch, W.T. (2015). Non-adjacent visual dependency learning in chimpanzees. *Animal Cognition* 18, 733–45.

Sorensen, A., Roebroeks, W. and van Gijn, A. (2014). Fire production in the deep past? The expedient strike-a-light model. *Journal of Archaeological Science* 42, 476–86.

Soressi, M. (2004). From the Mousterian of Acheulean tradition type A to type B: a change in technical tradition, raw material, task or settlement dynamics? In *Settlement Dynamics of the Middle Paleolithic and Middle Stone Age*, Volume II (ed. N.J. Conard), pp. 343–66. Tübingen: Kerns Verlag.

Soressi, M. and Roussel, M. (2014). European Middle to Upper Paleolithic transitional industries: Châtelperronian. In *Encyclopedia of Global Archaeology* (ed. C. Smith), pp. 2679–93. New York: Springer.

Soressi, M., McPherron, S.P. et al. and Texier, J.-P. (2013). Neandertals made the first specialized bone tools in Europe. *Proceedings of the National Academy of Sciences* 110, 14186–90.

Sousa, A.M.M., Zhu, Y. et al. and Sestan, N. (2017). Molecular and cellular reorganization of neural circuits in the human lineage. *Science* 358, 1027–32.

Speer, S.R. and Ito, K. (2009). Prosody in first language acquisition – acquiring intonation as a tool to organize information in conversation. *Language and Linguistics Compass* 3, 90–110.

Spelke, E. (1991). Physical knowledge in infancy: reflections on Piaget's theory. In *Epigenesis of Mind: Studies in Biology and Culture* (eds S. Carey and R. Gelman), pp. 133–69. Hillsdale, NJ: Erlbaum.

Spence, C. and Deroy, O. (2012). Crossmodal correspondences: innate or learned? *i-Perception* 3, 316–18.

Spinozzi, G. (1996). Categorization in monkeys and chimpanzees. *Behavioural Brain Research* 74, 17–24.

Spoor, F. and Zonneveld, F. (1998). Comparative review of the human bony labyrinth. *Yearbook of Physical Anthropology* 41, 211–51.

Sprinker, M. (1980). Gerard Manley Hopkins on the origin of language. *Journal of the History of Ideas* 41, 113–28.

Spunt, R.P. and Adolphs, R. (2017). A new look at domain specificity: insights from social neuroscience. *Nature Reviews Neuroscience* 18, 559–67.

Starowicz-Filip, A., Chrobak, A.A. et al. and Przewoźnik, D. (2017). The role of the cerebellum in the regulation of language functions. *Psychiatria Polska* 51, 661–71.

Steele, J., Clegg, M. and Martelli, S. (2013). Comparative morphology of the hominin and African ape hyoid bone, a possible marker of the evolution of speech. *Human Biology* 85, 639–72.

Steele, T.E., Álvarez-Fernández, E. and Hallett-Desguez, E. (2019). A review of shells as personal ornamentation during the African Middle Stone Age. *PaleoAnthropology* 2019, 24–51.

Steels, L. (1999). *The Talking Heads Experiment, Volume 1: Words and Meanings.* Antwerp: Laboratorium.

Steels, L. (2003). Evolving grounded communication for robots. *Trends in Cognitive Sciences* 7, 308–12.

Štekauer, P. (1998). *An Onomasiological Theory of English Word-Formation.* Amsterdam: John Benjamins.

Stepanchuk, V. (1993). Prolom II, a Middle Palaeolithic cave site in the eastern Crimea with non-utilitarian bone artefacts. *Proceedings of the Prehistoric Society* 59, 17–37.

Stevens, B.E. (2008). Symbolic language and indexical cries: a semiotic reading of Lucretius 5.1028–90. *American Journal of Philology* 129, 529–57.

Stockwell, R. (1978). Perseverance in the English vowel shift. In *Recent Developments in Historical Phonology* (ed. J. Fisiak), pp. 337–48. The Hague: Mouton.

Stockwell, R. (2002). How much shifting actually occurred in the historical English vowel shift? In *Studies in the History of the English Language: A Millennial Perspective* (eds D. Minkova and R. Stockwell), pp. 267–82. New York: De Gruyter Mouton.

Stoessel, A., David, R. et al. and Hublin J.-J. (2016). Morphology and function of Neandertal and modern human ear ossicles. *Proceedings of the National Academy of Sciences* 113, 11489–94.

Stout, D. (2011). Stone toolmaking and the evolution of human culture and cognition. *Philosophical Transactions of the Royal Society of London B* 366, 1050–9.

Stout, D. and Chaminade, T. (2012). Stone tools, language and the brain in human evolution. *Philosophical Transactions of the Royal Society of London B* 367, 75–87.

Stout, D., Toth, N., Schick, K. and Chaminade, T. (2008). Neural correlates of Early Stone Age toolmaking: technology, language and cognition in human evolution. *Philosophical Transactions of the Royal Society of London B* 363, 1939–49.

Stout, D., Semaw, S., Rogers, M.J. and Cauche, D. (2010). Technological variation in the earliest Oldowan from Gona, Afar, Ethiopia. *Journal of Human Evolution* 58, 474–91.

Stout, D., Apel, J., Commander, J. and Roberts, M. (2014). Late Acheulean technology and cognition at Boxgrove UK. *Journal of Archaeological Science* 41, 576–90.

Stringer, C. and Crété, L. (2022). Mapping interactions of *H. neanderthalensis* and *Homo sapiens* from the fossil and genetic records. *PaleoAnthropology* 2022(2), 401–12.

Svantesson, J.-O. (2017). Sound symbolism: the role of word sound in meaning. *Wiley Interdisciplinary Reviews – Cognitive Science* 8, 1441.

Tallerman, M. (ed.) (2005). *Language Origins: Perspectives on Evolution.* Oxford: Oxford University Press.

Tallerman, M. (2014). No syntax saltation in language evolution. *Language Sciences* 46, 207–19.

Tallerman, M. and Gibson, K.R. (eds) (2012). *The Oxford Handbook of Language Evolution.* Oxford: Oxford University Press.

Tanaka, M. (1995). Object sorting in chimpanzees (*Pan troglodytes*): classification based on physical identity, complementarity, and familiarity. *Journal of Comparative Psychology* 109, 151–61.

Tanaka, M. (2001). Discrimination and categorization of photographs of natural objects by chimpanzees (*Pan troglodytes*). *Animal Cognition* 4, 201–11.

Taylor, B. (2020). Nature and culture in the Epicurean theory of language. In *Lucretius and the Language of Nature* (ed. B. Taylor), pp. 15–42. Oxford: Oxford University Press.

Taylor, C. and Dewsbury, B.M. (2018). On the problem and promise of metaphor use in science and science communication. *Journal of Microbiology and Biology Education* 19, 19.1.46.

Tennie, C., Call, J. and Tomasello, M. (2009). Ratcheting up the ratchet: on the evolution of cumulative culture. *Philosophical Transactions of the Royal Society B* 364, 2405–15.

Thibodeau, P.H., Matlock, T. and Flusberg, S. (2019). The role of metaphor in communication and thought. *Language and Linguistics Compass* 13, e12327.

Thieme, H. (1997). Lower Palaeolithic hunting spears from Germany. *Nature* 385, 807–10.

Thiessen, E.D. and Saffran, J.R. (2003). When cues collide: use of stress and statistical cues to word boundaries by 7- to 9-month-old infants. *Developmental Psychology* 39, 706–16.

Thompson, P.D. and Estes, Z. (2011). Sound symbolic naming of novel objects is a graded function. *Quarterly Journal of Experimental Psychology* 64, 2392–404.

Thompson, R.L., Vinson, D.P., Woll, B. and Vigliocco, G. (2013). The road to language learning is iconic: evidence from British Sign Language. *Psychological Science* 23, 1443–8.

Timmermann, A., Yun, K.-S. et al. and Ganopolski, A. (2022). Climate effects on archaic human habitats and species successions. *Nature* 604, 495–501.

Tomasello, M. and Carpenter, M. (2005). The emergence of social cognition in three young chimpanzees. *Monographs of the Society for Research in Child Development* 70, 1–132.

Toro, J.M., Nespor, M., Mehler, J. and Bonatti, L.L. (2008). Finding words and rules in a speech stream: functional differences between vowels and consonants. *Psychological Science* 19, 137–44.

Toth, N. and Schick, K. (2009). The Oldowan: the tool making of early hominins and chimpanzees compared. *Annual Review of Anthropology* 38, 289–305.

Traugott, E.C. and Dasher, R.B. (2001). *Regularity in Semantic Change*. Cambridge: Cambridge University Press.

Trecca, F., Tylén, K., Højen, A. and Christiansen, M.H. (2021). Danish as a window onto language processing and learning. *Language Learning* 71, 799–833.

Trudgill, P. (2020). *Millennia of Language Change: Sociolinguistic Studies in Deep Historical Linguistics*. Cambridge: Cambridge University Press.

Tryon, C.A. and Faith, J.T. (2013). Variability in the Middle Stone Age of Eastern Africa. *Current Anthropology* 54, S234–54.

Turq, A., Roebroeks, W., Bourguignon, L. and Faivre, J.-P. (2013). The fragmented character of Middle Palaeolithic stone tool technology. *Journal of Human Evolution* 65, 641–55.

Tylén, K., Fusaroli, R. et al. and Lombard, M. (2020). The evolution of early symbolic behavior in *Homo sapiens*. *Proceedings of the National Academy of Sciences* 117, 4578–84.

Tyler, L.K. and Moss, H.E. (2001). Towards a distributed account of conceptual knowledge. *Trends in Cognitive Sciences* 5, 244–52.

Tzeng, C.Y., Nygaard, L.C. and Namy, L.L. (2016). The specificity of sound symbolic correspondences in spoken language. *Cognitive Science* 41, 2191–220.

Tzeng, C.Y., Nygaard, L.C. and Namy, L.L. (2017). Developmental change in children's sensitivity to sound symbolism. *Journal of Experimental Child Psychology* 160, 107–18.

Uomini, N.T. and Meyer, G.F. (2013). Shared brain lateralization patterns in language and Acheulean stone tool production: a functional transcranial Doppler ultrasound study. *PLoS One* 8, e72693.

Urban, M. (2014). Lexical semantic change and semantic reconstruction. In *The Routledge Handbook of Historical Linguistics* (eds C. Bowern and B. Evans), pp. 374–92. London: Routledge.

Usik, V., Rose, J.I. et al. and Marks, A.E. (2013). Nubian complex reduction strategies in Dhofar, southern Oman. *Quaternary International* 300, 244–66.

Vanhaeren, M., d'Errico, F. et al. and Erasmus, R.M. (2013). Thinking strings: additional evidence for personal ornament use in the Middle Stone Age at Blombos Cave, South Africa. *Journal of Human Evolution* 64, 500–17.

Van Herk, G. (2008). Fear of a Black phonology: the Northern Cities Shift as linguistic white flight. *University of Pennsylvania Working Papers in Linguistics* 14, 19.

van Peer, P. (1992). *The Levallois Reduction Strategy*. Monographs in World Archaeology 13. Madison, WI: Prehistory Press.

Villa, P. and Soriano, S. (2010). Hunting weapons of Neanderthals and early modern humans in South Africa: similarities and differences. *Journal of Anthropological Research* 66, 5–38.

Villa, P., Boscato, P., Ranaldo, F. and Ronchitelli, A. (2009a). Stone tools for the hunt: points with impact scars from a Middle Paleolithic site in southern Italy. *Journal of Archaeological Science* 36, 850–9.

Villa, P., Soressi, M., Henshilwood, C.S. and Mourre, V. (2009b). The Still Bay points of Blombos Cave (South Africa). *Journal of Archaeological Science* 36, 441–60.

Villanea, F.A. and Schraiber, J.G. (2019). Multiple episodes of interbreeding between Neanderthal and modern humans. *Nature Ecology and Evolution* 3, 39–44.

Villmoare, B., Kimbel, W.H. et al. and Reed, K.E. (2015). Early Homo at 2.8 Ma from Ledi-Geraru, Afar, Ethiopia. *Science* 347, 1352–5.

Vygotsky, L. (1934). *Thought and Language*. Cambridge, MA: MIT Press.

Wadley, L., Esteban, I. et al. and Sievers, C. (2020). Fire and grass-bedding construction 200 thousand years ago at Border Cave, South Africa. *PLoS One* 15, e0239359.

Walker, M.J., Anesin, D. et al. and Fernández-Jalvo, Y. (2016). Combustion at the late Early Pleistocene site of Cueva Negra del Estrecho del Río Quípar (Murcia, Spain). *Antiquity* 90, 571–89.

Warren, J.L.A., Ponce de León, M.S., Hopkins, W.D. and Zollikofer, C.P. (2019). Evidence for independent brain and neurocranial reorganization during hominin evolution. *Proceedings of the National Academy of Sciences* 116, 22115–21.

Warren, P. (2016). *Uptalk: The Phenomenon of Rising Intonation*. Cambridge: Cambridge University Press.

Wassmann, J. and Dasen, P.R. (1994). Yupno number system and counting. *Journal of Cross-Cultural Psychology* 25, 78–94.

Watson, S.K., Townsend, S.W. et al. and Slocombe, K.E. (2015). Vocal learning in the functionally referential food grunts of chimpanzees. *Current Biology* 25, 495–9.

Watson, S.K., Burkart, J.M. et al. and Townsend, S.W. (2020). Nonadjacent dependency processing in monkeys, apes, and humans. *Science Advances* 6, eabb0725.

Watts, I. (2002). Ochre in the Middle Stone Age of Southern Africa: ritualised display or hide preservative? *South African Archaeological Bulletin* 57, 1–14.

Watts, I. (2009). Red ochre, body-painting, and language: interpreting the Blombos ochre. In *The Cradle of Language* (eds R. Botha and C. Knight), pp. 62–92. Oxford: Oxford University Press.

Watts, I. (2010). The pigments from Pinnacle Point Cave 13B, Western Cape, South Africa. *Journal of Human Evolution* 59, 392–411.

Watts, I., Chazan, M. and Wilkins, J. (2016). Early evidence for brilliant ritualized display: specularite use in the Northern Cape (South Africa) between ~500 and ~300 ka. *Current Anthropology* 57, 287–310.

Weinreich, U., Labov, W. and Herzog, M.I. (1968). Empirical foundations for a theory of language change. In *Directions for Historical Linguistics: A Symposium* (eds W.P. Lehmann and Y. Malkiel), pp. 95–189. Austin, TX: University of Texas Press.

Welker, F., Hajdinjak, M. et al. and Hublin, J.-J. (2016). Palaeoproteomic evidence identifies archaic hominins associated with the Châtelperronian at the Grotte du Renne. *Proceedings of the National Academy of Sciences* 113, 1162–7.

Wellman, H.M. and Liu, D. (2004). Scaling of theory-of-mind tasks. *Child Development* 75, 523–41.

Wescott, R.W. (1971). Linguistic iconism. *Language* 47, 416–28.

Westermann, D. (1937). Laut und Sinn in einigen westafrikanischen Sprache. *Archiv für vergleichende Phonetik* 1, 154–72; 193–212.

White, R. (2001). Personal ornaments from the Grotte du Renne at Arcy-sur-Cure. *Athena Review* 2, 41–6.

White, R., Mensan, R. et al. and Chiotti, L. (2012). Context and dating of Aurignacian vulvar representations from Abri Castanet, France. *Proceedings of the National Academy of Sciences* 109, 8450–5.

White, R., Bosinski, G. et al. and Willis, M.D. (2020). Still no archaeological evidence that Neanderthals created Iberian cave art. *Journal of Human Evolution* 144, 102640.

White, S., Pope, M., Hillson, S. and Soligo, C. (2022). Geometric morphometric variability in the supraorbital and orbital region of Middle Pleistocene hominins: Implications for the taxonomy and evolution of later *Homo*. *Journal of Human Evolution* 162, 103095.

Whiten, A. (ed.) (1991). *Natural Theories of Mind: Evolution, Development and Simulation of Everyday Mindreading*. Oxford: Blackwell.

Whiten, A., Goodall, J. et al. and Boesch, C. (1999). Cultures in chimpanzees. *Nature* 399, 682–5.

Whorf, B.L. (1940). Science and linguistics. *MIT Technology Review* 42, 229–31. Reprinted in (ed.) J. B. Carroll (2011). *Language, Thought, and Reality: Selected Writings of Benjamin Lee Whorf*, pp. 207–219. Cambridge, MA: MIT Press.

Wierzbicka, A. (1986). Human emotions: universal or culture-specific? *American Anthropologist* 88, 584–94.

Wiessner, P.W. (2014). Embers of society: firelight talk among the Ju/'hoansi Bushmen. *Proceedings of the National Academy of Sciences* 111, 14027–35.

Wilkins, J., Schoville, B.J., Brown, K.S. and Chazan, M. (2012). Evidence for early hafted hunting technology. *Science* 338, 942–6.

Wilson, B., Spierings, M. et al. and Rey, A. (2020). Non-adjacent dependency learning in humans and other animals. *Topics in Cognitive Science* 12, 843–58.

Winter, B., Perlman, M., Perry, L. and Lupyan, G. (2017). Which words are most iconic? Iconicity in English sensory words. *Interaction Studies* 18, 433–54.

Wittkower, D.E. (2012). 'Friend' is a verb. *APA Newsletter on Philosophy and Computers* 12, 22–26.

Wragg Sykes, R. (2020). *Kindred: Neanderthal Life, Love, Death and Art*. London: Bloomsbury.

Wrangham, R. (2009). *Catching Fire: How Cooking Made Us Human*. New York: Basic Books.

Wray, A. (1998). Protolanguage as a holistic system for social interaction. *Language and Communication* 18, 47–67.

Wray, A. and Grace, G.W. (2007). The consequences of talking to strangers: evolutionary corollaries of socio-cultural influences on linguistic form. *Lingua* 117, 543–78.

Wurz, S. (2013). Technological trends in the Middle Stone Age of South Africa between MIS 7 and MIS 3. *Current Anthropology* 54, S305–19.

Xu, Y., Liu, E. and Regier, T. (2020). Numeral systems across languages support efficient communication: from approximate numerosity to recursion. *Open Mind: Discoveries in Cognitive Science* 4, 57–70.

Yellen, J.E., Brooks, A.S. et al. and Stewart, K. (1995). A Middle Stone Age worked bone industry from Katanda, Upper Semliki Valley, Zaire. *Science* 268, 553–6.

Young, D. (2005). The smell of greenness: cultural synaesthesia in the Western Desert. *Etnofoor* 18, 61–77.

Yu, C. and Smith, L.B. (2007). Rapid word learning under uncertainty via cross-situational statistics. *Psychological Science* 18, 414–20.

Zaccarella, E. and Friederici, A.D. (2015). Merge in the human brain: a sub-region based functional investigation in the left pars opercularis. *Frontiers in Psychology* 6, 1818.

Zentall, T.R., Wasserman, E.A. et al. and Rattermann, M.J. (2008). Concept learning in animals. *Comparative Cognition and Behavior Reviews* 3, 13–45.

Zhang, Y., Han, K., Worth, R. and Liu, Z. (2020). Connecting concepts in the brain by mapping cortical representations of semantic relations. *Nature Communications* 11, 1877.

Ziegler, M., Simon, M.H. et al. and Zahn, R. (2013). Development of Middle Stone Age innovation linked to rapid climate change. *Nature Communications* 4, 1905.

Zilhão, J. (2013). Neandertal–modern human contact in western Eurasia: issues of dating, taxonomy, and cultural associations. In *Dynamics of Learning in Neanderthals and Modern Humans*, Vol. 1 (eds T. Akazawa, Y. Nishiaki, and K. Aoki), pp. 21–57. Tokyo: Springer.

Zilhão, J., Angelucci, D.E. et al. and Zapata, J. (2010). Symbolic use of marine shells and mineral pigments by Iberian Neandertals. *Proceedings of the National Academy of Sciences* 107, 1023–8.

Zlatev, J. and Blomberg, J. (2015). Language may indeed influence thought. *Frontiers in Psychology* 6, 1631.

Zuberbühler, K. (2019). Syntax and compositionality in animal communication. *Philosophical Transactions of the Royal Society of London B* 375, 20190062.

Notes

1. Introduction: the puzzle of language

1   Determining how many words we know is not easy. This depends on how words are defined and will vary by the language spoken, age, social and educational background, and some argue gender. See Brysbaert et al. (2016).

2   See Rayner and Clifton (2009) for estimates of speaking, listening and reading speeds.

3   'Trequartista' derives from the Italian for three-quarters, *tre quarti*, with the -*ista* suffix denoting a person. See Oxford English Dictionary online: oed.com

4   Throughout this book we will be returning to the remarkable ability of babies, toddlers and children to acquire words. For a start, see Bergelson and Swingley (2012) for how six- to nine-month-olds learn the meaning of many common nouns.

5   The oldest DNA recovered from human remains dates to 450,000 years ago, coming from *Homo heidelbergensis* specimens from Sima de los Huesos, Spain (Meyer et al., 2016). In 2023, Madupe et al. (2023) announced the recovery of genetic material from fossils of *Paranthropus robustus*, dating to at least 2 million years old from East Africa – an achievement described as a potentially transformative breakthrough for palaeoanthropology (Callaway, 2023).

6   The Stone Age is part of the three-age system for prehistory of Stone, Bronze and Iron Ages, which was devised by the Danish antiquarian Christian Jürgensen Thomsen during the 1830s. He had noticed that stone artefacts always came before those of bronze which came before those of iron when ordering materials in the Danish National Museum. For a time when it was impossible to secure absolute dates on materials, this amounted to a scientific revolution, as important as radiocarbon dating in the 1950s. Rowley-Conwy (2007) provides an excellent study of how the three-age system was devised, became swiftly adopted in Scandinavia and eventually adopted in Britain after having been contested. Although the three ages are still used in a generic sense, they have less relevance today because we recognise that major transitions in social and economic behaviour are not associated with the change from stone to bronze or to iron. Moreover, the start and end of each of the periods vary by region. Formally the Stone Age

continued until the invention of metallurgy, which was about 5,500 years ago in Southwest Asia and in China. Also, please remember that throughout prehistory most tools and components of tools were likely made from organic materials, notably wood, bone, antler, plant fibres, sinews and hide.

7   See Mithen (2003) for a review of the origin of farming with some emphasis on climate change and its immediate consequences, and Mithen (2019) for a summary of the role of language in this process, making an argument that is developed within this book.

8   Morten Christiansen and Simon Kirby describe the origin of language as the hardest problem in science in their introductory chapter to their 2003 edited book on language evolution (Christiansen and Kirby, 2003).

9   Reviewing the history of research on the evolution of language would require a book in itself. Some of this history is embedded into my chapters as appropriate. For those wishing to have a review of the field as a whole, I recommend the *Oxford Handbook of Language Evolution*, edited by Maggie Tallerman and Kathleen Gibson (2012). The earlier collections edited by Christiansen and Kirby (2003), Tallerman (2005) and Botha and Knight (2009) are all immensely valuable and illustrate the large number of disciplines contributing to a subject of immense breadth and complexity.

10 See Berwick and Chomsky (2017) for a sudden appearance of language at 100,000 years ago, a position long argued by others. Bickerton (2009) has proposed a protolanguage consisting of strings of words without any grammar, while Wray (1998) argued that protolanguage consisted of holistic phrases. Fitch (2017) provides an excellent review of the study of language evolution and covers the various types of protolanguage that have been proposed: lexical, gestural, musical and mimetic. The origin and nature of protolanguage is further discussed in the edited collection of essays by Arbib and Bickerton (2010).

11 Dunbar (1997) argued that the need for social bonding provided the key selective pressure for language evolution, while Mithen (2005) provided a case for the role of musicality. Ferretti (2022) promotes the idea that the fundamental role of communication in both humans and animals is persuasion, and that hominins developed narrative forms of communication – storytelling – to enhance their persuasive abilities. He provides a compelling case and storytelling could feasibly provide the vehicle for transmitting persuasive information about social relationships, hunting and tool making by embedding these into a narrative framework.

12 The linguist Derek Bickerton argued that displacement is the hallmark of language in a succession of publications, e.g. Bickerton (2009). Noam Chomsky has long argued that recursion is the most critical feature of language. For a recent exposition of that view, see Berwick and Chomsky (2017).

13 In Dediu and Levinson's (2018) otherwise excellent review of existing
    evidence for language within the Neanderthals, they neglect to consider the
    archaeological evidence for Neanderthal behaviour, notably the stability
    in tool making and lack of flexibility to climate change, both of which are
    incompatible with a fully modern linguistic capability. Similarly, Dunbar's
    (1997) focus on the so-called social brain neglects the significance of
    technology as represented by stone artefacts of the archaeological record.

14 See Chater and Christiansen (2012, p. 631), who state that once language
    is viewed as an evolving cultural system transmitted from generation to
    generation 'it is natural to see language evolution as language-change writ
    large'.

15 This is the principle of uniformitarianism. The most compelling argument
    of this in linguistics has come from Trudgill (2020). He has used an
    understanding of the causes of linguistic diversity in the present to propose
    how the languages of Stone Age hunter-gatherers are likely to have differed
    from those of today. See especially Chapter 1 of his book: 'Prehistoric
    sociolinguistics and the uniformitarian hypothesis: what were Stone Age
    languages like?'

16 Chapter 2 follows a well-established understanding of human evolution,
    further details of which are readily accessible in a range of published
    accounts. I recommend Muehlenbein (ed., 2015). The dilemma with accounts
    of human evolution is that they rapidly go out of date as new discoveries
    are made and methods are applied to existing material. I write at a time of
    an ongoing genomic revolution which is transforming our understanding
    of human evolution. I will provide additional citations to the most recent
    discoveries, analyses and interpretations that are unlikely to have yet entered
    any textbook, and to any other articles that I find of particular significance.
    Further citations about the evolution of the brain, vocal tract, technology,
    fire use, and signs and symbols are provided in the later chapters which focus
    on these topics.

2. A brief history of humankind

1 Africa is the conventional source for the last common ancestor but has a
   scarcity of fossils within the critical time period of 8–6 million years ago.
   Europe has numerous ape fossils within this period, most recently a new
   genus called *Anadoluvius*, from the 8.7 mya site of Çorakyerler in central
   Anatolia (Sevim-Erol et al., 2023). This leads some to argue that the last
   common ancestor had evolved in Europe, with descendants dispersing into
   Africa.

2 See Blumenthal et al. (2017) for a consideration of aridity on the impact of
   hominin environments and evolution.

3 Villmoare et al. (2015) describe the earliest dated *Homo* specimen, part of a jaw from Afar, Ethiopia. This dates to between 2.80 and 2.75 mya.

4 See Lacruz et al. (2019) for the evolution of the flatter face in *Homo*, and how this strongly suggests that there are two distinct early *Homo* species, *H. habilis* and *H. rudolfensis*.

5 See Plummer et al. (2023) for the earliest known occurrence of Oldowan technology at 3.0 to 2.6 mya. See Harmand et al. (2015) for 3.3 mya stone tools from Lomekwi, Kenya, that lack the features of Oldowan technology.

6 A correlation between brain size, as measured by the ratio of the neocortex to the rest of the brain, and group size has been proposed by Dunbar (1992, 1995) and referred to as the social brain hypothesis. Aiello and Dunbar (1993) argued that the management of social relations in larger groups required enhanced vocalisations to supplement grooming, which they propose was the basis for language. Dunbar (1997) provides an extended account of their theory.

7 See Margari et al. (2023) for an extreme glacial event in Europe at 1.1 mya and its consequences for human occupation, and Lewis et al. (2019) for occupation at *c*.900,000 years ago at Happisburgh, Norfolk, UK.

8 See Dennell (2018) for a review of the Acheulean in Asia.

9 For the possible role of fire and cooking in the evolution of the brain, and by implication language, see Wrangham (2009).

10 See Harvati and Reyes-Centeno (2022) for a review of the evolution of *Homo* in the Middle and Late Pleistocene.

11 See White et al. (2022) for the problematic taxonomic status of *H. heidelbergensis* owing to the extent of morphological and geographical variability and poor chronological resolution of the Middle Pleistocene *Homo* fossil record.

12 Roksandic et al. (2022) proposed *Homo bodoensis* to resolve the 'muddle in the middle'.

13 Hu et al. (2023) used a new method of genomic analysis to project current human variation from 300 modern human genomes backwards in time to estimate the size of our ancestral population at specific points in the past. They identified that the severe population bottleneck had occurred between 930,000 and 813,000 years ago, and suggest this may have led to a speciation event. See the commentary on their arguments by Ashton and Stringer (2023).

14 See Ashton and Davis (2021) and Hosfield (2022) for hominin expansion in Europe indicative of behavioural change. See Hosfield (2020) for a review and interpretation of the earliest hominins in Europe, focusing on their adaptations to seasonal changes.

15 See Neubauer et al. (2018) for a study of the evolution of the modern human brain shape.

16 See Hublin et al. (2017) for the Jebel Irhoud fossils and discussion of their taxonomic status.

17 See Slimak et al. (2022) for a possible modern human incursion into France at 54,000 years ago. The case is largely dependent on the attribution of a tooth to *H. sapiens* rather than *H. neanderthalensis*, which remains to be fully validated.

18 See Froehle and Churchill (2009) for the energetic differences between Neanderthals and modern humans.

19 See Krause et al. (2007b) for the geographical extent of Neanderthals.

20 See Fabre et al. (2009) for the population distribution of Neanderthals.

21 The seminal publications for the Neanderthal and Denisovan genomes are Green et al. (2010), Reich et al. (2010) and Prüfer et al. (2014).

22 See Villanea and Schraiber (2019) for Neanderthal and modern human interbreeding.

23 See Ruan et al. (2023) for how climate change orchestrated interbreedings between the Neanderthals and Denisovans at *c.*120,000 and 78,000 years ago.

24 See Harvati et al. (2019) for *H. sapiens* at Apidima Cave, Greece, and Hershkovitz et al. (2018) for *H. sapiens* at Misliya Cave, Israel.

25 See Boaretto et al. (2021) for the earliest appearance of Upper Palaeolithic technology as documented at Boker Tachtit Cave, Israel, between 50,000 and 48,000 years ago.

26 See Aubert et al. (2014) for the cave art, which comes from Sulawesi, Indonesia. This was dated by uranium-series dating of overlying speleothems, which is methodologically challenging and the dates require independent validation.

27 The earliest date of modern humans in Australia has been particularly problematic. Clarkson et al. (2017) present what appears robust evidence from Madjedbebe rock shelter in northern Australia for a date of 65,000 years ago, which they adjust to 59,000 as the most conservative estimate.

28 Among others, Liu et al. (2015) argue that moderns humans were present in Fuyan Cave in southern China between 120,000 and 80,000 years ago, but their dating evidence has been challenged by Michel et al. (2016).

29 See Slimak et al. (2022) for the modern human occupation at *c.*54,000 years ago at Mandrin Cave, France. I would like to see further validation of the evidence for a modern human presence, notably the attribution of the tooth to *H. sapiens*.

30 The ongoing genomic revolution is providing new insights into the population movements within Europe during the Upper Palaeolithic period and the early Holocene, demonstrating an association between genomically defined populations and archaeologically defined cultures. See Posth et al. (2023) for the most recent publication at the time of writing – more will no doubt follow.

3. Words and language

1 For much of the remainder of this chapter, and often elsewhere in the book, I will refer to 'spoken' words simply for ease of writing. Please appreciate that the same descriptions and interpretations of words can be applied to signed and written words.

2 See MacNeilage (2010) for a comprehensive description of how speech is produced and a discussion of its origins.

3 The double disassociation between consonants and vowels was demonstrated by Caramazza et al. (2000) by testing the performance on vowel and consonant recognition and speaking of two Italian-speaking people with aphasia, one who had difficulty with vowels and the other with consonants. Consonants and vowels also play a different role in language acquisition: infants primarily use consonants to identify words, and vowels to acquire the rules with which words are put together within whatever language they are learning. See Toro et al. (2008) for the functional difference between vowels and consonants in language acquisition.

4 Brysbaert et al. (2013) used 4,000 participants to rate 37,058 English words and 2,896 two-word expressions on a scale from most abstract (1) to most concrete (5). The word *freedom* scored 2.54 and *democracy* scored 1.78. Other nouns were not far away: *fun* scored 1.97, *idea* 1.61 and *trouble* 2.25.

5 See Lupyan and Winter (2018) for a discussion of how we come to know the meaning of concrete and abstract words, and their significance for the role of iconic and arbitrary words in language.

6 During the 1950s, the linguist Charles Hockett defined 16 design features of language, six of which were unique to human communication (Hockett, 1958). One of these was arbitrariness between the word and its meaning, which is usually traced back to Ferdinand de Saussure (1857–1913).

7 Many onomatopoeias, however, seem quite arbitrary. Pinker (1994, p. 152) notes that while pigs go 'oink' in English, they go 'boo-boo' in Japanese.

8 Jackendoff (2002) refers to these expressions as 'defective' linguistic items. Their potential role as the first words in the evolution of language has been argued by Burling (2012) and Tallerman (2014).

9 For the seminal works on syntactical structures that emphasise the significance of hierarchical structures, see Chomsky (1957b) and Jackendoff (1972). For recent experimental research and discussion about the significance of hierarchical structure, see Coopmans et al. (2022).

10 For the significance of short pauses in speech, see University of Gothenburg, 'Pauses can make or break a conversation', *ScienceDaily*, 30 September 2015.

11 This does not necessarily mean that all languages are equally easy to learn. Trecca et al. (2021) argue that the phonetic structure of Danish delays Danish-learning children in several aspects of their language acquisition.

12 Chomsky's (1957a) ideas were initially published within a critical review of a

book by B.F. Skinner called *Verbal Behavior*. Skinner had argued that children learn language by associative learning, by reward and reinforcement, in just the same way that any animal might learn to undertake a task.

13 Cited from James (1890) and borrowed from Saffran and Kirkham (2018).

14 Pinker and Jackendoff (2009, p. 466) refer to Universal Grammar as a 'toolkit' in their response to Evans and Levinson's (2009) critique.

15 See Dąbrowska (2015) for a critique of Universal Grammar.

16 Evans and Levinson (2009) document the extent of linguistic diversity in the small fraction of languages recorded in the present – its diversity must have once been substantially higher. By doing so, they provide a critique of so-called language universals, and a proposed research programme for exploring the biocultural evolution of language capacity and languages without reliance on a theory of Universal Grammar. It is followed by extensive discussion by commentators, some of whom defend Universal Grammar, followed by Evans and Levinson's response.

17 Lupyan and Dale (2010) provide a figure of 6,912 languages taken from the Ethnologue, www.ethnologue.com/, an authoritative worldwide source of languages. By September 2023 it referred to 7,168 known living languages.

18 My figures for speakers of languages are taken from Marian (2023, p. 8).

19 For the last speaker of Resígaro, see 'Resígaro is the Amazonian language with only one remaining speaker', matadornetwork.com/read/resigaro-endangered-amazonian-language/ (accessed 11.10.2023)

20 See Isern and Fort (2014) for the rapid rate at which languages are disappearing.

21 Half a million languages is the figure given by Evans and Levinson (2009, p. 432), but this depends on how languages are differentiated from dialects, how language is defined and when it emerged, and a mass of other variables about which we have little idea.

22 There is a voluminous literature about colour terms, much of which concerns their impact on the perception of colour. I will address the key issues in Chapter 14. For now, the key work is Berlin and Kay (1969).

23 See Xu et al. (2020), Beller and Bender (2008) and Calude (2021) for descriptions of different number systems.

24 Lupyan and Dale (2016) provide examples of categories with varying lexicons between languages, along with citations.

25 See LeMaster and Monaghan (2004) for variability in sign languages.

26 See Maddieson (1984) for the challenges of defining the number of phonemes in a language. One could draw a distinction between phonemes and phones, the former being the mental representation of a sound within a language, and the latter being how it sounds. For my needs, I will use phoneme to cover both meanings.

27 See Evans and Levinson (2009) for variations in the number of phonemes in languages.

28 See MacNeilage (2010, Chapter 3) for the variants of vowels and consonants within languages.

29 See Nettle (2012) for the relationship between sound inventory and word length.

30 See Cummins et al. (1999) and Cruz and Frota (eds, 2022) for cross-linguistic studies of prosody.

31 See Kita (2009) for a cross-cultural study of speech-accompanying gestures.

32 See Evans and Levinson (2009) for how some languages appear to pack whole sentences into a single word.

33 See Lupyan and Dale (2010) for over-specification within languages, as in Yagua.

34 See Rijkhoff (2007) for languages lacking the four standard word classes of nouns, verbs, adjectives and adverbs.

35 See Evans and Levinson (2009) for word classes that are absent from English.

36 McWhorter (2001) covers the significance of dialects for linguistic diversity.

37 'A language is a dialect with an army and navy' is stated by Deutscher (2005, p. 55), who is citing an unnamed American linguist. The phrase is said to have been popularised by the sociologist Max Weinreich. Switzerland is an honourable exception to the relationship between national identity and language: a nation state that functions with four languages, Italian, French, German and Romansh.

38 'We reinvent language every time we speak' was said (in Spanish) by the linguist Eugenio Coșeriu.

39 This is not necessarily the case with written texts. When using these, one can feasibly keep track of many embedded clauses by having the written word and punctuation as a memory aid and for repeated reference to previous clauses. The capacity of language to enable such multiple embedding has been taken to be one of its key qualities, but this is more likely a consequence of the evolution of writing rather than of language itself.

40 See Akinnaso (1982) for an overall comparison between spoken and written language, focusing on English.

41 Here I am thinking about Chomsky, who long argued that recursion is the most critical feature of language, emphasising the ability to potentially embed an infinite number of clauses into one single expression. Yes, we can do that when writing, but no, that is never found in spoken language. For his most recent exposition, see Berwick and Chomsky (2017).

42 See Dediu and Ladd (2007) for the correlation between linguistic tone and genetic variability.

43 See Dediu and Moisik (2019) for variation in the articulation of /r/ and its potential impact on sound change and language evolution.

44 See Moisik and Dediu (2017) for the relationship between palette morphology and clicks.

45 Nettle (1998) provides a useful discussion about the distinction between a language and a dialect. He explains that one may find chains of dialects in which adjacent varieties are perfectly intelligible, but that those far from each other in the chain, and especially at the ends, are not. How then to draw boundaries to measure the number of languages being spoken?

46 See Mallory and Adams (2006) for a reconstruction of the Proto-Indo-European language with its implication for what is known about the Indo-European lifestyles. Once the idea of the Indo-Europeans was established in the early 18th century, there has been a persistent search for their homeland. In 1705, Gottfried Wilhelm Leibniz suggested this had been the central European steppe to the north of the Black Sea (now Ukraine and southern Russia). This idea was developed by the archaeologist Marija Gimbutas during the 1950s and 60s. She argued that the spread of the Bronze Age Kurgan culture from the Pontic Steppe led to the destruction of what she termed 'Old Europe', that of the Neolithic. Other possible homelands have been suggested, notably Schleswig-Holstein by Gustaf Kossinna in the 1920s, whose proposal was adopted by the Nazis to justify their annexations in Europe. In Colin Renfrew's 1987 book *Archaeology and Language*, he associated the spread of Proto-Indo-European with the dispersal of Neolithic farming from Southwest Asia and Anatolia from the eighth millennium BC. In 2015 a genomic study of ancient DNA from 69 Neolithic and Bronze Age Europeans provided support to the steppe as having been a source for the spread of Late Bronze Age communities and the likely source of Proto-Indo-European (Haak et al., 2015). Jean-Paul Demoule (2023) has provided a comprehensive history of the Indo-European question, covering all these works and much more. He provocatively challenges the tree-model for the evolution of Indo-European languages and the idea of an Indo-European people with their own homeland.

47 See Pagel et al. (2013) for the reconstruction of Eurasiatic, and Heggarty (2013) for a critique, which has my vote.

48 Nostratic was originally proposed in 1903 and has been championed by, among others, Merritt Ruhlen (1994).

49 These quotes are cited and the sources provided in Lupyan and Dale (2016, p. 651).

50 Cited from Boas (1911).

51 See Martin (1986) and Pullum (1991) for a critique of Boas (1911) regarding snow.

52 See Krupnik and Müller-Wille (2010) for a defence of Boas regarding snow.

53 See Regier et al. (2016) for Twitter traffic regarding snow.

54 See Conklin (1957) for the words for rice among the Hanunóo. That the Scots

have many different words for different types of rain comes from personal experience and needs to be academically validated. Consider, for instance, a *yillen* (a shower of rain, especially with wind), an *uplowsin* (heaving rain), a *smirr* (a fine rain drizzle), a *goselet* (a soaking, drenching, downpour) and *dreich* (an adjective meaning a blend of bleak, miserable, grey, depressing – and generally wet).

55 The following draws on Munroe et al. (2009).

56 See Munroe et al. (2009) for the proposed relationship between climate and consonants. The association between sonorous consonants and climate is not consistent. Voiced stops, such as /d/, the quietest of consonants, are more frequent in warmer climates, a finding that remains unexplained.

57 See Ember and Ember (2007) for arguments that topography and vegetation cover influence sonority in language, and Everett (2017) for humidity and language.

58 See Foster and Collard (2013) for the application of Bergmann's rule to modern humans.

59 See Gibson et al. (2017) for a study of colour naming and communication, and the proposal that this reflects need.

60 Majid et al. (2018) compared naming stimuli relating to the five basic senses across 20 diverse languages and concluded that 'the faculty of language does not constrain, due to intrinsic cognitive architecture, the degree to which different sensory domains are richly coded. Instead, the patterns we found suggest that the mapping of language onto senses is culturally relative.'

61 Majid and Kruspe (2018) compared the Semaq Beri hunter-gatherers and the Semelai swidden horticulturists within the Malaysian rainforest. Whether the greater linguistic ability regarding odour can be entirely accounted for by the hunter-gatherer lifestyle or is partly derived from cultural practices, such as the belief that people have their own personal odours, is unclear.

62 Majid and Burenhult (2014) report on the naming of odours by the Jahai hunter-gatherers of the Malay peninsula.

63 See Young (2005) for her discussion of the 'smell of greenness' among the Anangu hunter-gatherers of the Western Desert of Australia.

64 Blasi et al. (2019) describe changes in bite configuration and its impact on language phonology. They provide examples of skulls from the Palaeolithic that exhibit edge-to-edge bite configurations and from the Bronze Age with overbite, describe biomechanical models for the impact of bite configuration on producing labiodentals, and propose a model for the gradual increase of labiodentals within the Indo-European language family.

65 Hay and Bauer (2007) proposed a correlation between phoneme inventory and population size based on 216 languages, while the relationships between population size, phonetic inventories and word length are described by Nettle (2012). Moran et al. (2012) have challenged Hay and Bauer's (2007)

finding, arguing that when a larger sample (of 916 languages) is explored and various methodological issues addressed, the correlation between phoneme inventory and population size is lost.

66 See Lupyan and Dale (2010) for the association between population size and morphological complexity of language.

67 See Nettle (1998) for the relationship between environment and the distribution of languages.

68 Lupyan and Dale (2010) and Nettle (2012) emphasise the role of adult and especially second-language learners in creating linguistic differences between larger and smaller populations.

69 Wray and Grace (2007) have emphasised the role of labour and social differentiation.

70 Nettle (2012) proposes that 'heterogeneity in the learning set' may influence how language evolves within a community. Elsewhere he has argued that the turnover of linguistic change would be faster in smaller communities (Nettle, 1999), but that seems counter to the argument proposed here. Raviv et al. (2019) undertook experiments involving groups developing new languages in laboratory settings that suggest larger communities would develop more systematic languages by virtue of their size alone. Whether such experiments, in which the largest groups consisted of eight adult members and their language development was monitored over just one year via periodic bouts of interaction, can have any relevance to the real-world development of language remains contentious. I suspect they do.

71 See Nettle (2012) for an analogy between drift in language evolution and genetic drift in biology.

72 See Wray and Grace (2007) for the differences between esoteric and exoteric communication, and the likely causes.

4. Monkeys and apes

1   See Hayes and Hayes (1951) for an account of the development of Viki, a chimpanzee reared within a human household in a comparable way to a human child.

2   The figure of '300 or more' trials is cited from Premack (1990). Premack (1976) used plastic tokens to stand for spoken words in long-term experiments with a chimpanzee named Sarah, although his main aim was to gain insights into chimpanzee cognition rather than to teach language.

3   See Gardner and Gardner (1969) for a succinct account of teaching sign language to a chimpanzee, and Gardner et al. (eds, 1989) for a longer treatment. Washoe lived with the Gardners until she was five, and then moved to live under the care of Roger and Deborah Fouts, who provided their own published study (Fouts, 1998). Washoe died in 2007 at the age of 42.

4   Savage-Rumbaugh et al. (1986, 1998) developed a visual language based

on computer keyboards for chimpanzees named Lana and Matata. Kanzi, Matata's son, is said to have acquired symbols while watching his mother being taught. For a study of Kanzi, see Savage-Rumbaugh and Lewin (1994).

5 See Pinker (1994, pp. 334–42) for a damning review of all such chimpanzee language-learning experiments and their claimed outcomes. He asserts they are no more than trained animal acts (p. 339). Brakke and Savage-Rumbaugh (1995, p. 138) are (not surprisingly) more positive, arguing that Kanzi's performances 'indicate that many of the neural prerequisites for basic language acquisition were in place millions of years before anatomical changes . . . further neuronal growth and reorganization . . . and cultural growth facilitated their growth and elaboration'. For a discussion of these views that challenges those of Pinker, see Lloyd (2004). Schoenemann's (2022) recent analysis of Kanzi's performance in reversible sentences concludes that he understands word-order grammatical rules in English.

6 See Coski (2003) for Condillac's theories about the evolution of language and human thought in general.

7 See Condillac (1746) and Hewes (1973) for the early development of a gestural theory for the origin of language.

8 See Pollick and de Waal (2007) and Hobaiter and Byrne (2014) for two key publications regarding chimpanzee gestures.

9 Bullinger et al. (2011) classify ape gestures into 'intention-movements' and 'attention-getters'.

10 Notable recent proponents for a gestural origin of language are Corballis (2002) and Arbib (2005). Arbib et al. (2008) provide an evaluation of the likely roles of primate vocalisation and gestures for the evolution of human language. But note that article preceded the dramatic advances made in our understanding of ape vocalisations as described in this chapter. While it is implausible that gesture could have been the original mode for language, the co-evolution of gesture and speech is more credible. See Gillespie-Lynch et al. (2014) and Kendon (2017) for multimodal hypotheses for language origins.

11 Bohn et al. (2016) review the evidence for iconic gestures within wild and captive apes. Only a few claims have been made. One is of a male captive gorilla who swung his arm and tapped his genitals as a gesture towards a female that he wished her to engage in sexual activity, but this is open to other interpretations. Apes do not use pointing in the wild. Captive apes have learned to point but use this only to acquire something for themselves rather than to indicate where someone else should look, move to, or collect something for their own benefit. Bullinger et al. (2011) describe the different social motives in pointing by chimpanzees and human children.

12 Bohn et al. (2016) conducted an experiment to test whether chimpanzees can use iconic gestures by initially training chimpanzees in the required arm movements to operate three different types of apparatus, one by

pushing, one by pulling and one by turning a crank. It tested whether chimpanzees could use a gesture provided by a human that captured one of these movements to choose the correct apparatus from which to get a reward. The chimpanzees failed at this, in contrast to four-year-old children who had received the same training and spontaneously chose the correct apparatus. Neither the chimpanzees nor the children were able to quickly learn the association between an arbitrary gesture and a particular apparatus. Over time, the chimpanzees learned to pair the iconic gestures with the appropriate apparatus but continued to fail with the arbitrary gestures. See Bohn et al. (2016) for a full description of this experiment and comparison between human children and chimpanzees in their understanding of iconicity.

13 See Gillespie-Lynch et al. (2014) and Kendon (2017) for multimodal hypotheses for language origins.

14 The following draws on Cheney and Seyfarth (1990). Seyfarth et al. (1980) is a seminal paper in the study of non-primate vocal behaviour.

15 See Price et al. (2015) for a quantitative analysis of the vervet alarm calls.

16 Seyfarth et al. (1980, p. 1091) described the vervet alarm calls as being arbitrary because they did not 'resemble in physical contours what they denote'. I am not clear what that means.

17 See Price et al. (2015) for the multiple contexts in which the same vervet calls are made.

18 As Stevens (2008) explains, Lucretius, writing in *c*.60 BC, distinguishes between animal vocalisation (*uoces ciere*) and human language (*res uoce notare*): animal vocalisation is an automatic or involuntary signification of things or emotions immediately present, whereas human language is fully symbolic, i.e. a voluntary signification using arbitrary signs. See Driscoll (2015) for a comparison of the views of Rousseau, Herder and Kant on animal cries and their relationship to human language.

19 See Slocombe and Zuberbühler (2005) for a full description of testing whether different forms of rough grunts have different meanings. As with all the examples I describe in this chapter, my summaries do not cover the often sophisticated experimental designs and statistical analyses that are undertaken of the results.

20 The Sonso community and their use of rough grunts when finding food are discussed by Slocombe et al. (2010). They live in a moist, semi-tropical deciduous forest at an altitude of about 1,600 metres. In 2006, the community consisted of 72 individuals: 8 adult males, 21 adult females, 8 sub-adult males, 5 sub-adult females, 18 juveniles and 12 infants. Although the community has become habituated to humans since contact was made in 1990, there has been no provisioning and hence it can be considered a fully natural and wild group of chimpanzees.

21 See Crockford et al. (2012) for a full description of the experiments using an artificial viper. A further development of this study is described by Crockford et al. (2017).

22 See Lameira and Call (2018) for the study of displaced alarm calls by orangutans.

23 See Schel et al. (2013) for a full description of the experiments using an artificial python.

24 See Gruber and Zuberbühler (2013) for a study of chimpanzee travel-hoo calls made before travelling.

25 See Laporte and Zuberbühler (2010) for a study of pant-grunts and their social context.

26 Cited from Laporte and Zuberbühler (2010, p. 471).

27 See Lameira (2017) for a review of vocal learning in primates and other animals.

28 See Hopkins et al. (2007) for raspberry blowing by captive chimpanzees and Lameira (2017) for a review of their use in wild populations. Blowing raspberries has also been recorded in a few wild populations of chimpanzees and gorillas, also implying independent invention and for a purpose different to that of attracting human attention.

29 See Crockford et al. (2004) for a study of pant-hoots and vocal learning.

30 Kojima et al. (2003) describe a study using captive chimpanzees that shows how individuals were identified by other chimpanzees from their calls.

31 See Watson et al. (2015) for a study of vocal learning by comparing the development of the grunts made by the Edinburgh Zoo (ED) chimpanzees and Beekse Bergen Safari Park (BB) chimpanzees following cohabitation. Their interpretation that the grunt convergence indicates vocal learning has been challenged by Fischer et al. (2015) who argue there was insufficient control for the levels of arousal and the extent of acoustic overlap between the ED and BB calls at the start of the study.

32 See Lameira et al. (2022) for a study of orangutan vocal variability and population density.

33 See Hay and Bauer (2007), Lupyan and Dale (2010) and Nettle (2012) for the relationship between human population size and linguistic character.

34 See Clay et al. (2015) for a study of bonobo peep calls.

35 Zuberbühler (2019) reviews the evidence for syntax in animal communication.

36 See Ouattara et al. (2009) for a study of suffixation by Campbell monkeys.

37 See Schlenker et al. (2016) for a study of putty-nosed monkey calls.

38 See Miyagawa and Clarke (2019) for possible neurological constraints on the abilities of monkeys to combine more than two different types of call into a sequence.

39 See Fedurek et al. (2016) for an acoustic analysis of male chimpanzee pant-hoots.

40 See Clay and Zuberbühler (2011) for a study of how bonobos extract meaning from call sequences. Clay et al. (2015) note that peep calls might have a generalised social function relating to cohesion or spacing.

41 See Girard-Buttoz et al. (2022) for a full description, analysis and interpretation of the vocal sequences of the Taï National Park chimpanzees.

5. Speaking and hearing

1 See Carruthers (2017) for the consciousness, or otherwise, of thought. It is a tricky subject.

2 See MacNeilage (2010) for the process of speaking.

3 The double disassociation between consonants and vowels was demonstrated by Caramazza et al. (2000).

4 See Toro et al. (2008) for the functional difference between vowels and consonants in language acquisition.

5 In addition to these types of consonants, three more can be added. Affricates are combinations of plosives and fricatives, such as 'cheap' and 'jeep'. Liquids are when the tongue produces a partial closure in the mouth, resulting in a resonant, vowel-like consonant, such as 'late' and 'rate'. Glides are consonants that are phonetically similar to a vowel but function as a syllable boundary, rather than as the nucleus of a syllable, as in /j/ and /w/.

6 Ghazanfar and Takahashi (2014) emphasise the importance of rhythm in the production of speech.

7 Ghazanfar and Takahashi (2014) emphasise the significance of turn-taking in human speech. This is not found in any other primate other than the common marmoset, which they suggest is an example of convergent evolution.

8 See Girard-Buttoz et al. (2022) for the twelve chimpanzee calls made by communities in the Taï National Park. They are denoted by pant, grunt, hoo, bark, scream, panted-scream, panted-hoo, panted-bark, and panted-grunt, panted-roar, non-voiced sounds (lip-smacking, raspberry blowing) and whimper. For an example of monkey calls, see Cheney and Seyfarth (1990) for the calls of vervets. See further references in Chapter 4.

9 My description of the chimpanzee SVT and how it differs to that of modern humans follows that of Nishimura et al. (2006).

10 See Lacruz et al. (2019) for the evolutionary history of the human face.

11 See Pelaez et al. (2018) for a review of infant vocalisations and imitation.

12 Lieberman and Crelin (1971), Lieberman et al. (1972) and Laitman (1984) were most prominent in arguing that only the modern human SVT can produce the full range of vowels, notably /i/, /u/ and /a/, and claiming they are universal and required for human speech. Boë et al. (2002) undertook

computer simulations to conclude that the Neanderthal vowel space was
the same as that of modern humans. See Lieberman (2007), de Boer and
Fitch (2010) and Morley (2013) for reviews of the extensive literature and the
phonetic variability arising from different SVT configurations.

13 See Boë et al. (2017) for vowel-like production by a baboon.

14 See Lameira et al. (2015) for Tilda, and Ralls et al. (1985) for Hoover. De
Boer and Fitch (2010) state that any mammal can produce enough phonetic
distinctions to support a basic spoken language.

15 Morley (2013) provides a review of the history of research on the basicranium
regarding reconstructions of the vocal tract. Basicranial flexion was
prominently drawn on by Lieberman and Crelin (1971) in an early computer
simulation of a Neanderthal vocal tract. Lieberman continued to emphasise
the significance of the basicranium, although with developments in his
interpretations (as reviewed in Lieberman, 2007). Laitman et al. (1979)
claimed a direct relationship between the position of the larynx and
basicranial flexion. Laitman and Heimbuch (1982) and Laitman (1984)
examined a range of hominin skulls, arguing that the basicranial flexion in *H.
erectus* indicates that this species would have had a limited range of vowels.
Budil (1994) studied the basicrania of a large sample of modern humans,
non-human primates and hominin fossils to reconstruct vocal tracts. He
found that although that of *H. heidelbergensis* had a basicranium similar to the
modern form, that of later Neanderthals appeared archaic, a finding he was
unable to explain. It gradually became apparent that there are many factors
influencing the shape of the base of the skull (e.g. Spoor and Zonneveld,
1998; Arensburg et al., 1990; Aiello, 1996). Fitch (2009) provides a substantive
critique of the long history of research attempting to relate fossil basicranial
anatomy to vocal sound production.

16 See Steele et al. (2013) for the comparative morphology of human, African
ape and hominin hyoid bones.

17 See Alemseged et al. (2006) for the *A. afarensis* hyoid bone and other details
of this fossil's remains.

18 See Martínez et al. (2008) for the *H. heidelbergensis* hyoid bones from Sima
de los Huesos, Arensburg et al. (1990) for the Neanderthal specimen from
Kebara, and Rodríguez et al. (2003) for that from El Sidrón Cave.

19 See Arensburg et al. (1990) for the significance of the mylohyoid groove on
the mandible for reconstructing the muscle systems that supported the hyoid
bone and larynx. The date of *c*.450,000 for *Homo heidelbergensis* from Sima de
los Huesos is from Demuro et al. (2019) who provides 448 ± 15 ka.

20 See Nishimura et al. (2022) for the role of vocal membranes and significance
of their loss for the evolution of speech.

21 See Hewitt et al. (2002) and de Boer (2009) for the nature and possible
function of air sacs in primates.

22 The following text draws on the study by MacLarnon and Hewitt (1999).

23 Frayer and Nicolay (2000) challenged MacLarnon and Hewitt's (1999) conclusions. Morley (2013) weighs up their respective arguments, finding some value in both but that the conclusions drawn by MacLarnon and Hewitt remain valid.

24 See García-Martínez et al. (2018) for estimates of Neanderthal lung capacity.

25 See Lameira et al. (2017) for the analysis of orangutan kiss-squeaks and potential evolutionary significance.

26 See Lameira et al. (2015) for orangutan imitation of humans and Hopkins et al. (2007) for raspberry blowing by captive chimpanzees. Also, see Lameira (2017) for a review of raspberry blowing in wild populations.

27 See Fedurek et al. (2015) for the function of chimpanzee lip-smacking. Their study of chimpanzees in the Budongo Forest, Uganda, found that 'Lip-smacking at the beginning of grooming bouts was significantly more often followed by longer and reciprocated bouts than silent grooming initiations. Lip-smacks were more likely to be produced when the risk of termination of the interaction by the recipient was high, for instance when grooming vulnerable body parts. Groomers were also more likely to produce lip-smacks during face-to-face grooming where the visual aspect of the signal could be perceived.'

28 See Pereira et al. (2020) for the claimed significance of chimpanzee lip-smacks for the evolution of speech.

29 See Pereira et al. (2020) for references to speech-like rhythms in monkey lip-smacking and gibbon song. See Lameira et al. (2015) for Tilda's orangutan-speech like rhythms.

30 See Brodbeck et al. (2018) for the process of speech perception.

31 Quam et al. (2013a) describe the conservative nature of auditory anatomy and similarities and differences between human and chimpanzee ear ossicles.

32 Remodelling means the resorption of existing bone material and the deposition of new material. This is a means by which strong bones are constructed for adult use.

33 See Conde-Valverde et al. (2019) for the differences between the human and chimpanzee cochlea.

34 See Heffner (2004) and Quam et al. (2013b, 2015a) for differences in primate audiograms and their auditory implications.

35 Quam et al. (2015a) describe the significance of mid-range frequencies for short-range vocal communication.

36 Key publications documenting the discovery, measurement and interpretation of ear ossicles include Rak and Clarke (1979), Arensburg et al. (1981), Moggi-Cecchi and Collard (2002), Quam and Rak (2008), Quam et al. (2013a), Quam et al. (2013b) and Stoessel et al. (2016).

37 See Quam et al. (2013b) for the La Ferrassie ossicular chain and reference to those from Le Moustier and Darra-i-Kur.

38 See Beals et al. (2016), Quam et al. (2015b) and Conde-Valverde et al. (2019) for the use of CT scans.

39 See the annexes and supplementary materials in Quam et al. (2015b) and Conde-Valverde et al. (2021) for the methodology behind sound power transmission models and occupied bandwidth estimates.

40 See Rak and Clarke (1979), Moggi-Cecchi and Collard (2002) and Quam et al. (2013a) for descriptions and interpretations of early hominin ear ossicles, and Quam et al. (2015b) for their auditory capacities.

41 See Conde-Valverde et al. (2019, 2021) for the ossicles, cochlea and reconstructed auditory tracts and function of the Sima de los Huesos *H. heidelbergensis* specimens.

42 Stoessel et al. (2016) argue for a larger difference between Neanderthal and modern human ossicles than do Quam et al. (2013b) and Conde-Valverde et al. (2021), although they all agree on the functional equivalence of their auditory tracts. For an overview see Conde-Valverde et al. (2021).

43 See Beals et al. (2016) for the most detailed study of a Neanderthal cochlea, that coming from Krapina, Croatia, dating to 130,000 years ago.

44 See Quam and Rak (2008) for ossicles from Middle Palaeolithic sites in Southwest Asia, including the important sample from *H. sapiens* from Qafzeh Cave.

6. Iconic and arbitrary words

1 See Sidhu and Pexman (2018) for a review of various types of sound symbolism and the possible mechanisms by which non-arbitrary associations between meaning and sound might arise.

2 That sound-symbolic words are no more than a 'quaint curiosity' is stated by Pinker (1994, p. 166).

3 Sprinker (1980, p. 117) noted that Max Müller had refuted the 'bow-wow' and 'pooh-pooh' theories in *Lectures on the Science of Language*, 1st series (1861; reprinted New York, 1871), pp. 358–70. Despite Müller's disclaimer that the terms he coined 'were not intended to be disrespectful to those who hold the one or the other theory' (p. 358), Sprinker says it is hard to suppress one's natural amusement when discussing them under such labels.

4 The following draws on the *Cratylus*, a dialogue written by Plato in the fourth century BC. It draws on the interpretation provided by Keller (2000).

5 See Ross (1955) for a consideration of when the *Cratylus* was written.

6 See Mackey (2015) and Taylor (2020) for interpretations of the Epicurean theory for the origin of language.

7 See Lifschitz (2012, pp. 17–21) for a description and interpretation of the Epicurean theory for the origin of language. His summary is that Epicurus

proposed 'the same tree would have been denoted by different sounds if it had been first encountered in the desert, next to a waterfall, surrounded by sheep, or in the midst of a thunderstorm. After some time and due to social interaction, human beings grew used to such sounds as names for the corresponding objects. Only later was convention introduced into this process: after knowledge had accumulated, language became enriched by analogy, abstract terms, and additional categories (such as pronouns or prepositions).'

8   My account of thought about language during the Enlightenment draws on Avi Lifschitz's (2012) fascinating and superbly crafted account of the 18th-century debates about the nature of language, its role in thought and how it evolved centred on the Berlin Academy. I am grateful to Nicholas Mithen for sharing his knowledge about Enlightenment thinkers on language.

9   The quotation of the essay questions follows Lifschitz (2012).

10 'Having no answer to the question' arises from my own reading of Herder's essay and is the impression I gain from Lifschitz (2012). Other commentators such as Pan (2004) appear more positive.

11 See Herder (1986 [1771]) for Herder's essay.

12 Cited from Herder (1986 [1771], pp. 139–41).

13 See Seuren (2016) for Saussure's intellectual environment. As he explains, Saussure himself was reticent about his sources and intellectual pedigree. Seuren concludes that Saussure had aimed at integrating the 19th-century rationalist tradition in the sciences into a scientific theory of language.

14 Deutscher (2005, pp. 104–6) describes how in 1878 the young Ferdinand de Saussure 'proposed a revolutionary theory which in one stroke transformed the impenetrable complexity of the distribution of vowels in the daughter languages [of Proto-Indo-European] into a system of almost incredible simplicity'.

15 Svantesson (2017) notes the arbitrariness of language was widely held by linguists before Saussure. He also notes the marginal reference to sound symbolism in key textbooks of linguistics throughout the 20th century.

16 Jespersen (1919) published an article called 'The symbolic value of the vowel i', but his work is primarily known from his 1922 book. He was one of the most distinguished linguists of his generation, working on many aspects of language, but is best known for his studies of syntax and language development. The context of the cited quote is as follows: 'Here [in the work of Saussure] we see one of the characteristics of modern linguistic science: it is so preoccupied with etymology, with the origin of words, that it pays much more attention to what words have come from than to what they have come to be. If a word has not always been suggestive on account of its sound, then its actual suggestiveness is left out of the account and may even

be declared to be merely fanciful. I hope this chapter contains throughout what is psychologically a more true and linguistically a more fruitful view.' (Jespersen, 1922, p. 410).

17 Edward Sapir (1929b) combined linguistics with anthropology, having been a student of Franz Boas. His own student was Benjamin Lee Whorf, and their work is often collated into the Sapir–Whorf hypothesis that proposed language influenced perception and thought.

18 Wolfgang Köhler (1887–1967) was a German psychologist who contributed to the development of Gestalt psychology, which took a holistic approach to the human mind and body. He undertook research on the mentality of apes, believing this might provide insights into the human mind. His *maluma/ takete* experiment is described in Köhler (1929).

19 Westermann's (1937) study of sound symbolism in West African languages may have been the most significant contribution between 1920 and 1971.

20 Roger Wescott (1925–2000) was Professor of Linguistics and Professor of Anthropology at Drew University in Madison, New Jersey.

21 Murdock (1959) proposed a universal pattern of stop consonants for father terms and continuant consonants for mother terms.

22 Jakobson's (1960) proposal has been supported by a series of later studies. Brooks-Gunn and Lewis (1979), for instance, demonstrated that pre-speaking infants have a propensity to select the words 'mommy' and 'daddy' for their mother and father when presented with a range of images including those of strangers.

23 Svantesson (2017) describes the contrary use of /i/ in words for big and /a/ in words for small within in the Bahnar language.

24 This is the first of three hypotheses with which Wescott (1971) concluded his articles. The second is that new iconic words might be added to the language at the same rate as arbitrary words to maintain a constant ratio; third, that the ratio of iconic and arbitrary words fluctuates slowly but rhythmically. When paraphrasing Wescott, I have taken the liberty of using 'arbitrary word(s)' in place of his use of the word 'symbol(s)'.

25 See Perniss and Vigliocco (2014) and Ortega (2017) for a discussion and interpretation of the role of iconicity in sign language.

26 Perry et al. (2015) and Winter et al. (2017) rated 3,001 English words from 0 indicating arbitrariness to 5 indicating iconicity. Akita (2013) had previously developed a 'lexical iconicity hierarchy' (LIH) in which sound-symbolic words vary in their degree of iconicity. Those at the highest end, having absolute iconicity, were more often found across languages. Ortega (2017) reports that Akita found a clear order of acquisition with the most iconic words being mastered first and the less iconic words mastered at a later stage.

27 Phonoaesthemes are a problematic class of word, with the first element sometimes not considered to be a morpheme. Cuskley and Kirby (2013)

review phonoaesthemes in the context of sound symbolism and cross-modal thinking.

28 Cuskley and Kirby (2013) suggest that rather than having an iconic origin, *gl-* might arise from a particular branch of Proto-Indo-European words. Non-arbitrary associations between word form and their meaning category can arise by systematicity, without having any iconic role. See Dingemanse et al. (2015) for a description and comparison of the roles of arbitrariness, iconicity and systematicity in language.

29 *Sound Symbolism* was edited by L. Hinton, J. Nichols and J.J. Ohala (1994) from the University of California, and published by Cambridge University Press.

30 The chapters referred to are by Alpher (1994) concerning the use of ideophones by the Yir-Yoront, and Childs (1994) concerning their presence in African languages. See Dingemanse (2018) for a survey of the research history concerning ideophones and a synthesis of current research.

31 Ideophones are called 'expressives' in Southeast Asian languages and 'mimetics' in Japanese. Svantesson (2017) explains that the different terminology has arisen from independent research traditions in these different regions of the world.

32 Cited from Alpher (1994, p. 172).

33 Humphrey (2012) referred to synaesthesia as 'leakage' between cortical areas. Duncan Carmichael, the leading expert on synaesthesia, states that 'everybody potentially starts off as a synaesthete' (quoted in Bilby, E. 'Do we all have synaesthesia?', Horizon: *EU Research & Innovation Magazine*, 20 October 2015). For a study exploring the personality profile of child synaesthetes, see Rinaldi et al. (2020).

34 See Maurer and Mondloch (2006) for the lack of pruning of neuronal connections as the cause of synaesthesia. Kadosh et al. (2009) argue that cortical functional specialisation for cognitive functions emerges during human postnatal development for non-synaesthetes, but this is lacking for synaesthetes because pruning of neuronal connections between cortical areas does not occur. They note the increased quantities of white matter in the brains of those with synaesthesia. See Karmiloff-Smith (1992, 1998) for the development of domain-specific mentality.

35 See Bankieris and Simner (2015) for experimental studies involving sound symbolism that conclude synaesthesia is an exaggeration of the cross-modal perceptions that all people experience to varying extents, rather than being a qualitatively different phenomenon.

36 Ramachandran and Hubbard's 2001 synaesthetic account for how sound symbolism relating to shape arises has been challenged by Margiotoudi and Pulvermüller (2020). They argued that the difference between the articulatory gestures (the movements of the tongue and mouth) for making round-sounding and sharp-sounding phonemes is insufficient to support

Ramachandran and Hubbard's proposal. Based on an experimental study, they place more emphasis on the shapes and sounds created by body actions, such as that of a pen on paper when drawing a circle. They argue that correspondences between shapes and the sounds of hand movements when making such shapes are learned and provide the basis for sound symbolism. Margiotoudi and Pulvermüller's arguments are unpersuasive. They ignore the attention that Ramachandran and Hubbard had paid to the hand gestures and body movements that correlate and often occur simultaneously with making sounds. Moreover, they do not address the non-shape-related types of sound symbolism – the association of specific types of sounds with size, movement, light conditions and so forth, with these associations being dependent on articulatory gestures, such as enlarging the oral cavity when making a word for a large object and vice versa. Their emphasis on a learned association between body action and the sound it makes cannot account for the presence of shape-sound symbolism at an early stage of development before any such learning could take place.

37   Ghazanfar et al. (2007) describe experiments testing for cross-modal perception in rhesus monkeys.

38   See Ludwig et al. (2011) for experiments testing for cross-modal perception in chimpanzees. They concluded that their positive result indicates an innate rather than a learned capacity in the chimpanzee, but this was challenged by Spence and Deroy (2012). They suggested that in the chimpanzee world the main source of luminescence is the sky, and it is in the sky where small animals, notably birds, are found that make high-pitched noises. Large, low-pitched-sound animals such as elephants are found on the ground, which is darker. As such, the association that Ludwig et al. (2011) found might have been learned from experience.

39   See Ravignani and Sonnweber (2017) for a test of isomorphic mapping across sensory modalities in chimpanzees.

40   Margiotoudi et al. (2019) describe the *maluma/takete* test on chimpanzees.

41   Margiotoudi et al. (2022) describe the *maluma/takete* test using a bonobo participant. This was Kanzi, the most renowned language-learning ape.

42   Sidhu and Pexman (2018) describe how there had been 28 academic articles on sound symbolism published in 2001 and 193 in 2016.

43   Sidhu et al. (2021) examined whether the *maluma/takete* effect is attested in English across a sample of 1,757 objects for which participants were asked to specify their real names. This found that phonemes associated with roundness are indeed more common in words referring to round objects, and those referring to spikiness are more common in words referring to spiky objects.

44   Brent Berlin is an American anthropologist who undertook seminal work on colour terms with the linguist Paul Kay and wrote a classic work

on ethnobiological classification (Berlin, 1992). I draw on his insightful, thoughtful and entertaining article (Berlin, 2006).

45 Studies of large corpuses of words were undertaken by (among others) Berlin (2006), who drew on 17 Amazonian languages, Monaghan et al. (2014), who used word lists from English, and Blasi et al. (2016), who drew on word lists from 6,452 of the world's languages.

46 Large-scale experimental studies to explore the nature and role of iconic words in various aspects of language have been undertaken by (among others) Berlin (2006), Thompson and Estes (2011), Monaghan et al. (2012), Imai and Kita (2014) and Tzeng et al. (2016).

47 The *bouba-kiki* effect is not always found, but a meta-analysis by Fort et al. (2018) found there was sufficient evidence for this to be considered a consistent outcome.

48 See Imai and Kita (2014, pp. 4–5) for studies of children's sensitivity to sound symbolism.

49 See Tzeng et al. (2017) for developmental change in children's sensitivity to sound symbolism, and see the discussion of this finding in Fort et al. (2018).

50 Quoted from Tzeng et al. (2016, pp. 2210–11), who describe their set of experiments and collate evidence from those undertaken since the 1950s, of which the most significant are Imai et al. (2008) and Nygaard et al. (2009).

51 Blasi et al. (2016) undertook a global study of iconism in languages. Their data set consisted of 28–40 lexical items from 6,452-word lists with a subset of 328-word lists having up to 100 items.

52 Monaghan et al. (2014) describe how iconic words are learned earlier than arbitrary words. This ease of learning iconic words had already been noted, or at least implied, by the Classical writers. The study by Monaghan et al. was primarily concerned with non-arbitrary words, and hence the results reflect the role of both iconicity and systematicity in language acquisition.

53 See Fort et al. (2018) for a discussion of the different sensitivities to words associated with roundness and spikiness. Margiotoudi et al. (2022) note that a scenario in which sound-shape mappings are learned because of their systematic presence in human vocabulary is poorly supported by cross-linguistic studies.

54 Imai and Kita (2014) suggest the dense neural connectivity across different sensory brain regions may enable infants to spontaneously map perceptual experiences across different modalities onto speech sounds. They describe how evidence indicates that large-scale, synchronous neural oscillations play an important role in dynamically linking multiple brain regions in adults. Svantesson (2017) summarises recent EEG and brain imaging studies that demonstrate that sound-symbolic words are processed differently from other words in the brain. Of most interest is the study by Revill et al. (2014), concluding that MRI data 'provide support from cross-modal activation

as a mechanism by which sound symbolism facilitates word-to-meaning mappings. Heightened activation in left superior parietal lobe for sound symbolic relative to non-symbolic words suggests that sound symbolic foreign words engage cross-modal sensory integration processes to a greater extent than non-symbolic words.' Also see Sidhu and Pexman (2018) for a lengthy discussion about the possible mechanisms of sound-symbolic association.

55 Quine's *gavagai* problem is the potentially infinite number of possibilities for mapping between a word and its potential referent. It was promoted and is named after the philosopher Willard Van Orman Quine (1908–2000) and introduced in Quine (1960).

56 The child would also have been attracted to associate the word with the whole rabbit rather than one part of it because of the object bias they bring to learning the meaning of words, as described by Bloom (2000).

57 See Imai and Kita (2014, p. 7) for studies of how iconic words have a high prevalence in child-directed speech, and how carers modify their use with the child's age.

58 Thompson et al. (2013) demonstrated that iconicity facilitates the acquisition of signs by deaf children, while Ortega (2017) nuanced this by identifying action signs as the first ones acquired.

59 See Monaghan et al. (2012) for an experimental study that demonstrates iconic words are more effective for learning general categories rather than specific meanings. See also discussion of this issue and further experimental studies in Tzeng et al. (2017). Ortega (2017) identified that iconicity can hinder adults acquiring the exact phonological form of a sign when learning sign language as their second language.

60 Thompson and Estes (2011) undertook experimental studies that demonstrated the number of large or small phonemes within a nonsense word influences the relative size of an object that it is associated with. Words such as *wodolo*, with three large-sounding phonemes, are taken to refer to the largest object, and *kitete*, with three small-sounding phonemes, to the smallest. Svantesson (2017) provides real-world examples of graded sound-symbolic words from the Kammu language, spoken in Northern Laos, which refer to different sizes of fire and different loudness of noise made when drinking by animals of different size.

61 Gasser (2004) undertook computational modelling that demonstrated the need to introduce arbitrary words as a lexicon increases. This has been emphasised by Monaghan et al. (2014). Lupyan and Winter (2018) note how combining elements of iconicity and arbitrariness within single words is a means to expand the lexicon.

62 Imai and Kita (2014) refer to the relatively low frequency of iconic words for objects in Japanese. Perry et al. (2015) made the same finding for English.

63 Sidhu and Pexman (2018) undertook a statistical analysis of a word list from English to establish that iconic words are more frequent within what they described as categories of low semantic neighbourhood density.

64 Lupyan and Winter (2018) discuss the relationship between iconic words, arbitrary words and abstract concepts. They provide evidence that iconic words resist expressing abstractions.

65 Quoted from Cuskley and Kirby (2013, p. 21).

66 Quoted from Bankieris and Simner (2015, p. 193).

67 Quoted from Berlin (2006, p. 539).

68 Perniss and Vigliocco (2014) argue that iconicity might have played a key role in utterances concerning displacement that they view as a key development in the evolution of language. They also emphasise the role of iconicity in supporting referentiality (learning to map linguistic labels onto objects, events, etc., in the world), which is central to language acquisition.

69 Quoted from Perniss and Vigliocco (2014, p. 5).

70 Dellert et al. (2021) analysed sound change with Eurasian languages and concluded that iconic sound-meaning mappings are more likely to survive sound evolution and change compared with an average arbitrary connection between form and meaning.

7. Making tools

1 See Tennie et al. (2009) for a discussion of the ratchet effect regarding chimpanzee and human technology. While this and other work focus on the enhanced role of social learning and cooperation in human communities, there has been insufficient attention given to the role of language, in both enabling such cooperation and directly influencing technological change through time.

2 The first edition of *Man the Tool-maker* was published in 1949, but the 1972 edition is the one most frequently accessed (Oakley, 1972).

3 Goodall (1964) is the earliest published reference to chimpanzee tool use of which I am aware.

4 In a short note, Oakley (1969, p. 222) wrote: 'Are we then to abandon tool-making as a criterion of humanity? If we scrutinize any evolutionary process we become aware of the fallacy of the hard-and-fast line – in other words, we require a sense of proportion.' He went on to describe apes as occasional tool users and, in the case of chimpanzees, grading into incipient tool makers.

5 See Boesch and Boesch (1990) for a summary of tool use by chimpanzees and McGrew (1992) for an ecological approach to such tool use and its implications for human evolution.

6 See McGrew (1992), Sanz and Morgan (2013) and Grund et al. (2019) for

the relationship between chimpanzee tool use and ecological and social variables.

7 See Proffitt et al. (2022) for variability in the use of stone by chimpanzees in West Africa.

8 See Whiten et al. (1999) for social learning and patterns of cultural variation in chimpanzees, and Musgrave et al. (2020) for what they describe as teaching within chimpanzee communities.

9 See Toth and Schick (2009) for a comparison between Oldowan stone tools and those made by Kanzi. Roffman et al. (2012) provide the most recent account of Kanzi's stone tools. They claim that Kanzi can now make tools that fall into Oldowan tool categories, but they fail to provide a formal, quantitative comparison and one cannot make judgements from their published illustrations.

10 Stout (2011) provides an excellent summary for Oldowan and Acheulean technology.

11 See Key et al. (2020) for a study of raw material selection in the Oldowan.

12 See Stout et al. (2010) for a study and discussion of technological variation within the Oldowan.

13 Stout (2011) describes elaborate flake production as part of the Early Acheulean. It is illustrated by the Karari scrapers known from the basal Okote member at Koobi Fora, Kenya, dating to 1.6–1.5 mya.

14 The Acheulean is named after the French commune of Saint-Acheul where handaxes were first found within river gravels.

15 See Machin et al. (2007) for an experimental study to evaluate the contribution of handaxe symmetry to functional effectiveness for butchering carcasses.

16 See Kohn and Mithen (1999) for the case for handaxes as products of sexual selection. A remarkable example of such display was found from a 2021 excavation close to London when a handaxe 29.6 cm (11.7 inches) long was recovered, dating to around 320,000 years ago (Ingrey et al., 2023). The handaxe is beautifully symmetrical and so large that it is difficult to hold, and even more difficult to imagine how it could have been used for functional tasks such as animal butchery. It is a peacock's tail in stone.

17 See Ashton and Davis (2021) for the expansion of hominin range in Europe and enhanced behavioural complexity after 600,000 years ago.

18 See Rees (2000) for refitting knapping debris from Boxgrove, and García-Medrano et al. (2019) for the stone knapping methods at Boxgrove.

19 Cited from García-Medrano et al. (2019, p. 416).

20 Stout et al. (2014) emphasise the significance of platform preparation in Late Acheulean handaxe manufacture as illustrated at Boxgrove for increasing the cognitive complexity of tool making.

21 Dennell (2018, p. 209) was writing about the Asian record, but his summary is applicable to the Acheulean throughout the world.

22 The following summary draws on Stout et al. (2008).

23 See Uomini and Meyer (2013) for an experimental study of blood flow when making handaxes and silently generating words.

24 Cited from Stout et al. (2008, p. 1947).

25 See Stout and Chaminade (2012) for further discussion of the possible relationships between tool making and language in the brain.

26 Ohnuma et al. (1997) explored the role of verbal and non-verbal communication in the transmission of skills to make Levallois technology. Morgan et al. (2015) conducted an extensive experimental programme exploring five different conditions for transmitting technological knowledge (reverse engineering, i.e. inspecting cores and flakes; imitation and emulation; basic teaching in which experts could slow their actions or provide novices with a clear view; gestural teaching; gestural teaching with speech). Novices learned from experts and then taught another person, who taught another in a chain of cultural transmission. Lombao et al. (2017) undertook a similar programme to Morgan et al. (2015) but standardised the cores that were used to remove a major source of variability. Cataldo et al. (2018) separated spoken from gestural instruction, which had been conflated by Morgan et al. (2015).

27 See Morgan et al. (2015) for an experimental programme involving the transmission of knapping skills along a chain. There was no justification within the methodology for the extremely short times provided for a novice to learn a technique before having to pass it on to another.

28 Morgan et al. (2015) undertook the experimental programme involving the transmission of knapping skills along a chain.

29 See Cataldo et al. (2018) for experiments that involve speech without gesture.

30 The origin of Levallois technology, and prepared core technology in general, is one of the most important but under-researched issues in Palaeolithic archaeology. It remains unclear whether this was associated with one particular hominin that dispersed throughout the Old World, was spread by cultural transmission between hominin species, or was independently invented in different regions. The earliest date for the transition from Acheulean to prepared core technology designated as the Middle Stone Age in Africa is dated to *c*.320,000 (Deino et al., 2018). Middle Palaeolithic technology using Levallois technology has been dated to *c*.385,000 in South Asia (Akhilesh et al., 2018) and to *c*.335,000 in the southern Caucasus (Adler et al., 2014).

31 See van Peer (1992) for a comprehensive description of Levallois technology. When multiple flakes are removed from a Levallois core, the technique is called recurrent Levallois; when using two platforms it is called opposed core

Levallois. Flakes were sometimes removed from around the core, which is termed centripetal Levallois.

32 See Muller and Clarkson (2016) for an argument and experimental evidence that the change in stone tool technologies through time delivered a greater amount of cutting edge on flakes and blades from the same quantity of stone.

33 'Investigating the missing majority' is the subtitle for Hurcombe (2014), a book considering the perishable material culture from prehistory.

34 See Thieme (1997) for the discovery and archaeological context of the Schöningen spears. The specific dates of these spears remain unclear, but likely edge towards 300,000 years old (Richter and Krbetschek, 2015). The Clacton spear was found in 1911 and is described by Oakley et al. (1977). This was made from yew and likely part of a thrusting spear. See Barham et al. (2023) for two interlocking logs joined transversely together by an intentionally cut notch from the site of Kalambo Falls, Zambia. These are dated to at least 476,000 years ago and were found with an assortment of wooden tools dating to between 390,000 and 324,000 years ago.

35 Annemieke Milks (pers. comm) emphasises that such weapons were likely used in a variety of ways. Two throwing sticks from Schöningen that may have been used for hunting birds and small mammals such as hares.

36 See Milks et al. (2023) for a meticulous description and analysis using CT-scanning, 3D microscopy and Fourier transform infrared spectroscopy of one of the double-pointed sticks, and inferences for how it had been made and used. They concluded that 'the hominins selected a spruce branch which they then debarked and shaped into an aerodynamic and ergonomic tool. They likely seasoned the wood to avoid cracking and warping. After a long period of use, it was probably lost while hunting, and was then rapidly buried in mud' (p. 1). Their study sets a new standard for the analysis for ancient wooden artefacts.

37 See Haidle (2009) for the multi-stage manufacture of a Schöningen spear.

38 See Schoch et al. (2015) for behavioural implications of the Schöningen spears.

39 See Aranguren et al. (2018) for the Poggetti Vecchi wooden tools and reconstruction of their manufacturing process, and Revedin et al. (2020) for experiments in using fire as part of their production.

40 See Hardy et al. (2020) for a description of the twisted plant fibre from Abri du Maras. Before this find, further fragments of plant fibre had been found at Abri du Maras but were too fragmentary to identify positively (Hardy et al., 2013).

41 See Villa et al. (2009a), Villa and Soriano (2010), Wilkins et al. (2012) and Rots (2013) for evidence of hafted Middle Palaeolithic artefacts from Europe and Middle Stone Age artefacts from Africa.

42 Grünberg (2002) described two pieces of birch bark tar excavated from Königsaue, a Middle Palaeolithic site in Germany, one of which had a fingerprint and the negative impressions of bifacial tool. Mazza et al. (2006) describe birch bark tar partially covering three flakes dating to between 0.78 and 0.5 mya ago from Campitello, Italy.

43 Hardy et al. (2020) suggest that the fragment of fibre from Abri du Maras implies 'mathematical understanding of pairs, sets, and numbers'.

44 Kozowyk et al. (2017) proposed that Neanderthals were making birch bark tar using dry distillation methods, whereas Schmidt et al. (2019) showed that it could have been produced in a simple hearth when birch bark is burnt against stones, demonstrating that 'birch bark alone cannot indicate the presence of modern cognition and/or cultural behaviours in Neanderthals'.

45 The following consideration of tool making in Europe draws on de la Torre et al. (2013), Turq et al. (2013), Ruebens (2013) and Kuhn (1995, 2006). For eastern Africa, it draws on Tryon and Faith (2013); for northern Africa, Scerri (2017); and for southern Africa, Wurz (2013).

46 For the 'Mousterian of Acheulean Tradition', see Soressi (2004).

47 See Turq et al. (2013) for an outstanding study of Middle Palaeolithic stone tool technology when viewed from a landscape perspective, which draws on a number of superbly preserved sites that have allowed extensive refitting.

48 Tryon and Faith (2013), for instance, note the presence of impact damage on scrapers from the European Middle Palaeolithic, suggesting that they had been used in the same manner as points.

49 See Ruebens (2013) for an assessment of Middle Palaeolithic bifacial tool variability in western Europe.

50 Technological interstratification in Europe is exemplified in the sequences from Combe Grenal, France. Faivre et al. (2014) review the long history of analysis and interpretation of the stratified sequence from this site and provide their own study that concludes there is a temporal succession of Levallois, Quina and then discoid assemblages. My reading of their work suggests that these assemblage types are referring to the dominant tool-making method and consequently some degree of interstratification remains. De la Torre et al. (2013) describe several sites from Iberia with interstratification of discoid, Levallois and Quina flaking techniques. Steven Kuhn (2006, p. 110) stated that 'While the Mousterian may have changed over time, it was not going anywhere in particular.' Ignacio de la Torre et al. undertook a detailed study of the Mousterian in Iberia and concluded that although change through time exists, there is no patterning. He interpreted this as stochastic variation rather than directional change (de la Torre et al., 2013).

51 The period between 130,000 and 80,000 years ago was an interglacial, known as MIS 5. Open steppe was replaced with woodland of varying composition

and density. Gaudzinski-Windheuser and Roebroeks (2011) explain that the loss of the large mammals that had grazed on the steppe would have proved a challenge to the Neanderthals, being a top carnivore with large energetic requirements to fuel their physique and lifestyle. According to Defleur and Desclaux (2019), rather than adapt and thrive in such conditions by exploiting the new range of smaller game and plant food from the woodland, the Neanderthal population appears to have collapsed. As Gaudzinski-Windheuser and Roebroeks (2011) note, the record might be biased by low archaeological visibility of small Neanderthal groups.

52 Soressi et al. (2013) describe the supposed hide working tools, known as lissoirs, and summarise other bone tools made and used by Neanderthals in Europe. Wragg Sykes (2020) describes bone tools used as 'retouchers' from several sites, although, to my mind, exaggerates their significance.

53 See Berger and Trinkaus (1995) for Neanderthal injuries from close-encounter hunting. Beier et al. (2018) undertook a further study and found both similarities and differences in cranial trauma between Neanderthals and early Upper Palaeolithic humans, with the Neanderthals having had greater mortality risk.

54 See Lombard (2022) for the origin of spear throwers and bows and arrows in hunting technology, and the selective benefits these provide.

55 See McBrearty (2003) for the presence of Acheulean handaxes in Africa at 160,000 years ago and how these might be interpreted. They may either have been made by *Homo sapiens* or by relict populations of *Homo heidelbergensis*, sharing the same landscape.

56 Iovita (2011) notes that Aterian points have wear patterns that indicate their use as scrapers. Tryon and Faith (2013) state that points from East Africa were used as cutting tools rather than used on spears.

57 See McBrearty and Brooks (2000) for an extensive review of the Middle Stone Age in Africa.

58 See Iovita (2011) for the function of Aterian tanged points. Although resembling arrowheads, some were hafted for use as knives and scrapers. Dibble et al. (2013) argue that other than the presence of tanged points there is no difference between the Aterian and the preceding MSA industry, referred to as the Maghrebian Mousterian.

59 See Henshilwood et al. (2001) for excavations at Blombos Cave.

60 The backed bladelets from the MSA of South Africa are a defining feature of the Howiesons Poort culture. See Wurz (2013) for a consideration of the relationship between and chronology of the Still Bay and Howiesons Poort industries.

61 Tryon and Faith (2013) describe how the Kapthurin Formation in Kenya has sites with points and Levallois cores interstratified with those containing cleavers.

62 See Ziegler et al. (2013) for the correlation between environmental change and pulses of technological innovation in the Middle Stone Age of South Africa.

8. Lessons from an artificial language

1 See Smith (2012) for the case for why formal models are useful for evolutionary linguistics, and a counter to some of the common criticisms that are made.

2 For key publications by this Edinburgh group, see Kirby (2001, 2002), Brighton (2002) and Brighton et al. (2005).

3 As reviewed by Cangelosi (2012), the ILM is just one of what has become a large family of computer and mathematical models that have explored the evolution of language. Multi-agent simulation models have been used to explore the emergence of shared lexicons. In these models the agents communicate with each other while undertaking tasks such as finding food and differentiating between that which is edible and poisonous (e.g. Cangelosi and Parisi, 1998, 2002). See Gong et al. (2014) for further examples of simulation and formal models for language evolution. The evolution of shared words has been explored by using robots that interact with the world and each other through sensors, including microphones for sounds, cameras for images, and touch sensors (e.g. Steels, 1999, 2003). The robots play language games with each other that negotiate words for concepts to explore the minimal conditions required for shared lexicons to emerge.

4 The following description of the iterated learning model draws on Brighton et al. (2005).

5 See Appendix A of Brighton et al. (2005) for how compositionality is measured in the ILM.

6 See Pinker and Bloom (1990) for a biologically driven, incremental view for the evolution of language, and Clark (2013) for theories on the evolution of syntax.

7 See Kirby et al. (2008) for laboratory-based experiments using human participants.

8 See Brighton et al. (2005) for various experiments exploring learning processes, bottlenecks, compositionality and communicative accuracy.

9 See Kirby (2002) for learning, bottlenecks, and the evolution of recursive syntax.

10 Cited from Kirby (2001, p. 108).

11 See the arguments and multi-authored discussion in Christiansen and Chater (2008). The quote is from p. 489 of that article.

9. Finding and learning the meaning of words

1 Recapitulationist theories about language development are less frequently

considered now than in previous decades. For classic arguments see Lamendella (1976) and Parker and Gibson (1979).

2   I have taken the example of *ham* and *hamster* from Endress and Hauser (2010).

3   Partanen et al. (2013) undertook an experiment in which they played a pseudo-word, *tatata*, regularly to foetuses from 29 weeks to birth, including variations in the middle syllable and pitch changes. After birth, these infants showed neural response to this pseudo-word and its variations in a manner not found in infants who had not received the same stimulus when a foetus. The researchers argued that this demonstrated that the foetal brain has the same learning capacities as the infant brain – which seems a bit of a leap.

4   See Langus et al. (2017) for the use of prosody in first language acquisition, and Speer and Ito (2009) for the significance of rhythm in language learning. Languages have characteristic rhythmic structures. English, for instance, is a 'stress-timed' language because stressed syllables, which often have a longer duration than others, are spoken at regular intervals, whereas French and Italian are 'syllable-timed' languages because each syllable takes an equal time to say. Babies have already acquired some familiarity with the rhythm of their caregiver's language before birth and are highly sensitive to the prosody of what they hear.

5   See Saffran et al. (1996) for a seminal article that transformed our understanding of how infants find words, a first step in the acquisition of language.

6   Graf Estes et al. (2007) undertook experiments to test whether newly segmented words can be mapped onto meanings. See Endress and Hauser (2010) for a review of these and related experiments and alternative interpretations of the results. They place emphasis on language-specific learning mechanisms that used universal patterns of prosody in language to identify words. This appears, however, to be a retreat to a form of Universal Grammar and is unpersuasive because of the high degree of variability of prosody in languages that they acknowledge and attempt to counter in their article.

7   See Dahan (2015), who provides my cited examples and an excellent review of prosody and language comprehension.

8   See Shukla et al. (2007) for the relative ease of learning edge words and the interaction between prosody and statistics for the segmentation task.

9   See Matzinger et al. (2021) for experiments to test the relative impact of prosodic cues.

10 My text draws on Thiessen and Saffran (2003). Their study found that infants as young as six and a half months old can achieve word segmentation using transitional probabilities.

11 See Marian (2023, p. 90) for multilingual infants.

12 McMullen and Saffran (2004) explain that by seven and a half months old, infants are including talker-specific cues in their representations of spoken words. Before ten and a half months, they have difficulty in recognising words when spoken by voices other than their primary caregivers. After that age, talker-specific cues are no longer required.

13 See Romberg and Saffran (2010) for a review of the role of statistical learning in language acquisition, including how infants learn about the complexities of language.

14 See Pelucchi et al. (2009) for the backwards tracking of transitional probabilities, otherwise termed as 'learning in reverse'.

15 See Saffran and Wilson (2003) for how children move swiftly from syllables to syntax.

16 See Toro et al. (2008) for experiments that suggest that vowels and consonants play different roles in segmentation and acquiring grammar.

17 See Gómez (2002) for the learning of non-adjacent dependencies between words.

18 See Newport and Aslin (2004) for non-adjacent dependent learning experiments, and Wilson et al. (2020) for an excellent review of non-adjacent dependency learning.

19 See Endress and Hauser (2010) for how infants expect new words to begin after the end of words they already know.

20 Shoaib et al. (2018) showed that English-learning infants at 20 months old with relatively small vocabularies learned Italian words with high transitional probabilities as the labels for objects but did not learn the Italian words with low transitional probabilities. Infants with larger vocabularies also resisted learning the Italian words with high transitional probabilities, indicating that they are more resistant to words that do not fit the typical phonetic pattern of English.

21 See Romberg and Saffran (2010) for an emphasis on the role of the context in which language learning takes place.

22 Yu and Smith (2007) undertook experiments that showed that adults can use cross-situational statistics to acquire new word-object mappings. The experiments were repeated by Smith and Yu (2008) for 12- and 14-month-old infants with the same outcome.

23 McMullen and Saffran (2004) compared the acquisition of language and music, describing the use of statistical learning in both auditory domains. Saffran and Kirkham (2018) reviewed the evidence for statistical learning in the visual domain, noting how 'infants are sensitive to many different statistical regularities in the visual domain across both temporal and spatial input, enabling them to extract patterns for further processing' (p. 184). They conclude by citing William James (1890) by stating that statistical learning is a

rich and robust learning mechanism that allows infants to find structure (and meaning) in what James called a 'blooming' buzzing confusion' (p. 185).

24 Hauser et al. (2001) describe testing cotton-top tamarins for statistical learning. Once the tamarins had become familiarised with the sound stream, their responses to 'words' (syllable strings with high transitional probabilities, TPs) and non-words (those with low TPs) were tested by the extent of their response to hearing isolated 'words' and non-words. The experiment was also undertaken using part-words which consisted of the final syllable of one word and the first two syllables of another word. Their response to part-words was the same as to non-words. The tamarins performed as well as the infants, although they were allowed 20 minutes of listening time rather than the 8 minutes the human infants received. The tamarins were able to discriminate between speech syllables and track their transitional probabilities. They became familiar with recurrent patterns of syllables, thereby identifying what to humans are potential words. Newport et al. (2004) undertook further experiments to test whether tamarins were able to track non-adjacent syllables. Surprisingly, tamarins had no problem with this task, whatever sequence the syllables came in and despite having brains weighing a mere 10 g compared with the 500 g of a human infant. They likely achieved this feat by attending to a few syllables, tuning out before tuning in again to hear more syllables. By doing so, they didn't overload their brains with the mass of computations required for a continuous tracking of non-adjacent syllables, as happens when humans attempt the task. Experiments exploring non-adjacent statistical learning by South American capuchin monkeys using visual stimuli have had less success: the monkeys couldn't do it. That might reflect the specific experimental design or indicate that cotton-top tamarins had independently evolved this capacity long after we had shared a common ancestor, one that did not evolve in the lineage leading to present-day capuchin monkeys. Experiments by Sonnweber et al. (2015) explored whether chimpanzees use statistical learning by testing them with visual stimuli. By presenting different shapes in sequences that mimic the syllables in language-like sound streams, they found that chimpanzees can identify recurrent patterns from both adjacent and non-adjacent shapes. Watson et al. (2020) used sequences of different tones, testing marmosets (an Old World monkey), chimpanzees and humans on the same statistical learning task. All three species could process the sound streams to identify recurrent strings, leading the researchers to conclude that they were using the same statistical learning process, which must have evolved in the primate line at least 40 million years ago.

25 Apocope is the loss of one or more sounds or letters at the end of a word. Lenition is the process or result of palatalising a consonant.

26 Unless otherwise specified, the arguments and evidence in the following

text draw on Bloom's 2000 book *How Children Learn the Meanings of Words* and its précis (Bloom, 2001). That précis has a commentary from many psychologists, several of whom challenge Bloom's arguments and/or provide alternative viewpoints and further evidence. Their commentary is followed by Bloom's response, providing a thorough discussion of the many issues involved.

27 See Hirschfeld and Gelman (eds) (1994) for inherited, domain-specific knowledge. Spelke (1991) describes intuitive physics, arguing that babies are born with conceptual understanding (although not the words) for inertia (that inanimate objects cannot move by themselves) and solidity (that solid objects cannot pass through each other). Atran (1990) makes a similar case for intuitive biology.

28 See Gelman and Meyer (2011) for a review of competing theories for child categorisation. They discuss the role of 'generics', which are linguistic forms that refer to whole categories rather than individual members of a category. Semenza (2009) has discussed the difference between common names and proper names that fall outside of categories. He suggests the latter require different cognitive and neural pathways, possibly arising from different evolutionary pressures.

29 Gelman and Meyer (2011, p. 98) provide the example of learning the word for a table from a drawing of a table, and then applying it to both real tables and those in miniature.

30 See Fitch (2019) for how animal minds, especially those of apes, contain many concepts and how the majority have no expression within the signals they make. Spinozzi (1996) provides an overview of categorisation in monkeys and chimpanzees, and Zentall et al. (2008) review concept learning in animals.

31 See Fitch (2019) for the limited connection between animal signals and their mental concepts.

32 The essential protocol is to initially train chimpanzees with exemplars of two or more categories (Tanaka, 1995). They are then shown novel items and tested as to whether they can place them into the newly learned categories. One such experiment gave chimpanzees photographs of four types of natural objects: trees, flowers, weeds and ground surfaces (Tanaka, 2001). The chimpanzees were asked to match new photographs of these object types into the correct category. They did so with ease, which suggested that they already had mental concepts for these categories before the experiment. A more challenging experiment, undertaken by Gruber et al. (2019), tested both categorisation and social learning. It involved a direct comparison between adult chimpanzees and children aged between seven and 11 years old regarding their abilities to sort objects into tools and non-tools. Each participant watched someone attempt to open a box with four objects in

turn. Two of these were tools (a hammer and pizza slice) that could open the box, and two were non-tools (a spoon and a door wedge) that touched the box but could not open it. After the demonstration, the participant was presented with each object and asked to match it to one of the other objects. The four objects were different colours to reduce any perceptual similarities that might have been used to group the objects rather than their functional abilities to open the box. One of the chimpanzees, called Ai, and 40 per cent of the children spontaneously paired objects by their functional utility at a significantly higher frequency than non-functionally matched pairs – in other words, they placed tools and non-tools into separate categories. When the children were asked why they made these groups, they confirmed it was based on function. Although Ai was unable to say if she did likewise, the researchers decided that was the most reasonable explanation for her behaviour. Their only doubt came from the possibility that Ai and other chimpanzees relied more on perceptual similarities than the children, more frequently pairing a black tool with a dark green non-tool than would be expected by chance.

33 See Premack (1990) and Premack and Premack (1983) for an appearance bias in chimpanzee categorisation, and that of animals in general. Tanaka (2001) argues for limitations on superordinate categorisation by chimpanzees: they can form basic-level categories (because dogs look broadly similar) and subordinate-level categories (because so do terriers) but struggle with superordinate categories (because different types of animals look different). It appears, however, that chimpanzees can overcome this constraint when taught to associate arbitrary symbols with objects – otherwise known as language training. Premack (1983) suggested that language training provided chimpanzees with an ability to respond to abstract properties to explain the experimental results of Savage-Rumbaugh et al. (1980) in which three chimpanzees sorted identically shaped tokens with simplified pictures of different food items and tools into their correct categories of either food or tools. The chimpanzees – Sherman, Austin and Lana – had received several years of training in a language-learning programme. This appears to have enhanced their capability to develop categories based on abstract properties rather than appearance alone.

34 Cited in Driscoll (2015). Note that Rousseau had died in 1781 and *Essai sur l'origine des langues* was a posthumous publication.

35 See Wellman and Liu (2004) for these and other theory of mind tasks for measuring the extent of its development in children.

36 See Morales et al. (2000) for an experimental study exploring the relationship between joint attention and language acquisition.

37 Bloom (2001) refers to a 1973 study that had found 'what's that' words within the first vocabulary acquired by children.

38　See Begus and Southgate (2012) and Begus et al. (2014) for studies of pointing leading to self-guided learning.

39　Brosseau-Liard et al. (2015) describe the experiments that tested whether children can discriminate between reliable and unreliable informants when learning the meaning of words.

40　See Miller (2006) for an overview of the developmental relationship between language and theory of mind. She summarises an extensive literature on the relationship between theory of mind, language and communicative disorders. While an impaired theory of mind will inhibit language acquisition, the converse is also true: individuals who have impaired language arising from other developmental disorders may also have an inhibited theory of mind.

41　Premack and Woodruff (1978) kick-started the long debate about whether chimpanzees have a theory of mind. They had required lone chimpanzees to watch video clips of humans attempting to complete tasks and solve problems that were quite unnatural to them. It was not until around 2000 that experimental designs were used that provided confident insights into the chimpanzee mind, these primarily involving social interaction and competition for food. There has been a long sequence of such experiments. An important set was undertaken by Daniel Povinelli and his colleagues, concluding that chimpanzees (and other apes) are more likely to rely on associative learning rather than inferring the intentions and thoughts of others, known as being a behaviorist rather than having a theory of mind. (e.g. Povinelli et al., 1996; Povinelli and Vonk, 2003). Another set has been undertaken by Michael Tomasello, Josep Call, Brian Hare, Fumihiro Kano and colleagues. They have been more persuaded that chimpanzees can at least understand that others have knowledge, goals and intentions. Their experiments made greater attempts to devise procedures that mapped onto the situations that chimpanzees would experience in the wild. These involved exploring the use of deception by chimpanzees (e.g. Hare et al., 2006), and often made explicit comparisons with human infants (e.g. Call and Tomasello, 1999; Tomasello and Carpenter, 2005). As an example, Hare et al. (2001) required a subordinate and dominant individual to compete for food that was placed in various locations on the subordinate's side of an opaque barrier. In some conditions both the subordinate and the dominant saw where the food was placed. In others, only the subordinate observed this, or how the food was later moved to another location unseen by the dominant. If the subordinate was sensitive to what the dominant had or had not seen during the experimental set-up, and hence what knowledge the dominant possessed, they should preferentially approach the food only when the dominant had not seen its placement or later movement. That is what occurred, indicating that in some situations at least, chimpanzees know what other chimpanzees know or do not know, and adjust their behaviour accordingly.

42 Call and Tomasello (2008) reviewed the state of knowledge regarding chimpanzee theory of mind.

43 That conclusion has been recently challenged by Kano et al. (2019), who provide a further innovation in experimental design known as the goggles test – although the chimpanzee version does not actually use goggles. The essence of this test is to enable chimpanzees to use their self-experience of either seeing or not seeing through a barrier to infer what an actor would see and hence believe about the world. In one experiment a group of chimpanzees was split into two sets. They were all given experience with a barrier that for one set was opaque and for the other was see-through. Next, both sets watched a video in which a human actor saw an object being hidden. The actor was then placed behind a barrier that was identical to that which each group of chimpanzees had experienced. The chimpanzees watched the object being removed and then the barrier was taken away. The eye movements of the chimpanzees were then tracked as the actor began to look for the object. If chimpanzees can entertain the idea of false beliefs, those who had experienced the barrier as opaque should anticipate (by their eye movements) that the actor would look for the object in the original position, whereas those who had experienced the barrier as see-through would have no such focus. If there was no difference in the eye movements between the two groups, this would support the proposal that chimpanzees cannot attribute a false belief to another individual. The results indicated that chimpanzees could understand and attribute false beliefs: the eye movements of those who had experienced the barrier as opaque were biased towards looking to where the object had first been hidden, anticipating that would be where the actor searched for it despite their own knowledge that it was no longer there. The chimpanzees who had experienced the barrier as see-through had no such bias. As the experimenters admitted when discussing their results, the possibility of a non-mentalistic explanation remains. Rather than attributing a false belief to the actor, the chimpanzee could have been following a behavioural rule such as 'agents search for objects in the last location where they established a line of sight'. As such, the final verdict on whether chimpanzees can understand false beliefs has yet to be made. The most reasonable stance is that although there may be some capacity for understanding false beliefs, this capacity remains significantly below that of a four-year-old human and never comes to pervade thought and action as it does when children grow up and become adept users of language.

44 See Markman (1990) for the role of the mutual exclusivity principle as a constraint on the possible meanings of words. She provides the example of the round, red apple.

45 'John said that Fred went shopping' and 'Lucy thinks the moon is made of green cheese' are drawn from Hale and Tager-Flusberg (2003) and Miller (2006) respectively. Miller (2006) explains that only certain verbs allow sentential complements: those about mental states, such as *think*, *guess* and *believe*, and those about communication, such as *say* and *tell*.

46 Hale and Tager-Flusberg (2003) describe experiments exploring the impact of learning sentential complements on the development of theory of mind. De Villiers (2000) discusses the relationship between language and theory of mind, noting the significance of sentential complement phrases. Hale and Tager-Flusberg (2003, p. 348) cite de Villiers (2000, p. 90): 'The complement structure invites us to enter a different world . . . and suspend our usual procedures of checking truth as we know it. In this way language captures the contents of minds, and the relativity of beliefs and knowledge states. These sentence forms also invite us to entertain the possible worlds of other minds, by a means that is unavailable without embedded propositions.'

47 See Marian (2023, pp. 87–9) for bilingualism and theory of mind.

48 See Bloom (2001, p. 1129) for a discussion of the difference between category-recognition and category-assembly as proposed by Michael Maratsos within a comment published following Bloom's article (pp. 1111–1112).

49 Bloom (2001, p. 1130) concludes his response to commentors on his précis of *How Children Learn the Meanings of Words* by stating that 'while many of the capacities that underlie word learning might be innate adaptations, word learning itself is not. It is a lucky accident. The mind is heterogeneous in its nature and development, and while some very interesting aspects of human nature may well be innate, modular, and adapted, some other very interesting aspects of human nature, such as the ability to learn the meanings of words, may not be.'

10. Fire

1 See Wiessner (2014) for the difference between daytime and night-time talk by hunter-gatherers.

2 Cited from Wiessner (2014, p. 14027).

3 Cited from Wiessner (2014, p. 14023).

4 Cited from Lynn (2014, p. 983).

5 See Wrangham (2009) for the potential role of cooking in human evolution.

6 See Gowlett et al. (1981) for the evidence from Chesowanja, and Hlubik et al. (2019) for that from FxJj20. Interpretation of such early traces is inevitably challenging. Fires arising from lightning strikes and other types of natural combustion are not infrequent in eastern Africa. Gowlett (2016) notes that many lightning strikes lead to bush and forest fires, especially just before the rainy season when vegetation is dry and susceptible to fire.

7 See Berna et al. (2012) for use of fire at Wonderwerk Cave, South Africa.

Their evidence consists of the microscopic identification of ashed plant material and burnt bone, indicating combustion at 30 metres from the cave entrance. See Goren-Inbar et al. (2004) and Alperson-Afil (2008) for fire use at Gesher Benot Ya'aqov. This is an open site adjacent to a lake. Hearths are claimed to have been present by the spatial distribution of burnt artefacts; because no charred material survives these have been referred to as 'phantom' hearths. Fire appears to have repeatedly occurred through the sequence of deposits, causing Alperson-Afil (2008, p. 1733) to argue that the hominins 'had a profound knowledge of fire-making, enabling them to make fire at will'. This, however, is not accepted by most other researchers on the history of fire use, largely because of the absence of fire at other sites of the same period.

8   See Walker et al. (2016) for combustion at Cueva Negra.

9   See Roebroeks and Villa (2011) for a review of the earliest evidence for the habitual use of fire in Europe. Gowlett (2016) notes that natural fires can be frequent in northern environments, with more than 100 fires burning in Alaska every year.

10 There has been a long history of debate about the evidence for use of fire at Zhoukoudian, China. Gao et al. (2017) review the evidence to argue that fire was first used in layer 4, locality 1, dating to between 292,000 and 312,000 years ago.

11 Cited from Lee (1979, p. 248).

12 See Sorensen et al. (2014) for methods of fire production and experiments using strike-a-lights.

13 See Shimelmitz et al. (2014) for the evidence of fire use at Tabun Cave.

14 See Karkanas et al. (2007) for the use of microscopic analysis of sediments to identify the habitual use of fire at Qesem Cave, Israel, and Shahack-Gross et al. (2014) for the identification of a central hearth at this site.

15 See Preece et al. (2006) for fire use at Beeches Pit, and Sanz et al. (2020) for that at Aroeira.

16 See Fernández Peris et al. (2012) for the evidence of fire use at Bolomor Cave. They focus on the hearths dating to c.225,000–240,000 years ago.

17 See Sandgathe et al. (2011) for the evidence of hearths from Pech-de-l'Azé and Roc de Marsal, and Marcazzan et al. (2023) for those from Grotta di Fumane.

18 See Heyes et al. (2016) for the likely use of manganese dioxide as part of the tinder to make fire by Neanderthals.

19 Sandgathe et al. (2011) describe the marked absence of fireplaces at several Neanderthal sites, including some dated to especially cold periods. Murphree and Aldeias (2022) also note a relatively low density of Upper Palaeolithic fireplaces during the colder periods of the last glacial cycle, suggesting their rarity may reflect an absence of woody fuel to burn. Brittingham et al. (2019) devised a means to distinguish between the residues left by wildfires and by

domestic, human-controlled fires. Applying this to multiple samples from Lusakert Cave, Armenia, led them to conclude that Middle Palaeolithic fire use was regionally differentiated, with some Neanderthal groups either having lost the capacity to make fire or never having acquired it.

20 See Albert et al. (2012) for fire use at Kebara Cave. They provide an impressive study of phytoliths, micro-stratigraphy and the geochemistry of site sediments to gain insights into the Neanderthal use of fire.

21 See Jaubert et al. (2016) for the Neanderthal construction in Bruniquel Cave.

22 See Wadley et al. (2020) for the use of fire and ashes in Border Cave,

23 See Haaland et al. (2021) for a geoarchaeological study of Blombos Cave that describes the hearths. They use these and other data to infer changes in human mobility patterns.

24 See Murphree and Aldeias (2022) for a review of Upper Palaeolithic fire use in Europe.

25 See Marcazzan et al. (2023) for the sequence of hearth types in Grotta di Fumane and its interpretation. The change in hearth construction between the Middle and Upper Palaeolithic has been interpreted as reflecting changes in mobility, with modern humans occupying the cave for longer periods of time.

11. Language and the brain

1 The function of glial cells has long been debated. See Jäkel and Dimou (2017) for a review.

2 See Kennedy (2016) for how synaptic plasticity leads to learning and memory.

3 Electroencephalography involves attaching sensors to the brain scalp that pick up electrical signals produced by the brain. Magnetoencephalography measures the magnetic fields produced by the brain's electrical circuits.

4 A left lateralisation of language was appealing because it seemed to match the asymmetry of the brain, as did the predominance of right-handedness in modern populations and conditions that involved language disorders, such as autism which is more prominent in those with more symmetrical brains (Herbert et al., 2004). See Poeppel et al. (2012) for how our understanding of the neurobiology of language had changed within the decade before 2012.

5 See Zaccarella and Friederici (2015) and Friederici (2017) for the role of Broca's area in processing hierarchical phrase structures.

6 See Mariën et al. (2014) and Starowicz-Filip et al. (2017) for the multiple roles of the cerebellum in language.

7 See Crosson et al. (2007) and Bohsali and Crosson (2016) for the role of the basal ganglia in language.

8 Friederici (2011, p. 1357) provides a summary of evidence (as of 2011) for where a sentence is parsed within the brain: 'Different brain regions in the left and right hemisphere have been identified to support particular

language functions. Networks involving the temporal cortex and the inferior frontal cortex with a clear left lateralization were shown to support syntactic processes, whereas less lateralized temporo-frontal networks subserve semantic processes . . . Suprasegmental prosodic information overtly available in the acoustic language input is processed predominantly in a temporo-frontal network in the right hemisphere . . . [The] corpus callosum . . . plays a crucial role in the interaction of syntactic and prosodic information during language processing.' The sentence being parsed by this network of connections might be either spoken or signed because despite the original signals from these being the auditory and visual cortices respectively, their processing to derive meaning uses the same language network.

9 See Huth et al. (2016) and Zhang et al. (2020) for where the brain stores words.

10 Huth et al. (2016) noted that narrative-related words were more likely located within the right hemisphere, while Zhang et al. (2020) suggested a preference for abstract words in this region. A limitation of the initial research by Huth et al. (2016) was that in real life we process continuous streams of speech or written text rather than listen to isolated words. Consequently, in the research by Zhang et al. (2020), subjects listened to two hours of stories while the changes in their cortical blood flow were monitored at the most detailed level possible, known as voxels which are 3D units of the brain equivalent to those of pixels in a digital photograph. Following sophisticated statistical analysis of the vast quantities of data, involving placing meaning-related words (such as *week* and *month*) into 985 semantic domains and relating each domain to sets of voxels, a complete 'semantic atlas' for the brain was derived. In one study, striking similarities were found in the concept networks found in different subjects. It could not be determined whether such similarities arose from an innate condition of the brain for where concepts will be located or from the similar backgrounds of the subjects – they were all university-educated English speakers.

11 Cited from Zhang et al. (2020, pp. 6–7).

12 Poeppel et al. (2012) refer to the multiple sub-units of Broca's area. See Fedorenko et al. (2012) for the identification of language-specific and domain-general functions of Broca's area.

13 See Koziol et al. (2014) for the multifunctional roles of the cerebellum.

14 See Knecht et al. (2000) for individuals with a right-hemisphere lateralisation of language and Golestani (2014) for co-variation between brain structure and linguistic processing.

15 See Paulesu et al. (2000) for a comparison between reading Italian and English on the brain.

16 See Kovelman et al. (2008) for a comparison between bilingual and monolingual brains. Mamiya et al. (2016) describe how learning a second

language changes the quantity and distribution of both grey and white matter in the brain.

17 Marian (2023, pp. 24–8) describes how bilingual and multilingual people process languages in parallel, characterising their co-activation as a multidimensional ripple effect.

18 Herculano-Houzel (2012) places the human brain into context not only with those of other primates but also with other large animals in general, some of which have brains much larger than humans, notably elephants and various whale species. The human brain ranks first in the relative size of the cerebral cortex (but only just) and in the overall number of neurons, 86 billion. These are found at lower densities in other large-brained animals. Their number in the human brain is what we would expect for a primate of our body size. Consequently, she describes the human brain as remarkable but not extraordinary.

19 Schoenemann (2006) compares the human and chimpanzee brain.

20 Dogs, for instance, with brains of no more than 100 cm³ have a better sense of smell than humans by anything between 1,000 and 10,000 times, even being able to detect the odour of cancer growing inside a human body. See Kokocińska-Kusiak et al. (2021).

21 See Schenker et al. (2010) for the homologue of Broca's area in the chimpanzee. In both humans and chimpanzees, the relevant area is Brodmann areas 44 and 45.

22 See Mora-Bermúdez et al. (2016) for an attempt to find out how human brains develop a larger number of brain cells by growing miniature human and chimpanzee brains in laboratory petri dishes. These are known as brain organoids and are grown by reprogramming stem cells to act as brain cells. Mora-Bermúdez and his colleagues found that while the human and chimpanzee brain organoids are strikingly similar in their mix and arrangement of cell types, there is a vital difference in how the specialised cells that form the cerebral cortex develop. They arise from general progenitor cells which were found to spend 50 per cent more time in a stage of cell division called the metaphase within the human than in the chimpanzee organoids. The metaphase is when cells make sure that the chromosomes they contain can be separated and equally distributed between their two daughter cells. How and why the length of the metaphase comes to influence cortical cell production remains to be established, but this research is tracking down how the human brain becomes so much larger than that of the chimpanzee.

23 See Schoenemann et al. (2005) for the differences in the proportion of white and grey matter between human and chimpanzee brains.

24 Building connectomes involves the integration of several advanced neuroscience methods for brain imaging, modelling and statistical analysis.

Ardesch et al. (2019) describe the methods within the supplementary materials to their article, but these are not for the faint-hearted non-specialist who is not prepared to spend time googling to gain even the basic understanding of the terms and techniques they use. There is a helpful introduction at the website of Omniscient Neurotechnology, a private healthcare company specialising in brain imaging. Its website has beautiful images of human connectomes: www.o8t.com

25 Ardesch et al. (2019) built their chimpanzee and human connectomes by combing data derived from multiple subjects (57 human and 20 chimpanzees).

26 Friederici (2017) also argues for stronger connections between Broca's area and the temporal lobe in humans than between the homologous Broca's area and temporal lobe in the chimpanzee.

27 See Sansalone et al. (2023) for changing degrees of cerebral cortex integration during development and evolution of the *Homo* genus. This study measured the degree of integration by the extent of co-variation between cortical areas into adulthood.

28 Caramazza and Shelton (1998) focus on the animate–inanimate distinction. See Semenza (2009) for a further type of double disassociation, that between proper names and common names.

29 See Capitani et al. (2003) for a systematic review of the clinical evidence for semantic category-specific deficits. We must note here that there are two competing theories (or clusters of theories) about how category-specific deficits arise. The view proposed by Caramazza and his colleagues is that domain-specific conceptual categories, such as inanimate versus animate, are served by distinct neural mechanisms. The alternative view is that the categorical nature of the deficits is a result of selective damage to non-categorically organised sensory or functional semantic subsystems, referred to as the sensory-functional argument. For the needs of this book and the latter arguments about domain-specific knowledge and cognitive fluidity, the precise neural mechanism for category-specific deficits is of limited consequence. To follow the debate, see Caramazza and Shelton (1998) and Capitani et al. (2003). See Mahon and Caramazza (2003, 2011) for category-specific neural mechanisms and Tyler and Moss (2001) and Moss and Tyler (2003) for the sensory-functional argument. See Spunt and Adolphs (2017) for a broader reflection on the nature of domain-specificity.

30 Cited from Mahon and Caramazza (2011, pp. 97–8).

31 Please note that Mahon and Caramazza (2011, p. 99) refer to the fusiform gyrus, which is an alternative name for the occipitotemporal gyrus.

32 See Karmiloff-Smith (2015) for a development perspective on domain-specific knowledge. She argues that brains start out with a number of basic-level biases to the processing of certain kinds of input over others but become

domain-specific over time. See Hirschfeld and Gelman (eds) (1994) for domain-specific knowledge in babies. Spelke (1991) makes a case for intuitive physics and Atran (1990) does likewise for intuitive biology. The significance of this intuitive conceptual knowledge is considered in Chapter 14.

33 This draws on Karmiloff-Smith (2015), who in turn draws on Dehaene (2009).

34 Cited from Karmiloff-Smith (2015, p. 96). The removed piece of the quote (the . . .) is where she attributes the term 'global workspace' to Dehaene (2009).

35 Warren et al. (2019) argue that changes in neurocranial structure do not reflect cortical reorganisation but constraints related to encephalisation (increase in brain size) and obligate bipedalism.

36 See Schoenemann (2006) and Herculano-Houzel (2012) for changes in brain size, numbers of neurons and the costs of having a large brain.

37 DeSilva et al. (2021) argue that the modern human brain size has decreased only during the last 3,000 years. They attribute this to the externalisation of knowledge, distributed cognition, the storage and sharing of information, and group decision making.

38 See Du et al. (2018) and Du and Wood (2020) for the most authoritative current account of changes in brain size of *Homo* through time. The figures for brain size provided in my text are approximate and rounded. Variations will be found in the literature reflecting how different specimens are sometimes placed into different taxons and different samples were available for study. My figures do not cover the extreme cases of either very small or very large brains which arise from pathological conditions.

39 Several partial endocasts have been derived from the *Homo naledi* specimens coming from southern Africa, which together provide an almost complete cortical coverage (Holloway et al., 2018). These show a mix of ancestral traits shared with australopiths and derived human-like traits, including Brodmann areas 44 and 45. Despite the small size of the brain, *H. naledi* may have possessed a human-like pattern of brain organisation. It remains unclear whether the small size of the brain had been retained from the earliest *Homo* or arose from a reduction in the brain size of *H. erectus* because of selective pressures. Similarly, study of an endocast from *Homo floresienses* suggests that its small brain had human-like derived features in the frontal and temporal lobes and a lunate sulcus in a derived position, suggesting higher cognitive processing (Falk et al., 2005). There has been controversy regarding whether *H. floresiensis* is a separately evolved species or microcephalic *Homo sapiens*, which appears to have been resolved to the former (Falk et al., 2005).

40 See Gunz et al. (2020) for interpretation of australopith endocasts.

41 See Ponce de Léon et al. (2021) for the evolution of the frontal lobe area within early *Homo*.

42 See Roach and Richmond (2015) for skeletal evidence for throwing by *H. erectus*.

43 See Holloway (1981) and Melchionna et al. (2020) for hominin brain asymmetry.

44 See Green et al. (2010) for estimates of genetic and population divergence times between Neanderthals and modern humans. Both rely on estimates for the time of divergence between humans and chimpanzees, about which there is some uncertainty.

45 See Gunz et al. (2012) for a comparison between chimpanzee, Neanderthal and modern human brain shapes at birth and their pattern of growth. Their study involves the cranial reconstruction of a Neanderthal newborn from the site of Mezmaiskaya Cave in the North Caucasus.

46 See Neubauer et al. (2010) for a comparison of the development of the brain in humans and chimpanzees.

47 See Kochiyama et al. (2018) for the digital reconstruction of the Neanderthal brain – my summary of how this was achieved is a gross simplification. Their method used digital models from more than 1,000 living humans to account for the large degree of variability in the sulcal and gyral patterns in the modern human brain, constructed an average 3D model and then fitted this to fossil endocasts. Their article provides striking images for the similarities and differences in the structures of the Neanderthal and modern human brain.

48 See Pearce et al. (2013) for the larger visual cortex in the Neanderthal brain than with that of *Homo sapiens* and how this may have impacted on Neanderthal group size.

49 More specifically, Kochiyama et al. (2018) observed differences in the cerebellar hemisphere, which is more inferiorly projected in modern humans than in Neanderthals. The cerebellar hemisphere is partitioned into two hemispheres, separated by the mid-section of the cerebellum, called the vermis. Barton and Venditti (2014) had noted the potential significance of the cerebellum for the evolution of ape and human cognition before differences in the cerebellar hemispheres of Neanderthals and modern humans were recognised.

50 Kochiyama et al. (2018) summarise the cognitive implications of variation in the size of the cerebellum in modern humans. See also Paradiso et al. (1997).

51 The greater importance of connectivity within the brain rather than its shape has been challenged by Pang et al. (2023). They argued that the geometry of the surface of the brain provides a better explanation of brain activity than models that attempt to capture the intricate patterns of connectivity between different brain regions. Their study is highly technical and their methods impenetrable to non-specialists in neuroscience but a

commentary is provided by Castelvecchi (2023) and a short accessible account of their findings by Pang and Fornito (2023).

12. The genetics of language

1   Fisher (2017) authoritatively explains why single gene mutations for the appearance of language are not feasible. Examples of 'single gene mutations' arguments for the appearance of language are provided by Chomsky (2011), Crow (1997) and Klein and Edgar (2002).

2   A nucleotide consists of a sugar molecule, which is deoxyribose in DNA, attached to a phosphate group and a nitrogen-containing base, which in DNA is either adenine (A), cytosine (C), guanine (G) or thymine (T).

3   Cases of linguistically deprived children are fortunately extremely rare and often subject to much controversy because of emotive factors. They can be informative about the extent to which the brain can repair itself once stimuli are present. Simon (1978) considers the case of Kaspar Hauser (1812–33), who is said to have been kept in a dungeon until the age of 17, with minimal social contact and language input. Once released he was able to acquire some language capability. Cheng et al. (2019) explore the effects of language deprivation on brain connectivity.

4   Genomic research moves forward at a remarkable pace. Consequently, any description of how specific genes operate and how they influence human physiology, behaviour or cognition, will inevitably be outdated when published in a book for the general reader. Moreover, any summary I provide can only be a simplification of the incredibly complex manner in how genes function and interact and neural networks develop. For these reasons, I was advised to avoid any description of specific genes and remain at a conceptual level for brain and language evolution. That struck me as inappropriate: however difficult, we all need to learn about genes because they are frequently discussed and often misrepresented in the popular media, and genomic research is coming to dominate so many aspects of our lives.

5   See Fisher and Vernes (2015), Galaburda et al. (1985) and Carrion-Castillo et al. (2013) for the genetics of dyslexia.

6   See Fisher and Vernes (2015) for the genetic basis of language disorders.

7   For the impact of mutations on *NRXN1* on language and other developmental disorders, see Ching et al. (2010).

8   See Mamiya et al. (2016) for this example of how genetic variation between individuals can influence their degree of language proficiency.

9   Key publications regarding *FOXP2* include the original study by Enard et al. (2002), which was picked up by the media and led to the designation of *FOXP2* as 'the language gene'; Krause et al. (2007a) identified that the modern variant of *FOXP2* was also shared with the Neanderthals; Enard et al. (2009) explored the impact of the human variant of *FOXP2* within mice; and

Atkinson et al. (2018) repeated the 2002 study with a larger sample population and found no evidence for a selective sweep by the new variant that had been identified by Enard et al. (2009).

10 Fisher (2017) notes that multiple genes may influence the presence of verbal dyspraxia.

11 See Benítez-Burraco et al. (2022) for *FOXP2* in modern humans, Neanderthals and Denisovans, with an argument for differences in the manner in which this gene is regulated.

12 The 1000 Genomes Project ran between 2008 and 2015, creating the largest public catalogue of human variation and genotype data. See www.internationalgenome.org/about

13 For variation in the regulation patterns of *FOXP2*, see Benítez-Burraco et al. (2022).

14 Green et al. (2010), Reich et al. (2010) and Prüfer et al. (2014) are seminal publications for the Neanderthal and Denisovan genomes.

15 See Melchionna et al. (2020) for the evolution of brain asymmetry within *Homo*.

16 See Villanea and Schraiber (2019) for Neanderthal and modern human interbreeding.

17 See Sankararaman et al. (2014) and McCoy et al. (2017) for the impact and distribution of Neanderthal-introgressed DNA.

18 See Green et al. (2010), Reich et al. (2010) and Prüfer et al. (2014) for the first Neanderthal and Denisovan genomes.

19 The catalogue of SNCs (single nucleotide changes) is continually changing as further archaic genomes are derived and samples of modern humans increase, with initial values provided in Green et al. (2010), Reich et al. (2010) and Prüfer et al. (2014). The counts provided by Kuhlwilm and Boeckx (2019) are significantly lower than previous estimates and can only be considered as an interim statement.

20 Kuhlwilm and Boeckx (2019) provide the 2019 catalogue of genetic changes that distinguish modern humans from archaic hominins. See Moriano and Boeckx (2020) for genetic changes in regularity regions of the human genome.

21 See Schaefer et al. (2021) for an estimate of how much of the human genome is unique to *Homo sapiens*. They explain that people today carry Neanderthal DNA for two reasons. One is by admixture, i.e. from interbreeding. The other is from the common ancestor. Because there was original genetic variation in that (proposed) ancestor, *H. heidelbergensis*, a different range of genes will have been inherited by different people today. This is referred to as incomplete lineage sorting (ILS).

22 See Shi et al. (2017) for the impact of *CASC5* on grey matter volumes. Its

variability within living modern humans accounts for the larger volumes of grey matter in East Asians.

23 See Smith et al. (2010) for differences in the rate of Neanderthal and modern human development. Neanderthal children are thought to have grown up more quickly because their ages for dental eruption were younger than in modern humans. As Hublin et al. (2015) explain, however, it is not simply the rate of development that differs but also the specific pattern of brain development.

24 Cited from Kuhlwilm and Boeckx (2019, p. 7).

25 See Buisan et al. (2022) for the significance of brain areas outside the cerebral neocortex that underwent positive selection within what is now a Neanderthal-introgression desert.

26 Pinson et al. (2022) describe the impact of variants of *TKTL1* on brain development. Pinson and her team inserted either the ancestral or the human variant of *TKTL1* into the brains of mouse and ferret embryos. Those with the human gene developed significantly more neural progenitor cells and a different shape to their brain – suggesting that *TKTL1* may contribute towards globularisation of the modern human brain. A further experiment was undertaken by growing brain organoids in the laboratory – mini-brains made from human stem cells. When the ancestral version of the *TKTL1* gene was inserted into an organoid, fewer neural progenitor cells were produced than when the organoid received the modern human version of *TKTL1*. Moreover, Pinson and her team found higher quantities of the mRNA produced by *TKTL1* in the frontal than in the occipital lobe of modern humans, suggesting that the modern variant of this gene played a role in the expansion of the frontal lobe. Although further tests are required, it appears that the mutation on *TKTL1* within the modern human lineage had profound consequences on the development of the brain by producing a greater number of neurons during foetal development. Although how this fed into the development and function of neural networks remains unknown, an impact on the acquisition of language seems likely.

27 See Gunz et al. (2019) for a ground-breaking study that uses modern human genetic variation and cranial shape to make inferences about Neanderthal brain development.

28 One of the 11 SNPs that Gunz et al. (2019) identified down-regulates the *UBR4* gene that controls the formation of neurons as the cortex is being formed and promotes their migration to designated parts of the developing brain. Another of the SNPs up-regulates the *PHLPP1* gene. This gene encodes a negative regulator of the AKT protein that drives myelination. Carriers of the Neanderthal allele would cause the expression of *PHLPP1* to be higher in the cerebellum, which would reduce the AKT-driven myelination, with potential implications for cognition and language.

29   See Pang et al. (2023) for the case that shape is the primary determinant of brain function. Pang and Fornito (2023), who the comparison between ripples in a pond and telecommunications networks (p. 4). Castelvecchi (2023) quotes David Van Essen, a neuroscientist leading a connectome project, saying that the comparison Pang et al. draw between the shape-based models and connectivity models is 'not fair'.

30   According to Reich et al. (2010), the Neanderthal population bottleneck occurred after the divergence of the Denisovans.

31   See Harris and Nielsen (2016) for an estimation of Neanderthal inclusive fitness based on inferred levels of inbreeding and distribution of mutation fitness effects.

32   See Gicqueau et al. (2023) for the anatomically modern human at Grotte du Renne, with a series of alternative interpretations.

33   See Harris and Neilsen (2016) and Simonti et al. (2016) for the positive and negative impacts of Neanderthal introgression.

34   See Sankararaman et al. (2014) for how the DNA of modern humans in the presence of introgressed Neanderthal alleles is sometimes enriched with genes indicating the Neanderthal alleles had helped humans adapt to non-African environments.

35   See Slimak et al. (2022) for evidence of modern humans in Europe at *c*.54,000 years ago, at least 10,000 years before their permanent arrival. This might be described as a 'failed' colonisation event. Stringer and Crété (2022) review evidence for later pre-Aurignacian dispersals into Europe that date between 45,000 and 41,000 years ago, with evidence from sites such as Zlatý kůň Cave, Czechia, Bacho Kiro Cave, Bulgaria, and Grotta del Cavallo, Italy. These incursions would have also provided opportunities for Neanderthal–modern human interactions, including interbreeding. Stringer and Crété suggest the absorption of Neanderthals into modern human communities may have contributed to the demise of the Neanderthal population.

36   See Green et al. (2010) for the lack of evidence for modern human alleles within Neanderthal genomes and an explanation for their counterintuitive finding.

37   See Harris and Nielsen (2016) and Simonti et al. (2016) for the 'genetic cost of Neanderthal introgession'.

38   See Kuhlwilm and Boeckx (2019) and Sankararaman et al. (2014) for introgression deserts in the modern human genome and their interpretation.

39   See Sankararaman et al. (2014) and Harris and Nielsen (2016) for introgression deserts on the X chromosome.

40   See McCoy et al. (2017) and Kuhlwilm and Boeckx (2019) for the rarity of Neanderthal alleles and their down-regulation in brain-related regions of the modern human genome.

13. Words keep changing

1   See Lifschitz (2012, pp. 17–21) for a description and interpretation of the Epicurean theory for the origin of language.

2   See Urban (2014) for an overview of semantic change that is drawn on throughout this chapter.

3   The changing meaning of 'woke' is described in the *Economist*, 30 July 2021, and is covered at length in a five-part BBC Radio 4 podcast, 'Woke: The Journey of a Word', by Matthew Syed, originally broadcast in February and March 2023.

4   Cited from Haiman (1998, p. 166).

5   See Clark and Clark (1979) for a lengthy and excellent study of verbification.

6   See McWhorter (2016, p. 180) for sound backshift that converts a noun to a verb. The reason for this sound change might be to ensure that the hearer knows whether the verb or the noun is being used, although this is usually evident from other words and the context of the utterance.

7   Deutscher (2005, pp. 175–7) provides several examples of backformation.

8   Traugott and Dasher (2001) provide a framework for the process of semantic change.

9   Traugott and Dasher (2001, p. 35) call the adoption of a peripheral or a misunderstood meaning of a word as the main meaning of the word as 'the semanticization of pragmatics'.

10 See Regalado, A., 'Who coined "cloud computing"?' *MIT Technology Review*, 31 October 2011, for the coining of the term 'cloud computing'.

11 See Robinson (2012) for the study of the semantic change of 'skinny'.

12 Robinson's (2012) method was to ask her participants to identify something that is skinny, such as their dog, and then ask them to justify why they called it skinny. By so doing, she was able to elicit a more natural and real-world usage of the word than if she had simply asked them to define the meaning of the word.

13 Traugott and Dasher (2001) suggest that because creating new word meanings presupposes knowledge of discourse-pragmatic rules, children are unlikely to be responsible for innovations in semantics. I interpret 'discourse-pragmatic rules' as those that influence how the meanings of spoken words are related to the social context in which they are employed.

14 Blank (1999) emphasises the desire for efficient and effective communication as a driver for semantic change, and provides the examples cited in my text.

15 IDK = I don't know; NBD = No big deal; GOAT = Greatest of all time.

16 See Garber (2013) for Shakespeare's use of friend as a verb in *Henry V*: 'Disorder, that hath spoil'd us, friend us now!' See Wittkower (2012) for a consideration of friendship via Facebook as an activity.

17 See Skinner (2014) for changes to the meaning of the word 'cool'.

18 Google was likely derived from the number googol, which is a 1, followed

by 100 zeros. As such, it merely represents an example of semantic, morphological and sound change, rather than a case of word creation.

19 *Fahrenheit* from Daniel Gabriel Fahrenheit, who derived this measure of temperature; *Caesar salad* from the chef Caesar Cardini who invented it; America from the Italian explorer Amerigo Vespucci; *nicotine* from Jean Nicot, who sent powered tobacco leaves and seeds to France from Portugal.

20 Word formation can be categorised in other ways, making either more or fewer than seven categories. Moreover, each of the categories I provide has been subdivided at a much finer level of description. See Table 1.1 in Schmid (2015) for the types of information used at that level.

21 McWhorter (2016, pp. 178–8) describes such changes in pronunciation and provides my examples.

22 Deutscher (2005, pp. 175–7) describes backformation and provides these examples.

23 The extent and complexity of word formation study can be appreciated from Müller et al. (2015), the five-volume handbook of word formation, which contains 207 articles and yet covers only European languages.

24 See Štekauer (1998) for his approach to onomasiology.

25 See Fernández-Domínguez (2019) for a summary of the history of onomasiology as an introduction to his review of the field.

26 See Grzega (2015) for his CoSMOS model.

27 You may have already guessed the animal is a bullfrog. In 2000 more than 7,000 giant American bullfrogs were found living in a pond in Sussex, seemingly released as tadpoles brought from the United States. These can grow to twice the size of normal frogs and have been known to eat fish, birds and baby alligators (although not necessarily in Sussex).

28 Körtvélyessy et al. (2021) explored the role of individual creativity in the naming process within an experimental study by measuring the creativity of participants and exploring how this relates to the two aspects of word formation that are often in tension, economy of expression (preferred by speakers) and semantic transparency (preferred by hearers). Using 309 university undergraduates as their participants, they measured the creativity of each participant by using the Torrance tests of creative thinking (TTCT), which use measures described as fluency, flexibility, originality and elaboration. The participants were given various tests that required them to coin new words, the character of which were then compared with their creativity. The researchers found a correlation between creativity and a tendency towards economy of expression. This is perhaps why so many of the words we have come across in linguistics, coined by clever and creative academics, are semantically opaque – onomasiology being a case in point.

29 See Fontaine (2017) for a study of the semantics of the word 'Brexit'. She finishes her analysis by stating that 'If the UK does exit the EU, perhaps the

term will be restricted to history books. At the time of writing, "Brexit" is on the tip of everyone's tongue in Europe.'

30   The blog was by Peter Wilding, who was a political advisor to the then prime minister David Cameron.

31   Addressing supporters in Birmingham on 11 July 2016, two days before she became prime minister, Theresa May said: 'First, our country needs strong, proven leadership to steer us through this time of economic and political uncertainty and to negotiate the best deal for Britain as we leave the EU and forge a new role for ourselves in the world. Because Brexit means Brexit, and we're going to make a success of it.'

32   See Marian (2023, p. 161) for changing pronouns in response to changing notions of gender.

33   See Salmons (2021) and Garrett (2014) for reviews of the field of sound change, which are drawn on throughout the following section.

34   The relative significance of speaking (production) and hearing (perception) to sound change is debated. Ohala (1981, 2003) prioritised the listener's role arguing that the listener reconstructs what they believe the speaker was intending to say by using various cognitive methods. Mistakes might arise. Ohala uses the terms 'hypocorrection' to refer to instances when a listener mishears a sound and fails to correct it, and 'hypercorrection' when a sound is correctly heard but is altered in the listener's mind by applying those cognitive methods.

35   Weinreich et al. (1968) promoted the view of language as a social object, as summarised by Garrett (2014, p. 242).

36   See Mallory and Adams (2006) for a full consideration of how Proto-Indo-European words are reconstructed.

37   Deutscher (2005) provides an excellent coverage of sound change and several of the examples used in this chapter. The change of leading consonants became enshrined in historical linguistics as Grimm's Law after Jacob Grimm, who first described it in 1822. The word 'law' is significant because the 19th-century linguists (and many who followed) believed these and other sound changes to be regular and to occur without exception – they automatically spread throughout the whole language. Hence every /p/ in Indo-European would have become /f/ in Germanic. This was an attractive idea: by defining laws and treating language as an object prone to evolution, it made linguistics appear more science-like. It is not a view that survived, however, because many instances of irregular and sporadic sound change were soon recognised. Another view – one I much prefer – is summarised in the adage that 'every word has its own history'. See Salmons (2021) for an extended discussion about the attitudes lying behind Grimm's Law and the other so-called laws developed in the 19th and early 20th centuries by linguists who became known as the Neogrammarians. Salmons (2021, p. 23)

cites the motto *'Chaque mot a son histoire'* ('Every word has its own history'), which he attributes to Jules Gilliéron, a French dialectologist writing in 1918.

38   See Salmons (2021) for an academic and Deutscher (2005) for a more accessible discussion of least-effort theories for sound change.

39   McWhorter (2001, pp. 8–21) describes sound erosion with reference to French verbs.

40   The erosion of word endings can cause the meaning of a word to be distinguished solely by the tone of a remaining vowel. John McWhorter (2001, 195–197) provides an example from Vietnamese, now a part-tonal language but once entirely reliant on syllables to differentiate words. He suggests that at about 2,000 years ago the words for *big toe, nostalgic* and *shitty little monkey* were spelt *da, da'* and *dah* respectively, with the ' of the syllable *do'* indicating a 'catch in the throat' sound and the *h* of *dah* pronounced lightly, as in 'holly', but without the 'olly'. These endings made subtle changes in the sound of the 'a' in the preceding da: it was slightly raised in tone by the ending, and slightly lowered in tone by the h ending, although the words were primarily distinguished by the sounds of their endings. Those endings became eroded, leaving all three words as da, and distinguished only by their tones. Whether sound erosion is the cause of tonal languages in general, as McWhorter suggests, remains unclear. These form a major component of the world's languages and have not received the attention from western linguists they deserve. Cantonese Chinese, for instance, has no word endings to indicate how the words should be understood regarding tense or the object of a verb, having an entire reliance on variations in tone. Some words can have six different levels indicating six entirely different meanings.

41   Salmons (2021, pp. 4–5) provides several examples of the umlaut effect, including that for how the plural of mouse is mice rather than mouses.

42   See Giancarlo (2001) and Stockwell (2002) for alternative interpretations of the Great Vowel Shift.

43   See accounts in Crystal (ed.) (2003), Millward and Hayes (2012) and Nevalainen and Traugott (eds) (2012) for accounts of the Great Vowel Shift.

44   As stated by the linguist Robert Stockwell, 'This kind of vowel shifting is a pervasive and persevering characteristic of vowel systems of a certain type.' Quoted from Stockwell (1978, p. 337). This is also cited in Salmons (2021, p. 99).

45   For Labov's seminal contributions to sociolinguistics, see Labov (1994, 2001) and Labov et al. (2006).

46   See Labov (2007) for dialect mixing as a cause for the Northern Cities Shift.

47   See Van Herk (2008) for the Northern Cities Shift that arose from white northerners seeking to differentiate themselves from the Black population.

48 See Labov (2001, p. 447) for the role of women as the drivers of sound change.

49 See Eckert (1988) for the study of social class and sound change in Detroit adolescents.

50 See Warren (2016) for a coverage of upspeak. McWhorter (2016, pp. 36–7) interprets upspeak as a means of acknowledging the listener.

51 Quoted from McWhorter (2016, p. 3).

52 See Narrog and Heine (2012) for a review of the central issues and debates concerning grammaticalisation. Many definitions and schemes for grammaticalisation have arisen, along with debates about what should be covered by the term, whether grammaticalisation should have independent status as a linguistic process or be seen as a collation of others such as semantic and sound change, and whether it is unidirectional or can be reversed.

53 Bernd Heine is a specialist on African languages, many of which provide examples in *The Genesis of Grammar* (Heine and Kuteva, 2007). Heine and Kuteva (2002) provide the standard reference book for grammaticalisation – a remarkable compendium appropriately called the *World Lexicon of Grammaticalization*.

54 Smith (2012) noted the extensive use of diverse body parts in adpositions.

55 Deutscher (2005, pp. 229–30) describes how demonstratives map onto pronouns. Latin has a fourth demonstrative, *is* (for male nouns), but this lacks any specific spatial or temporal associations like the other demonstrative pronouns.

56 Linguists are divided in their opinion about whether syntax evolved incrementally through time or in one event. Berwick and Chomsky (2017) appear to adopt a saltationist view for the origin of spoken syntax, suggesting it emerged fully formed by the occurrence of a genetic mutation at around 100,000 years ago. They are, however, largely isolated, and arguments for a gradually incremental development are more persuasive. See Tallerman (2014) for a powerful critique of Berwick and Chomsky's views, and Clark (2013) for an excellent discussion of the relationship between theories of syntax and contrasting evolutionary accounts.

57 For the emergence of syntax in artificial languages, see Kirby (2012) and Chapter 8.

58 Aronoff et al. (2008) describe the linguistic organisation of Al-Sayyid Bedouin Sign Language that arose about 80 years ago in a small, isolated community with a high frequency of deafness. This has some of the syntactical elements but not others that one would expect with a Bedouin language, causing Aronoff et al. (2008, p. 149) to conclude that 'the existence of certain syntactic mechanisms and the lack of others suggests that language does not appear all at once, but rather develops incrementally. Even syntax is not an

"indecomposable bloc"; instead, it builds up over time.' This is also cited in Clark (2013).

59 Jim Hurford is a linguist based at Edinburgh University who has made major contributions to our understanding about the evolution of language, notably in his 2012 book *The Origins of Grammar* (Hurford, 2012). Ray Jackendoff is a linguist and professor of philosopher at Tufts University, Massachusetts, who has written extensively about the nature and evolution of language. I recommend Jackendoff (1999, 2002, 2011).

60 See Bybee (2012) for the role of domain-general processes as the basis for grammar, and de Boer (2012) for how cultural evolution may lead to self-organisation in linguistic systems.

61 See Hopper and Traugott (2003) for the notion of a cline.

62 The phrase 'I'll see you later' has already frequently become front-end-eroded into 'See you later'.

63 Hodge (1970) introduced the idea of a linguistic cycle, providing the example from Ancient Egyptian. This was nicely captured in a remark by Talmy Givón (1971, p. 413): 'If today's bound morphemes are yesterday's lexical words, then today's morphology is yesterday's syntax.'

64 Narrog and Heine (2012, p. 16) refer to the time frame for grammaticalisation in pidgins and creoles. Nevalainen and Palander-Collin (2012) provide the sample of grammaticalisation in Toronto English. The study of Akkadian was by Deutscher (2012). He defined 'speech introducing clauses' (such as 'she definitely said . . .') as the source of the grammaticalisation, rather than any single word within the clause.

65 Nevalainen and Palander-Collin (2012) provide a study of the *ne* deletion in Parisian French, comparing its frequencies using studies published in 1975 and 1990.

66 Heine and Narrog (2015, p. 402) describe the motivation for grammatical-isation as arising from 'linguistic forms for meanings that are concrete, easily accessible and/or clearly delineated to also express less concrete, less easily accessible and less clearly delineated meanings. To this end, lexical or less grammaticalized linguistic expressions are pressed into service for the expression of more grammaticalized functions.'

14. Language, perception and thought

1 Marian (2023, p. 101) states that when 1,000 bilinguals were asked if they feel like a different person when using a different language, two-thirds said yes. My own anecdotal observations confirm this is the case, primarily based on my father-in-law, Eric Orton. He was a quiet, softly spoken, reserved Englishman, who spent his career teaching French in Leicester. Each summer he took his family for holidays in France. After marrying his daughter, I joined them on several occasions to find that Eric was transformed when

speaking French, in France with his French friends: a jovial, expressive, fun-loving extrovert. This wasn't because he was on holiday, it was because his personality changed when speaking French. Whether it had done the same in a Leicester schoolroom I never had the chance to find out.

2   See Pavlenko (2014) and Leavitt (2011) for the origin and history of linguistic relativism.

3   Quoted from Boas (1920, reprinted in 1966, p. 289).

4   See Lakoff (1987) and Mervis and Rosch (1981) for definitions of concepts drawn on in my text.

5   See Hirschfeld and Gelman (eds) (1994) for inherited domain-specific knowledge.

6   See Spelke (1991) for intuitive physics. Inertia: that inanimate objects cannot move by themselves. Solidity: that solid objects cannot pass through each other.

7   See Atran (1990) for intuitive biology.

8   See Keil (1994) for why horses in striped pyjamas are horses and not zebras.

9   See Whiten (ed.) (1991) for intuitive psychology, sometimes characterised as mind reading.

10   See Ljubicic et al. (2018) for alternative categorisation of caribou herds.

11   Quoted from Sapir (1929b, reprinted in 1949, p. 162).

12   Quoted from Whorf (1940, reprinted in 2012, pp. 213–14).

13   Pavlenko (2014) describes how the extensive, nuanced but often convoluted writings of Sapir and Whorf were transformed by later writers into what became known as the Sapir–Whorf hypothesis of linguistic determinism. This was a radically simplified and vulgarised version of their recognition of linguistic relativism, one that became a bête noire of the academic establishment.

14   Quoted from Pinker (1994, p. 54).

15   Bloom (2001, p. 1128–29) provides a critique of Whorf's view that the grammatical structure of language influences perception.

16   See Zlatev and Blomberg (2015) for a summary of how recent interdisciplinary research has provided a substantial degree of support for linguistic relativity and a rebuttal of the arguments made against the influence of language on thought. One of their most intriguing suggestions is that the language-based conceptual framework that Steven Pinker brings to his work, using terms such as 'modules' and 'information processing', effectively predetermine his negative view of linguistic relativism – ironically providing evidence that language does indeed influence thought.

17   See Deutscher (2010) for an excellent review of research up to 2010 on the impact of language on thought, placing this into a long-term historical perspective.

18   Stephen Levinson (2003b) described three frames of reference for talking

about space, which can be referred to as linguistic strategies. European speakers are most familiar with the 'relative' (or egocentric) frame of reference in which terms such as front, back, left and right are used for talking about the locations and directions. An 'intrinsic' (or object-centred) frame of reference is also frequently used, specifying a relationship to a reference object, such as 'behind the mountain'. Third is an 'absolute' (or geocentric) frame of reference that uses the cardinal directions of north, south, east and west. While speakers of European languages make use of this when speaking about long distances, such as the relationship between Edinburgh and London, other languages use it as their primary frame of reference. A well-described example is the Australian language of Guugu Yimithirr that would, for instance, refer to people sitting at a table as to the east or west of each other.

19 Haun et al. (2011) undertook experiments to test whether the way in which one talks about space influences non-linguistic spatial cognition, providing a detailed description of their methodology and results.

20 See Boroditsky (2000, 2001) for experiments to test whether the differences in which English and Mandarin speakers talk about time affect the way they think about time.

21 See Boroditsky et al. (2003) for experiments to test whether the gender of an object influences the way it is conceived. See Mickan et al. (2014) for a failure to replicate some of these experiments, questioning their validity.

22 Paul Ekman has many publications that can be easily accessed via the web. For a start, I recommend Ekman and Davidson (eds) (1994).

23 See Lindquist et al. (2015) for how we only make sense of our sensations once they are attached to words.

24 See Wierzbicka (1986) and Barrett (2017) for the relationship between language and emotion.

25 See Crozier (2014) for the difficulties of defining shame.

26 See Li et al. (2004) for a study of Chinese shame concepts.

27 See Malt et al. (2011) for words and concepts about locomotion.

28 It is neither necessary nor possible for me to review the debate about colour words and the perception of colour – it can be easily accessed via the internet with many academic and even more popular publications. The debate was kicked off by the seminal publication by Berlin and Kay (1969). Paul Kay maintained a relentless study of this topic. He and his co-authors defined two key questions: (1) Are colour categories determined by largely arbitrary convention? and (2) Do colour terms affect colour perception? Whereas a universalist would answer 'no' to both, a relativist would answer 'yes'. Kay's recent work suggests that neither question should have a simple yes or no answer. Key publications in the debate include Kay and Kempton (1984), Kay and Regier (2003), Regier et al. (2005) and Roberson et al. (2000, 2005).

29   See Regier and Kay (2009) for Whorf being half right about colour.

30   Quoted from Regier and Kay (2009, p. 442).

31   See Rosch Heider (1975) for Dani colour categorisation.

32   See Borghi et al. (2018) for a study of how abstract concepts are acquired by language and social interaction.

33   See Berk (1994) and Diaz and Berk (eds) (1992) for studies on private speech in children.

34   Carruthers (1996) believes that speech itself constitutes a type of thought.

35   See Lupyan (2015) and Clark (1998) for overviews of 'language-augmented thought'. As Clarke explains, ideas about language-augmented thought have a long history, with pioneering work having been undertaken by the Soviet psychologist Lev Vygotsky (1934).

36   See Dennett (1994) and Lupyan (2012) for the significance of labelling items.

37   See Boutonnet and Lupyan (2015), Lupyan and Bergen (2015), Edmiston and Lupyan (2015), Lupyan and Clark (2015) and Lupyan and Thompson-Schill (2012) for experimental studies exploring the impact of labels.

38   The following draws on Marian (2023) for 'corners' and Boroditsky (2011) for 'accidents'.

39   Words for weather and food are frequently cited examples of those that vary between speech communities, as noted in Chapter 3 regarding snow, rice and rain.

40   Quoted from Levinson (2003a, p. 36).

41   See Lupyan and Winter (2018) for a discussion of how abstract ideas are defined by words alone.

42   See Borghi et al. (2018) for the significance of talking about abstract ideas,

43   See Lancy (1983) for number systems in languages of the island of New Guinea. See Wassmann and Dasen (1994) for a fascinating study of one such number system.

44   Xu et al. (2020) argue that cross-language variation in number systems may be understood in terms of shared functional need to communicate precisely while using minimal cognitive resources.

45   See Beller and Bender (2008) and Calude (2021) for reflections on the evolution of number systems.

46   See Dehaene (1997) for a review and interpretation of numerical cognition.

47   See Thibodeau et al. (2019) for a review of the role of metaphor in communication and thought.

48   Lakoff and Johnson (1980) is a seminal work in the understanding of language and metaphor.

49   Deutscher (2005) provides a chapter entitled 'A reef of dead metaphors', which I have drawn on in my text.

50   Thibodeau et al. (2019) discuss the 'drain the swamp' metaphor, noting that it goes back more than a century. They use this example to demonstrate that

metaphors typically only carry some aspects of the source domain into the target domain, in this case leaving the positive aspects of swamps – locations of flourishing biodiversity – behind.

51 See Hoffman (2018) for a review of the role of metaphor in science. Although metaphor is essential and pervasive, Taylor and Dewsbury (2018) explain that it might also be restraining.

52 See Leivada et al. (2021) for bilinguals being more able than monolinguals to detect manipulative discourse.

53 The following draws on Marian (2023).

54 Cited from Marian (2023, p. 62).

15. Signs and symbols

1 Louis Liebenberg (1990) provides an account of hunter-gatherer tracking. Both he and the philosopher Peter Carruthers (2002) compare this process to how scientists develop theories and test hypotheses, suggesting it might lie at the origin of science.

2 See Mithen (1988) for the depiction of hoofprints in Upper Palaeolithic art. Those in the Aurignacian are often misinterpreted as vulvas (e.g. White et al., 2012), although there is a possibility of deliberate ambiguity in their depiction.

3 See Botha (2020, pp. 104–5) for differences between linguistic signs and cultural symbols.

4 Hodgskiss (2020) provides a review for ochre use in the Middle Stone Age of Africa. The earliest evidence is described by Watts et al. (2016).

5 See Watts (2009) for interpretation of the Blombos Cave ochre; Watts (2010) for that from Pinnacle Point Cave; and Hodgskiss (2012) for that from Sibudu Cave.

6 See Hovers et al. (2003) for ochre in Qafzeh Cave.

7 See Roebroeks et al. (2012) for the earliest use of red ochre in Europe.

8 See d'Errico and Soressi (2002) and d'Errico (2008) for the use of manganese pigments at Pech-de-l'Azé.

9 See Zilhão et al. (2010) for description of shells with possible pigment from the coastal site of Cueva de los Aviones, Spain. Three pierced shells were found alongside lumps of red and yellow mineral; residues suggestive of red pigment were found inside a *Spondylus* shell. A perforated *Pecten* shell said to be painted with orange pigment was found at Cueva Antón, an inland site, located 80 km from Cueva de los Aviones. Both sites were dated to *c.*50,000 years ago. I remain unpersuaded that the shells had been deliberately pierced. The shells from Cueva de los Aviones have been re-dated to between 115,000 and 120,000 years ago by Hoffmann et al. (2018a).

10 Both d'Errico and Zilhão have many publications on these themes, usually co-authored with other colleagues. On the archaeological evidence, they

provide meticulous detail, but I often find their interpretations to be wanting, seeming to carry considerable bias to finding evidence for Neanderthal symbolism and language and with inadequate, if any, definition of these terms. Useful reviews, although now rather out of date, are provided by d'Errico et al. (2003) and Zilhão et al. (2010).

11 'Evidence for symbolic thinking' is cited from d'Errico and Soressi (2002) and 'body decoration can be regarded as a proxy for language abilities' from d'Errico et al. (2009, p. 25). These are also cited, along with further such claims, by Botha (2020).

12 Rudolf Botha's critique of the claims for a necessarily symbolic role of pigment use as body painting, symbolism and implicating language are found in Botha (2020). This book has an equally insightful and comprehensive (up to 2020 publications) analysis of other claims for Neanderthal language based on archaeological evidence.

13 See Heyes et al. (2016) for the use of manganese dioxide in fire use by Neanderthals.

14 Hodgskiss (2020) describes utilitarian uses for ochre with supporting evidence and citations. See Watts (2002) for a consideration of whether ochre had been used for hide processing or ritualised display.

15 Cited from Humphrey (1976, p. 97).

16 See Knight et al. (1995) for an association between the use of red ochre and menstruation in recent Khoisan San communities of southern Africa and the projection of this into the Middle Stone Age of Africa.

17 See Hoffmann et al. (2018b) for U-Th series dates interpreted as indicating that the painting of red pigments on the walls of Iberian caves was undertaken by Neanderthals. Pike et al. (2012), using similar methods, had suggested a similar conclusion based on the study of paintings in El Castillo Cave. In their case, because only one out of 54 dates gave a date (40,800 years ago) that could potentially exclude modern humans, their interpretation was more cautious than that of Hoffmann et al. (2018b): 'It cannot be ruled out that the earliest paintings were symbolic expressions of the Neanderthals' (Pike et al., 2012, p. 1412).

18 See Pearce and Bonneau (2018), Aubert et al. (2018) and White et al. (2020) for a critique of Hoffmann et al.'s (2018b) methods and interpretations. Note that White et al. (2020) has over 40 authors, including many of the authorities on cave art.

19 See Hoffmann et al. (2020) for a defence of their methods and conclusions.

20 See Slimak et al. (2022) for the modern human occupation at c.54,000 years ago of Mandrin Cave, France.

21 See Metz et al. (2023) for the use of bows and arrows by modern humans at Mandrin Cave. I am largely persuaded by the evidence for modern humans

at Mandrin Cave presented by Slimak et al. (2022), but further analysis is required for full validation.

22 See Steele et al. (2019) for a comprehensive review of shell beads from the African Middle Stone Age. Unless otherwise stated, the following text draws on this source.

23 See d'Errico et al. (2005) for the first description of shell beads from Blombos Cave. Vanhaeren et al. (2013) provide a further study that included additional beads.

24 See d'Errico and Blackwell (2016) for the *Conus* shells from Border Cave.

25 See Sehasseh et al. (2021) for the shell beads from Bizmoune Cave, Morocco.

26 See Dibble et al. (2012) for excavations at Contrebandiers Cave and description of the shell beads.

27 See Bouzouggar et al. (2007) for a description of the beads from Taforalt (Grotte des Pigeons) and their archaeological context.

28 See Bar-Yosef Mayer et al. (2009) for the shells from Qafzeh.

29 See Bar-Yosef Mayer et al. (2020) for her evolutionary scenario of shell selection and use. She places considerable emphasis on the invention of string as transforming the means of object display.

30 See Gicqueau et al. (2023) for the most recent publication regarding the archaeology at Grotte du Renne and debates about its interpretation.

31 As Gicqueau et al. (2023) describe, there have been seven radiocarbon dating programmes attempting to resolve the chronology of Grotte du Renne. A large number of dates for Grotte du Renne were published by Higham et al. (2010). These were challenged by Caron et al. (2011) and Hublin et al. (2012), to which Higham et al. (2012) responded. My figures follow those of Hublin et al. (2012).

32 Slimak (2023) argues the Châtelperronian was made by a second wave of *H. sapiens* entering Europe at *c*.45,000 years ago.

33 The debate about body ornaments from Arcy-sur-Cure is too lengthy to cover in a footnote. A useful summary is provided by Botha (2020, pp. 35–40). Key articles that promote their manufacture by Neanderthals are by d'Errico et al. (1998), Caron et al. (2011) and Zilhão (2013). Those doubting the stratigraphic integrity of the site include White (2001), Bar-Yosef and Bordes (2010) and several commentators published within the d'Errico et al. (1998) publication, notably Mellars and White. Higham et al. (2010) undertook a dating programme that demonstrated stratigraphic mixing, its conclusions supported by Mellars (2010), but this was in turn challenged by Hublin et al. (2012). The debate about the ornaments from Grotte du Renne has sometimes become confused with whether Neanderthals or modern humans were responsible for the Châtelperronian tools. The weight of biomolecular evidence suggests it was the Neanderthals (Welker et al., 2016), although a Neanderthal–Châtelperronian association has been ruled out at the site

of Saint-Césaire which also has a Middle Palaeolithic–Châtelperronian–Aurignacian sequence (Gravina et al., 2018). At that site a meticulous study of stratigraphic mixing using the refitting of stone artefacts showed that all Neanderthal skeletal remains had derived from the underlying Middle Palaeolithic layer.

34 The following draws on Gicqueau et al. (2023).

35 Zilhão et al. (2010) and Hoffmann et al. (2018a) describe the shells from Cueva de los Aviones.

36 See Peresani et al. (2013) for the ochred marine shell from Grotta di Fumane.

37 See Radovčić et al. (2015) for evidence of using talons for pendants at Krapina; Morin and Laroulandie (2012) for their use at Combe Grenal and Les Fieux; and Romandini et al. (2014) for similar evidence from Rio Secco Cave in Italy and Mandrin Cave in France.

38 See Peresani et al. (2011) for the intentional removal of feathers at Grotta di Fumane, and Finlayson et al. (2012) for processing of large birds at Gorham's Cave, although with less specific evidence for feather removal.

39 See Pritchard (2013) for a comprehensive study of the use of bird skins, talons and feathers by shamans within Indigenous North American communities. All birds, not just raptors, are imbued with meanings, captured in his phrase 'No bird is boring'.

40 See Henshilwood et al. (2009) for a meticulous description of the engraved pieces of ochre from Blombos Cave.

41 See Morphy (1999) for the geometric art of Indigenous central Australian peoples and how this contrasts with their representational art.

42 See Tylén et al. (2020) for the experimental study of the salience, memorability and reproducibility of engravings from Blombos Cave and Diepkloof Rock Shelter.

43 See Prévost et al. (2021) for the incised bone from Nesher Ramla, Israel.

44 See Shaham et al. (2019) for the incised bone from Quneitra.

45 For the engraved objects from Crimea, see Majkić et al. (2018) for incised cortex from Kiik Koba; Stepanchuk (1993) for potentially non-utilitarian bone objects from Prolom II; and Majkić et al. (2017) for the decorated raven bone from the Zaskalnaya VI site.

46 See Leder et al. (2021) for the engraved bone from Einhornhöhle.

47 See Rodríguez-Vidal et al. (2014) for the rock engraving from Gorham's Cave, Gibraltar.

48 Botha (2020) summarises the questions that have been raised about interpreting the rock engraving in Gorham's Cave as having been made by Neanderthals and being designated as symbolic.

49 See Marquet et al. (2023) for a detailed description of the finger-flutings attributed to the Neanderthals from La Roche-Cotard, France.

INDEX

Note: The index covers the main text but not the Notes or Bibliography. *Italic* page references indicate a relevant illustration on that page.

Steven Mithen is professor of early prehistory at the University of Reading and is one of the most esteemed archaeologists working today. He has authored over two hundred academic articles and books, including *After the Ice* and *The Singing Neanderthals,* and was elected as a fellow of the British Academy in 2004. He lives in Reading, England.